Theory and Methods for Distributed Data Fusion Applications

Related titles on radar:

Advances in Bistatic Radar Willis and Griffiths
Airborne Early Warning System Concepts, 3rd Edition Long
Bistatic Radar, 2nd Edition Willis
Design of Multi-Frequency CW Radars Jankiraman
Digital Techniques for Wideband Receivers, 2nd Edition Tsui
Electronic Warfare Pocket Guide Adamy
Foliage Penetration Radar: Detection and characterisation of objects under trees Davis
Fundamentals of Ground Radar for ATC Engineers and Technicians Bouwman
Fundamentals of Systems Engineering and Defense Systems Applications Jeffrey
Introduction to Electronic Warfare Modeling and Simulation Adamy
Introduction to Electronic Defense Systems Neri
Introduction to Sensors for Ranging and Imaging Brooker
Microwave Passive Direction Finding Lipsky
Microwave Receivers with Electronic Warfare Applications Tsui
Phased-Array Radar Design: Application of radar fundamentals Jeffrey
Pocket Radar Guide: Key facts, equations, and data Curry
Principles of Modern Radar, Volume 1: Basic principles Richards, Scheer and Holm
Principles of Modern Radar, Volume 2: Advanced techniques Melvin and Scheer
Principles of Modern Radar, Volume 3: Applications Scheer and Melvin
Principles of Waveform Diversity and Design Wicks *et al.*
Pulse Doppler Radar Alabaster
Radar Cross Section Measurements Knott
Radar Cross Section, 2nd Edition Knott *et al.*
Radar Design Principles: Signal processing and the environment, 2nd Edition Nathanson *et al.*
Radar Detection DiFranco and Rubin
Radar Essentials: A concise handbook for radar design and performance Curry
Radar Foundations for Imaging and Advanced Concepts Sullivan
Radar Principles for the Non-Specialist, 3rd Edition Toomay and Hannan
Test and Evaluation of Aircraft Avionics and Weapons Systems McShea
Understanding Radar Systems Kingsley and Quegan
Understanding Synthetic Aperture Radar Images Oliver and Quegan
Radar and Electronic Warfare Principles for the Non-specialist, 4th Edition Hannen
Inverse Synthetic Aperture Radar Imaging: Principles, algorithms and applications Chen and Martorella
Stimson's Introduction to Airborne Radar, 3rd Edition Baker, Griffiths and Adamy
Test and Evaluation of Avionics and Weapon Systems, 2nd Edition McShea
Angle-of-Arrival Estimation Using Radar Interferometry: Methods and applications Holder
Biologically-Inspired Radar and Sonar: Lessons from Nature Balleri, Griffiths and Baker
The Impact of Cognition on Radar Technology Farina, De Maio and Haykin
Novel Radar Techniques and Applications, Volume 1: Real aperture array radar, imaging radar, and passive and multistatic radar Klemm, Nickel, Gierull, Lombardo, Griffiths and Koch
Novel Radar Techniques and Applications, Volume 2: Waveform diversity and cognitive radar, and target tracking and data fusion Klemm, Nickel, Gierull, Lombardo, Griffiths and Koch
Radar and Communication Spectrum Sharing Blunt and Perrins
Systems Engineering for Ethical Autonomous Systems Gillespie
Shadowing Function from Randomly Rough Surfaces: Derivation and applications Bourlier and Li
Photo for Radar Networks and Electronic Warfare Systems Bogoni, Laghezza and Ghelfi
Multidimensional Radar Imaging Martorella
Radar Waveform Design Based on Optimization Theory Cui, De Maio, Farina and Li
Micro-Doppler Radar and Its Applications Fioranelli, Griffiths, Ritchie and Balleri
Maritime Surveillance with Synthetic Aperture Radar Di Martino and Antonio Iodice
Electronic Scanned Array Design Williams
Advanced Sparsity-Driven Models and Methods for Radar Applications Li
Deep Neural Network Design for Radar Applications Gurbuz
New Methodologies for Understanding Radar Data Mishra and Brüggenwirth
Radar Countermeasures for Unmanned Aerial Vehicles Clemente, Fioranelli, Colone and Li
Holographic Staring Radar Oswald and Baker
Polarimetric Radar Signal Processing Augusto Aubry, Antonio De Maio and Alfonso Farina
Fundamentals of Inertial Navigation Systems and Aiding M. Braasch

Theory and Methods for Distributed Data Fusion Applications

Felix Govaers

The Institution of Engineering and Technology

Published by SciTech Publishing, an imprint of The Institution of Engineering and Technology, London, United Kingdom

The Institution of Engineering and Technology is registered as a Charity in England & Wales (no. 211014) and Scotland (no. SC038698).

First published 2023

The Institution of Engineering and Technology
Futures Place
Six Hills Way, Stevenage
Herts SG1 2AU, United Kingdom

www.theiet.org

British Library Cataloguing in Publication Data
A catalogue record for this product is available from the British Library

ISBN 978-1-83953-439-3 (hardback)
ISBN 978-1-83953-440-9 (PDF)

Typeset in India by MPS Limited

Cover Image: Global Connection Network (World Map Credit to NASA);
Yuichiro Chino Moment via Getty Images

Contents

About the author vii

1 Introduction 1

2 Stochastic motion 7
2.1 Stochastic processes 8
2.1.1 Brownian motion and the diffusion equation 8
2.2 Wiener process 16
2.3 Stochastic differential equation 19
2.3.1 Evolution of the mean and covariance 23
2.4 Path integral formulation 24
2.4.1 Path integral in quantum physics 25
2.4.2 Path integral for stochastic processes 27

3 Bayesian state estimation and target tracking 29
3.1 Fundamental equations of linear estimation 32
3.2 Kalman filter 35
3.3 Least squares 38
3.4 Kalman filter as a least squares solution 40
3.5 Kalman filter as an orthogonal projection 43
3.6 Data association 52
3.6.1 Nearest neighbor 53
3.6.2 Global nearest neighbor 54
3.6.3 Probabilistic data association 55
3.7 Probabilistic data association filter 59
3.7.1 Moment matching 60
3.8 Multi-hypotheses-tracker 63
3.9 Interacting multiple model filter 69
3.10 Sequential likelihood ratio test 72
3.11 Conclusion 77

4 Trajectory estimation 79
4.1 Retrodiction 79
4.2 Notion of accumulated state densities 81
4.3 ASD posterior 82
4.4 Recursive measurement fusion using ASDs 87
4.5 Generalized smoothing and out-of-sequence processing 92

4.5.1 Batch processing 96
4.5.2 Block-line combined estimator and smoother 98
4.6 Quantum physical interpretation 103
4.7 The free quantum particle 103
4.7.1 State estimation for macroscopic objects with path integrals 105
4.7.2 The free dynamic particle 106
4.7.3 Linear-Gaussian measurements 109
4.8 Application on a Gaussian density 110

5 Track-to-track fusion with known correlations 113
5.1 Bar-Shalom–Campo formula 115
5.2 T2TF for an arbitrary number of sensors with known cross-covariances 118
5.3 Fusion with arbitrary gains 119
5.4 Iterative calculation of cross-covariances 121
5.5 Distributed calculation of the cross-covariances using the square root decomposition 123

6 Track-to-track fusion with unknown correlations 127
6.1 Naïve fusion 127
6.2 Covariance intersection 130
6.3 Inverse covariance intersection 142
6.4 Tracklet fusion 145
6.5 Safe fusion 148
6.6 Conclusion 151

7 Distributed Kalman filter 153
7.1 Local error covariance globalization 154
7.2 DKF prediction 157
7.3 DKF filtering 159
7.4 Federated Kalman filter 161
7.5 DKF sum formulation 163
7.6 Consensus of agents in graphs 165
7.7 Distributed accumulated state density filter 168

8 Track-to-track association and distributed detection 177
8.1 Track-to-track association 177
8.2 Exact distributed sequential likelihood ratio test 180
8.3 Approximative distributed sequential likelihood ratio test 185

Appendix A 189
Appendix B 191
References 195
Index 201

About the author

Dr. Felix Govaers received his Diploma in Mathematics and his Ph.D. with the title "Advanced Data Fusion in Distributed Sensor Applications" in Computer Science, both at the University of Bonn, Germany. Since 2009 he has worked at Fraunhofer FKIE in the Sensor Data Fusion and Information Processing department where he led the research group "Distributed Systems" for three years. Since 2017 he is the deputy head of the department. The research of Felix Govaers is focused on track-to-track fusion, state estimation in non-linear applications, and quantum algorithms for information fusion. This includes quantum optimization methods, tensor-decomposition techniques, processing of delayed measurements, the Distributed Kalman filter, and track correlation mitigation. He is also interested in advances in state estimation such as particle flow and homotopy filters and the point process theory approaches.

Felix Govaers is an active member of the IEEE Aerospace and Electronic Systems Society (AESS) where he serves as an Associate Editor and is the financial chair of the Germany Section. He also is an AESS Distinguished Lecturer and Short Course Instructor. As an elected member of the Board of Directors (BoD) for the International Society for Information Fusion (ISIF), he is actively contributing to the future of the information fusion community. At the University of Bonn, he serves as a lecturer since 2014 with seminars, labs, and lectures on distributed data fusion systems and quantum algorithms for data fusion.

References

[1] Airbus. Online Presentation of the Future Combat Air System FCAS. Available from: https://www.airbus.com/en/products-services/defence/multi-domain-superiority/future-combat-air-system-fcas.

[2] Harley M. Mercedes Benz Level 3 Testing; 1999. Available from: www.forbes.com/sites/michaelharley/2022/08/02/testing-and-trusting-mercedes-benz-level-3-drive-pilot-in-germany/.

[3] Duraisamy B, Schwarz T, and Ludwig S. Influence of the sensor local track covariance on the track-to-track sensor fusion. In: *2015 IEEE 18th International Conference on Intelligent Transportation Systems*, 2015. p. 2643–2650.

[4] Jovanoska S, Brötje M, and Koch W. Multisensor data fusion for UAV detection and tracking. In: *2018 19th International Radar Symposium (IRS)*, 2018. p. 1–10.

[5] Castrillo VU, Manco A, Pascarella D, *et al.* A review of counter-uas technologies for cooperative defensive teams of drones. *Drones*. 2022;6(3):65. Available from: https://www.mdpi.com/2504-446X/6/3/65.

[6] Kalman RE. A new approach to linear filtering and prediction problems. *Transactions of the ASME—Journal of Basic Engineering*. 1960;82(Series D): 35–45.

[7] Chong CY, Chang KC, and Mori S. A review of forty years of distributed estimation. In: *2018 21st International Conference on Information Fusion (FUSION)*, 2018. p. 1–8.

[8] Chong CY. Hierarchical estimation. In: *2nd MIT/ONR Workshop on Distributed Information and Decision Systems*, 1979.

[9] Bar-Shalom Y. On the track-to-track correlation problem. *IEEE Transactions on Automatic Control*. 1981;26(2):571–572.

[10] Chong CY. Forty years of distributed estimation: a review of noteworthy developments. In: *2017 Sensor Data Fusion: Trends, Solutions, Applications (SDF)*, 2017. p. 1–10.

[11] Hall D, Chong CY, Llinas J, *et al.*, editors. *Distributed Data Fusion for Network-Centric Operations*. London: CRC Press, 2014.

[12] Benacquista MJ and Romano JD. Classical mechanics. In: *Undergraduate Lecture Notes in Physics*. New York, NY: Springer International Publishing, 2018. Available from: https://books.google.de/books?id=l2BODwAAQBAJ.

[13] Merzbacher E. *Quantum Mechanics*. New York, NY: Wiley, 1998. Available from: https://books.google.de/books?id=6Ja_QgAACAAJ.
[14] Stein DJ. *Problems of Living: Perspectives from Philosophy, Psychiatry, and Cognitive-Affective Science*. New York, NY: Elsevier Science, 2021. Available from: https://books.google.de/books?id=8bQjEAAAQBAJ.
[15] Laskey K. Agents with free will: a theory grounded in quantum physics. *Perspectives on Information Fusion*. 2022;5(1).
[16] Pavliotis GA. Stochastic processes and applications: diffusion processes, the Fokker–Planck and Langevin equations. In: *Texts in Applied Mathematics*. New York, NY: Springer New York, 2016. Available from: https://books.google.de/books?id=jXAsvgAACAAJ.
[17] Dirac P. The physical interpretation of the quantum dynamics. In: *Proceedings of the Royal Society of London*, 1927.
[18] Mörters P and Peres Y. Brownian motion. In: *Cambridge Series in Statistical and Probabilistic Mathematics*. Cambridge: Cambridge University Press, 2010. Available from: https://books.google.de/books?id=5BjknQEACAAJ.
[19] Stroock DW. An introduction to Markov processes. In: *Graduate Texts in Mathematics*. New York, NY: Springer Berlin Heidelberg, 2005. Available from: https://books.google.de/books?id=SP9Vzq_xctgC.
[20] Chandra TK. The Borel–Cantelli lemma. In: *Springer Briefs in Statistics*. India: Springer, 2012. Available from: https://books.google.de/books?id=heepk_FGPQ0C.
[21] Lipcer RŠ, Liptser RS, Širâev AN, *et al*. *Statistics of Random processes: II. Applications of Mathematics Stochastic Modelling and Applied Probability Series*. Springer; 2001. Available from: https://books.google.de/books?id=7An21SYEATsC.
[22] Johnson O. *Information Theory and the Central Limit Theorem*. Imperial College Press, 2004. Available from: https://books.google.de/books?id=r5XI8a0lYykC.
[23] Prochorov JV, Prokhorov IUV, Prokhorov YV, *et al*. Probability theory III: stochastic calculus. In: *Encyclopaedia of Mathematical Sciences*. New York, NY: Springer, 1998. Available from: https://books.google.de/books?id=3bbTgvGFuDkC.
[24] Csiszár I and Michaletzky G. Stochastic differential and difference equations. In: *Progress in Systems and Control Theory*. New York, NY: Springer, 1997. Available from: https://books.google.de/books?id=BDnWFJNOnRgC.
[25] Risken H and Frank T. The Fokker–Planck equation: methods of solution and applications. In: *Springer Series in Synergetics*. Springer Berlin Heidelberg, 1996. Available from: https://books.google.de/books?id=MG2V9vTgSgEC.
[26] Feynman R, Leighton R, and Sands M. *The Feynman Lectures on Physics*, vol. 1, 2nd ed. Boston, MA: Addison-Wesley, 1963.
[27] Kac M. On some connections between probability theory and differential equations. In: *Proceedings of the 2nd Berkeley Symposium on Mathematical Statistics and Probability*, 1950.

[28] Tarasov VE. *Applications in Physics, Part B*. De Gruyter Reference. De Gruyter, 2019. Available from: https://books.google.de/books?id=L3acDwAAQBAJ.

[29] Dirac P. The Lagrangian in quantum mechanics. *Physikalische Zeitschrift der Sowjetunion* 1933;3:64–72.

[30] Bar-Shalom Y, Li XR, and Kirubarajan T. *Estimation with Applications to Tracking and Navigation*. New York, NY: Wiley-Interscience, 2001.

[31] Mardia KV and Marshall RJ. Maximum likelihood estimation of models for residual covariance in spatial regression. *Biometrika*. 1984 04;71(1):135–146. Available from: https://doi.org/10.1093/biomet/71.1.135.

[32] Blackman SS and Populi R. *Design and Analysis of Modern Tracking Systems*. Artech House, 1999.

[33] Stone LD, Streit RL, Corwin TL, *et al*. Bayesian multiple target tracking. In: *Radar/Remote Sensing*, 2nd ed. Artech House, 2013. Available from: https://books.google.de/books?id=fJhQAgAAQBAJ.

[34] Koch W. *Tracking and Sensor Data Fusion*. New York, NY: Springer, 2014.

[35] Thomas P, Barr J, Hiscocks S, *et al*. Stone-soup: an open-source framework for tracking and state estimation. In: ISIF Perspectives on Information Fusion. 2019 March;2(1).

[36] Streit R. *Poisson Point Processes: Imaging, Tracking, and Sensing*. New York, NY: Springer, 2010.

[37] Streit R, Angle B, and Efe M. *Analytic Combinatorics for Multiple Object Tracking*. New York, NY: Springer, 2021.

[38] Reid D. An algorithm for tracking multiple targets. *IEEE Transactions on Automatic Control*. 1979;24(6):843–854.

[39] Koch W. *Tracking and Sensor Data Fusion: Methodological Framework and Selected Applications*. New York, NY: Springer, 2014.

[40] Blackman SS. Multiple hypothesis tracking for multiple target tracking. *IEEE Transactions on Aerospace and Electronic Systems*. 2004;19(1):5–18.

[41] Yu Z, Patnaik S, Wang J, *et al*. *Advancements in Mechatronics and Intelligent Robotics: Proceedings of ICMIR 2020. Advances in Intelligent Systems and Computing*. Springer Singapore, 2021. Available from: https://books.google.de/books?id=Z_E5EAAAQBAJ.

[42] Blair WD and Bar-Shalom T. Tracking maneuvering targets with multiple sensors: does more data always mean better estimates? *IEEE Transactions on Aerospace and Electronic Systems*. 1996;32(1):450–456.

[43] Rong Li X and Jilkov VP. Survey of maneuvering target tracking. Part V. *Multiple-Model Methods*. 2005;41(4):1255–1321.

[44] Ghosh BK. *Sequential Tests of Statistical Hypotheses*. Addison-Wesley Publishing Company, 1970. Available from: https://books.google.de/books?id=Wgk_AAAAIAAJ.

[45] Challa S, Morelande MR, Mušicki D, *et al*. *Fundamentals of Object Tracking*. Cambridge Books Online. Cambridge, MA: Cambridge University Press, 2011. Available from: https://books.google.de/books?id=DOKJGnNh9HgC.

[46] Musicki D and Evans R. Joint integrated probabilistic data association: JIPDA. *IEEE Transactions on Aerospace and Electronic Systems*. 2004;40(3): 1093–1099.

[47] Musicki D and Evans RJ. Multiscan multitarget tracking in clutter with integrated track splitting filter. *IEEE Transactions on Aerospace and Electronic Systems*. 2009;45(4):1432–1447.

[48] Rauch HE, Tung F, and Striebel CT. Maximum likelihood estimates of linear dynamic systems. *American Institute of Aeronautics and Astronautics (AIAA) Journal*. 1965;3:1445–1450.

[49] Koch W and Govaers F. On accumulated state densities with applications to out-of-sequence measurement processing. *IEEE Transactions on Aerospace and Electronic Systems*. 2011;47(4):2766–2778.

[50] Koch W, Koller J, and Ulmke M. Ground target tracking and road map extraction. *ISPRS Journal of Photogrammetry and Remote Sensing*. 2006;61(3–4): 197–208.

[51] Del Moral P. *Feynman-Kac Formulae*, 1st ed. Springer, 2004.

[52] Heisenberg W. Encounters with Einstein: and other essays on people, places, and particles. In: *Princeton Science Library*. Princeton University Press, 1989. Available from: https://books.google.de/books?id=FI9KD4jLPckC.

[53] Olfati-Saber R. Distributed Kalman filtering for sensor networks. In: *Proceedings of the 46th Conference on Decision and Control*, 2007. p. 5492–5498.

[54] Radtke S, Noack B, and Hanebeck UD. Distributed estimation using square root decompositions of dependent information. In: *2019 22th International Conference on Information Fusion (FUSION)*, 2019. p. 1–8.

[55] Zhang F. The Schur complement and its applications. In: *Numerical Methods and Algorithms*. New York, NY: Springer US, 2006. Available from: https://books.google.de/books?id=EMEyg8NcuskC.

[56] Khan W. Circumscribing an ellipsoid about the intersection of two ellipsoids. *Canadian Math Bulletin* 1968;11.

[57] Reinhardt M, Noack B, Arambel PO, *et al.* Minimum covariance bounds for the fusion under unknown correlations. *IEEE Signal Processing Letters*. 2015;22(9):1210–1214. Available from: http://dx.doi.org/10.1109/LSP.2015.2390417.

[58] Noack B, Sijs J, Reinhardt M, *et al.* Decentralized data fusion with inverse covariance intersection. *Automatica*. 2017;79:35–41. Available from: http://dx.doi.org/10.1016/j.automatica.2017.01.019.

[59] Carlen EA. Trace inequalities and quantum entropy: an introductory course. In: *Entropy and the Quantum*, Amer. Math. Soc., 2009.

[60] Sijs J, Lazar M, and Bosch PPJvd. State fusion with unknown correlation: ellipsoidal intersection. In: *Proceedings of the 2010 American Control Conference*, 2010. p. 3992–3997.

[61] Nygårds J, Deleskog V, and Hendeby G. Safe fusion compared to established distributed fusion methods. In: *2016 IEEE International Conference on Multisensor Fusion and Integration for Intelligent Systems (MFI)*, 2016. p. 265–271.

[62] Hogg R and Craig AT. *Introduction to Mathematical Statistics*. Prentice Hall, 1995.

[63] Olfati-Saber R. Distributed Kalman filter with embedded consensus filters. In: *Proceedings of the 44th IEEE Conference on Decision and Control*, 2005. p. 8179–8184.

[64] Braca P, Marano S, Matta V, *et al.* Quickest distributed detection via running consensus. In: *2011 19th European Signal Processing Conference*, 2011. p. 417–421.

Chapter 1
Introduction

With the rise of digitalization, artificial intelligence and modern data communication, engineering is increasingly challenged by the complexity of systems-of-systems, in which perception sensors have their own processing mechanism, information is merged at some centric entity, and cross-platform exchange of data is conceptualized in a holistic approach. Therefore, distributed architectures for data fusion, state estimation, and multi-target tracking have been becoming increasingly important. In both, the industry and the military domain, systems comprised of heterogeneous multi sensor data fusion nodes can be found for various applications such as for instance air and ground surveillance, autonomous driving and driver assistance systems, guidance and control, robotics, manufacturing processes, and automated medical diagnosis. The use of multiple sensors to perceive and compute the situational awareness picture leads to the generation of vast amounts of data, which need to be filtered, enriched, interpreted, and evaluated. This becomes infeasible without the notion of distributed algorithms, which are condensing information close to the sensor and on the platform level such that the computational burden is shared among multiple individual instances and communication bandwidth is spared. Such an hierarchical fusion scheme is illustrated in Figure 1.1, where at each level, the data from the subjacent fusion algorithms is combined and condensed for an extended, more precise, and more complete situational awareness picture.

Example applications of track-to-track fusion and distributed fusion

- The largest and maybe most important current European defense project is the *Future Combat Air System* (FCAS) [1]. FCAS is a designated next-generation fighter platform with state-of-the-art technology as part of a large system-of-systems, where unmanned remote carriers and backing legacy fighters collaborate with each other as well as share joint information with *Airborne Early Warning and Control System* (AWACS) aircrafts, sensors from interconnected satellites and others. In the case that information on joint entities in the field of interest is shared via a direct transmission or a cloud infrastructure, it is of high importance that sensor measurements have been processed such that data from clutter and false alarms has been reduced and information has been condensed. This is the most effective way for a consistent ontology of entities and a successful association of pre-existing tracks of potential targets. It is well known that particular in the field

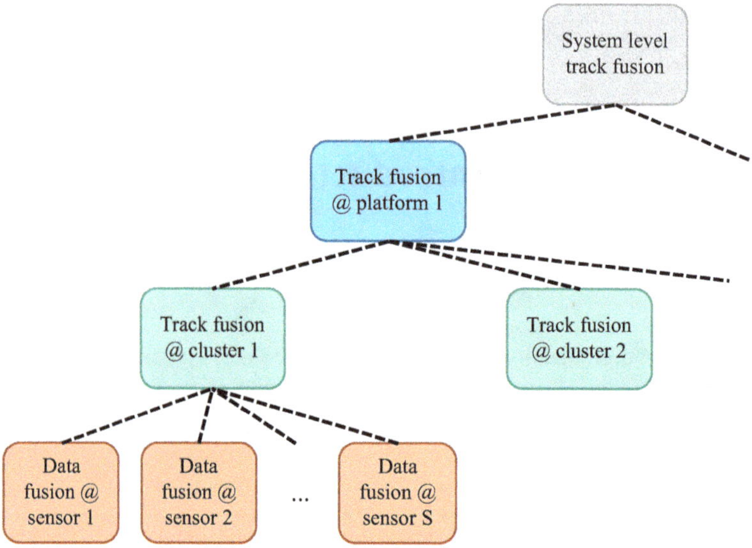

Figure 1.1 Schematic figure for an hierarchical fusion architecture

of reconnaissance spatially distributed sensors contribute in a highly effective manner due to their divers perspectives.

- The technology for *Autonomous Driving* (AD) and *Advanced Driver Assistance Systems* (ADAS) is maturing continuously [2]. Despite the clash of concepts with respect to the question how far machine learning-based methods can and should solve the tremendous challenges for AD in the infinite versatile environment of road traffic, it has become obvious that a mix of multiple sensors such as electrooptical cameras, lidars, and radars covering an all-round view is required. All of which gather a ton of data at each scan with the consequence that a central processing node receiving all observations would be immediately and heavily overloaded. Thus, even if the communication of raw data is possible, there are situations where a distributed computation is necessary in order to make the system work in real time. The pre-processed tracks then need to be combined by lightweight track-to-track fusion algorithms in order to compute a global estimate, an error covariance matrix and potentially additional parameters such as track existence or an agility index [3].
- Another example of distributed fusion is the sensor-based detection and tracking of drones for *Counter Unmanned Air Systems* (C-UAS). Due to the advanced capabilities of small and light drones to fly and even navigate autonomously, defense systems need to engage a fleet of spatially distributed and heterogenous sensors in order to optimize the information gain, detection probability, and tracking accuracy [4]. Various challenges imply that there is no single sensor system, which can cover all kinds of drones in various scenarios. Here, both the transmission of

information over long ranges and the amount of collected data make the use of distributed fusion algorithms inevitable [5].

It is well-known that the state estimation based on noise corrupted data is solved optimally under specific conditions* by the *Kalman filter*. As it will be derived later in this book, the Kalman filter uses a closed-form solution to compute the posterior probability density function of a stochastic *state variable* conditioned on the measurements. The key point for the derivation is the famous *Bayes formula*, which provides the computation of a posterior density by means of a prior and a *likelihood function*, which describes the statistical distribution of the sensor data given that the true state is known. A two-step recursive formula which consists of the *prediction* to compute the prior and the *filtering* for processing the latest data eventually allows a continuous computation and handling of incoming information to adapt the estimate based on the full set of measurements. The algorithm was invented in the 1960s of the past century by multiple researchers in parallel but Rudolph Kalman eventually was eponymous [6].

The problem of decentralized parameter estimation for noise corrupted measurements in control applications were already addressed in the late 1970s [7]. The first solution optimally solved a distributed computation of a multi-sensor state estimate which nowadays is known as the so-called *tracklet fusion* with full communication [8]. The general track-to-track fusion with arbitrary communication was mainly focused in the 1980s, initialized by Yaakov Bar-Shalom [9], who proved that tracks referring to the same target have actually correlated estimation errors such that the fusion approach to treat them as independent yields sub-optimal and too optimistic solutions. This insight actually found a new research area. For the interested reader regarding the history of distributed estimation, it is highly recommended to consider the publication of Chee Y. Chong in [10] on "forty years of distributed estimation." It is our aspiration to provide the full derivations for the most important methods which came up during those decades to give a solution to the track-to-track fusion problem. Some of them are not named explicitly, since they can be considered special cases of more general algorithms. The constellations in different scenarios, with different bandwidth and connectivity between the nodes in a network, and different processing capabilities are countless as described in detail for instance in [11]. However, a profound understanding of the approaches and algorithms enables the reader to easily find the optimal hierarchy and methods for a distributed estimation challenge.

This book is designed to provide state-of-the-art approaches and algorithms to the interested engineer for data fusion in distributed systems such that a profound understanding of the constraints, assumptions, benefits, and drawbacks are becoming evident to the reader. To this end, the derivation of the each method will be given in great detail without omitting auxiliary calculations or intermediate steps. Since this comes at the expense of readability to some degree, having the correct orientation in mind as while navigating the contents of this book is important. To this end, the structure of this book is as follows.

*Linear Gaussian models, given track existence and perfect data association.

Structure of this book

The upcoming chapter states the fundamentals of stochastic motion, which constitutes the underlying models in target tracking and parameter estimation. After the introduction of the stochastic process based on a formal definition, the focus will be on Brownian motion as itself and as part of a compound state. Within the context of this book, the computation of a mean vector and a covariance matrix is of particular interest and will coin the term "track." It will be shown that the time evolution of an Îto stochastic process is a linear flow with respect to the mean and covariance matrix. The chapter will close with an excursion to the path integrals as typically used in quantum theory. Since there is a close relation between quantum theory and state estimation, other chapters will come back to this formulation.

Chapter 3 exposes the most important basic target tracking and state estimation algorithms (without considering distributed multi sensor application explicitly). Quite naturally, the Kalman filter will be the starting point. We will present three equivalent derivations of the Kalman filter, for two reasons: This provides a profound insight into the mechanism of the Kalman filter by providing complementary views on the same result and on the other hand many of the tools required to reach this objective will also be used in the derivations of track-to-track fusion methods later on. The strict assumptions of the Kalman filter will be relaxed in extensions on data association and track extraction for ambiguous sensor readings. Tracking expert readers may well omit this chapter and come back whenever necessary.

Those methods are extended to the estimation of full trajectories with states from multiple time stamps in Chapter 4. First, the Rauch–Tung–Striebel equations are derived, since these are well-known as "smoothing" or "retrodiction" for updating past states with current information. Then, the *Accumulated State Densities* are introduced. They provide the optimal means for trajectory estimation including cross-covariances of individual states. This fact will be used later on for track-to-track fusion methods and for online processing of out-of-sequence methods, which are a regular challenge in distributed multi sensor applications. The chapter ends with aspects on the duality between quantum theory and state estimation based on Accumulated State Densities.

Chapter 5 then introduces the problem of distributed fusion and presents first solutions. Since tracks from distributed sensors are often correlated, the cross-covariances cannot be ignored. This chapter reveals the nature of correlations in distributed state estimation and target tracking by providing a recursive computation. Optimal methods will be derived, where the full knowledge of the cross-covariances is assumed. This, however, as we will see is infeasible in most applications. Still, it is important to understand the approach of exact fusion with known cross-covariances in order to derive methods, which relax this harsh assumption. The chapter closes with a distributed fusion method, where the cross-covariances are reconstructed in the fusion center by means of a transmission of additional parameters.

Then, Chapter 6 will be on track fusion methods with unknown cross-covariances. Most of them are approximations to the statistically optimal fusion, but these are often easy to compute and perform well in practical applications. A simplistic approach is to ignore the cross-covariance matrices, that is, to assume that they are zero. This results in an optimistic fused error covariance matrix, which does not cover the actual expected error. A more precise fused error covariance can be obtained by approaches, which are based on geometrical results on the intersection of ellipses. This allows to derive the smallest error covariance matrix, which still covers the optimal fusion for all possible cross-covariances. The last section of the chapter is on a method, which uses joint diagonalization in order to fuse individual parameters of the state based on the maximum information among the sensors.

A whole chapter is dedicated to the distributed Kalman filter and its variants in Chapter 7. It is shown that the Kalman filter recursion with its prediction and filtering for multiple sensors can be fully parallelized and distributed in the case of the standard Kalman assumptions, which are linear Gaussian models for the sensor likelihoods and transition model. The distributed computation is achieved based on the notion of global information, that is, all sensors have knowledge on the full amount of information in the system without any exchange of data between sensors. Since these assumptions are often not given, the closely related federated Kalman filter is introduced. Then, it is shown that optimal fusion can also be achieved based on the Accumulated State Densities which are derived in Chapter 4. This requires transmissions of cross-covariance matrices of multiple states in time, but allows to compute the exact fusion result without the assumptions of the distributed Kalman filter.

In Chapter 8, methods for track-to-track association are addressed. While in the previous chapters, it had been assumed that all sensors observe the same target, this assumption might not be given in multi-target scenarios. If targets are closely spaced and a fusion center receives a set of tracks from each sensor, sophisticated information metrics have to be used in order to avoid miss-associations. Then, the focus will be on track detection and confirmation. The sequential likelihood ratio test provides the means to decide on track existence in a stochastically optimal way, if all measurements are available. In this chapter, it will be shown that the likelihood ratio score can well be reconstructed based on transmitted scalar values from each sensor in a distributed computation. The method is based on the information representation of the distributed Kalman Filter and has the same assumptions. However, we will show that this approach can be well extended to the general case of distributed filtering in ambiguous measurement scenarios, if the unnormalized weights from the data association step are available. While this leads in general to an approximation of the exact method, it works astonishing well in real-world applications. This is mainly due to the fact that the likelihood ratio may well be estimated in an approximative manner in order to provide the correct decision on track existence.

Chapter 2
Stochastic motion

The theory of stochastic modeling of dynamic systems is highly appreciated among heterogeneous scientific disciplines, which emphasizes the importance of a consistent mathematical handling of phenomena, which may be described as "random" in their nature – at least in our perspective as an observer. It is not unlikely that mankind will never be able to truly understand, if the underlying principles of nature are deterministic as in classical physics [12] or random as in quantum physics [13], and philosophical argumentation on the topic quite fast end up in basic questions on the existence of free will and predefined destiny [14,15]. However, it is evident that the theory of stochastic events is a diamond among the pyrites of scientific achievements, not at least because it provides the means to formalize what we *know* and what we *don't know* about the state of a system.

For the theory of distributed state estimation unfolded in this book, we are particularly interested in modeling *dynamic* systems, which evolve over time in the sense that their state may not be static but may experience some inherent random effects. Those random effects may well be the consequence of a causal process, but due to the complexity of the environment, it is impossible and infeasible to be aware of all related parameters for a deterministic prediction.* Likewise they may also be implicated by inherent random noise, for instance from chaotic systems or quantum effects on microscopic bodies.†

This chapter introduces the basic notion of random processes in terms of the required mathematical framework to code the knowledge on the state of such a system.‡ Moreover, stochastic motion also is affected or even dominated by deterministic processes as provided by Newton's laws of motion for example. Therefore, we will

*Examples for this are the number of cars coming up at a traffic light in a given time period, a pilot's maneuver in a fighter jet, or the evolution of the spreading of a contagious disease.

†This can be observed as thermal or electronic noise of an electrical conductor as a consequence of the thermal energy acting according to quantum mechanics. But also chaotic systems such rolling a dice or drawing the numbers of a lottery can be considered to be inherently random.

‡Actually, the mathematical theory has a more distinct notion of the "random variable itself" and its "realization." Throughout this book, we will refrain from this distinction, since the notational overhead is tremendous while its gain with respect to the considered methods is limited. Whenever necessary, we will provide additional explanations to compensate for the loss of technical precision. For instance, this may happen whenever we "observe" a random variable.

also look at the combination or “superposition” of deterministic and random processes [16]. Moreover, the derivation of crucial statistics like the mean and covariance for such processes will be important in the upcoming chapters and methods. For the interested reader, the last part of this chapter will inspect the relation between the random processes considered in data fusion and target tracking and quantum mechanics as considered in the “transformation theory” of Paul Dirac [17].

2.1 Stochastic processes

A stochastic process [16] is a collection of random variables $\{\mathbf{x}_t\}_{t\geq 0}$ or $\{\mathbf{x}_k\}_{k\in\mathbb{N}}$, which usually is indexed by a continuous time parameter t or discrete time index $k = 1, 2, \ldots$, respectively. It is important to note that, for instance, $\mathbf{x}_{t_1}$ and $\mathbf{x}_{t_2}$ for some given time variables t_1 and t_2 are two different random variables for all $t_1 \neq t_2$, though they may depend on each other or be *correlated*, in particular if $t_1 \approx t_2$. One may use stochastic processes to model the evolution of systems by gathering all parameters of interest into a “state” variable $\mathbf{x}_k$ or $\mathbf{x}_t$ and providing a conditional transition density $p(\mathbf{x}_k|\mathbf{x}_{k-1})$ for instance. This allows to derive a prediction of a future state $\mathbf{x}_k$ based on a current state $\mathbf{x}_{k-1}$ by marginalization:

$$p(\mathbf{x}_k) = \int \mathrm{d}\mathbf{x}_{k-1}\, p(\mathbf{x}_k|\mathbf{x}_{k-1})\, p(\mathbf{x}_{k-1}) \tag{2.1}$$

where the integral is over the total state space of $\mathbf{x}_{k-1}$. The integral either has to be solved analytically[§] or numerically.

This section considers stochastic processes, which yield Gaussian probability density functions. It turns out that these models are general enough to cover a wide variety of applications and can be handled well analytically.

2.1.1 Brownian motion and the diffusion equation

A simple but descriptive example is the discrete, one-dimensional *random walk* [18]. Consider a particle on the reel axis at $\mathbf{x}_k \in \mathbb{R}$. At each time step k, it jumps a distance of h either left or right with no preference, that is with probability $1/2$, independent of the previous trajectory. This process can be expressed in terms of a Markovian[‖] *transition kernel*:

$$p(\mathbf{x}_k|\mathbf{x}_{k-1}) = \begin{cases} \frac{1}{2}, & \text{if } |\mathbf{x}_k - \mathbf{x}_{k-1}| = h \\ 0, & \text{else.} \end{cases} \tag{2.2}$$

An example trajectory of a such a random process is given in Figure 2.1, where the x-axis represents the time steps and the y-axis the 1D state of the system. Since the transition probability is independent of the past history of states, it is called a *Markov Chain*.

[§] Actually, analytical solutions are only available for a low number of parametrized probability density families.

[‖] The Markov assumption states that the current transition is independent from the past trajectory [19].

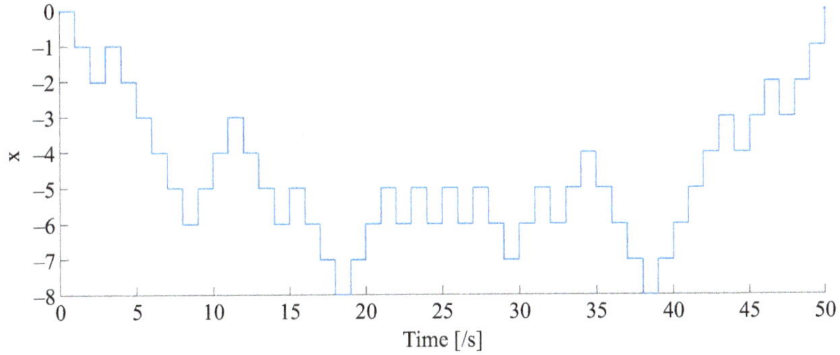

Figure 2.1 Example trajectory in 1D of a discrete Brownian motion

Due to our transition model in (2.2), the distribution at time t_{k+1} can be computed recursively as

$$p(\mathbf{x}_{k+1}) = \frac{1}{2}p(\mathbf{x}_k + h) + \frac{1}{2}p(\mathbf{x}_k - h). \tag{2.3}$$

Introducing the distribution of the state at time t_k yields

$$p(\mathbf{x}_{k+1}) - p(\mathbf{x}_k) = \frac{1}{2}\Big(p(\mathbf{x}_k + h) - 2p(\mathbf{x}_k) + p(\mathbf{x}_k - h)\Big). \tag{2.4}$$

By inserting the step size h and the discrete time increment $\tau = t_{k-1} - t_k$ into this equation, it can be rewritten as

$$\frac{p(\mathbf{x}_{k+1}) - p(\mathbf{x}_k)}{\tau} = \frac{h^2}{2\tau}\frac{p(\mathbf{x}_k + h) - 2p(\mathbf{x}_k) + p(\mathbf{x}_k - h)}{h^2} \tag{2.5}$$

By taking the limit $h \to 0$ and $\tau \to 0$ while keeping the *diffusion constant* $D = \frac{h^2}{2\tau}$ fixed, the following differential equation is obtained:

$$\partial_t p(\mathbf{x}_t) = D\partial_x^2 p(\mathbf{x}_t), \tag{2.6}$$

which is the well-known *diffusion equation*.

Taking the limit for h and τ to zero can also be seen as the transition from a discrete to a continuous stochastic process. It is well-known that the diffusion equation is solved by the fundamental solution for $D = \frac{1}{2}$, which is given by

$$f(\mathbf{x}_t) = \frac{1}{\sqrt{t}}e^{-\frac{1}{2}\frac{\mathbf{x}_t^2}{t}} \tag{2.7}$$

for $\mathbf{x}_0 = 0 \in \mathbb{R}$. This is easily verified by

$$\partial_t f = D\partial_x^2 f = \frac{1}{2}\left(\frac{\mathbf{x}_t^2}{t^2} - \frac{1}{t}\right)f. \tag{2.8}$$

The function f from above, however, is not a density function due to the lack of normalization. Using the fact that

$$\int \mathrm{d}x\, \mathrm{e}^{-x^2} = \sqrt{\pi}, \tag{2.9}$$

the correct normalization is achieved:

$$p(\mathbf{x}_t) = \frac{1}{\sqrt{2\pi t}} e^{-\frac{1}{2}\frac{\mathbf{x}_t^2}{t}} \tag{2.10}$$

$$= \mathcal{N}(\mathbf{x}_t;\, 0,\, t), \tag{2.11}$$

where $\mathcal{N}(x;\, \mu,\, \sigma^2)$ is the Gaussian normal density function in the random variable x with mean μ and variance σ^2.

This implies that the diffusion spreads the probability mass of a density function $p(\mathbf{x})$ spatially over the state space over time. A Gaussian example is shown in Figure 2.2 whereas some example realizations of a continuous Brownian motions are depicted in Figure 2.3. The transition from the discrete motion to the continuous limit results in a curve, which has some interesting properties:

Nowhere differentiable: By construction the curve is continuous, however, it is nowhere differentiable. This can be seen by means of the Borell–Cantelli lemma [20].

Self-similarity: In addition, it is possible to prove the self-similarity of the Brownian motion, that is $\mathbf{x}'_t := \mathbf{x}_{\lambda t}$ for an arbitrary λ is also a Brownian motion with the same properties.

We can now well generalize this result to a multivariate walk in $\mathbb{R}^n$, where the particle chooses randomly a direction from $\{\pm\mathbf{e}_i\}_{i=1,\ldots,n}$ with probability $\frac{1}{2n}$ and length h, where $\mathbf{e}_i$ is the ith unit vector. Example trajectories for 2D and 3D walks, respectively, are shown in Figure 2.3. The recursive formulation of the distribution of the position is then given by

$$p(\mathbf{x}_{k+1}) = \frac{1}{2n}\sum_{i=1}^{n}\left(p(\mathbf{x}_k + h\mathbf{e}_i) + p(\mathbf{x}_k - h\mathbf{e}_i)\right). \tag{2.12}$$

Analogously to the one dimensional case, this can be rewritten as

$$\frac{p(\mathbf{x}_{k+1}) - p(\mathbf{x}_k)}{\tau} = \frac{h^2}{2n\tau}\sum_{i=1}^{n}\frac{p(\mathbf{x}_k + h\mathbf{e}_i) - 2p(\mathbf{x}_k) + p(\mathbf{x}_k - h\mathbf{e}_i)}{h^2} \tag{2.13}$$

By means of the general diffusion constant $D = \frac{h^2}{2\tau n}$ and the Laplace operator $\Delta_x = \sum_{i=1}^{n}\partial_{x_i}^2$ we directly obtain the diffusion equation in $\mathbb{R}^n$:

$$\partial_t p(\mathbf{x}_k) = D\Delta_x p(\mathbf{x}_k). \tag{2.14}$$

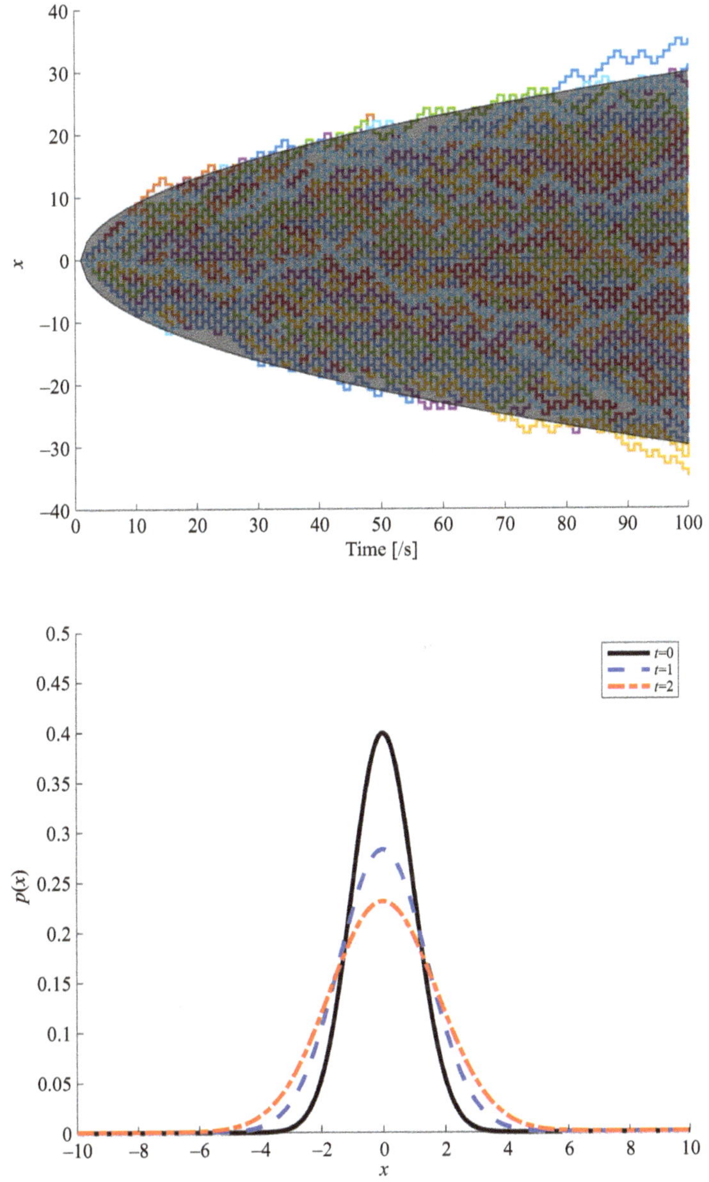

Figure 2.2 *The top figure shows different realizations of a discrete Brownian motion process each indicated by different colors. The gray-shaded area in the top diagram denotes the increasing variance. Bottom: example evolution of the probability density function of a diffusion process for three different time steps.*

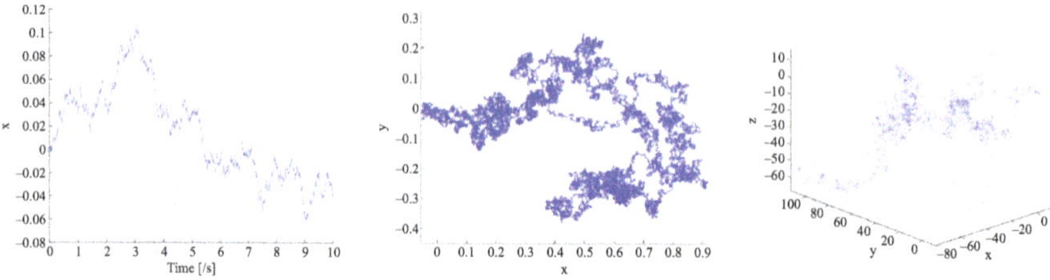

Figure 2.3 Example trajectory in 1D (left), 2D (center), and 3D (right) of a continuous Brownian motion

Until now, it was assumed that the system represented by the state does not prefer any of its possible directions. Going back to the 1D case, we may modify our model to a *Bernoulli* process, which assigns probabilities u and v such that $u + v = 1$ for the options of going up or down, respectively. Then, the recursive computation of the probability density becomes:

$$p(\mathbf{x}_{k+1}) = up(\mathbf{x}_k + h) + vp(\mathbf{x}_k - h). \tag{2.15}$$

It should be noted that this is equivalent to

$$p(\mathbf{x}_{k+1}|\mathbf{x}_k) = \begin{cases} u, & \text{if } \mathbf{x}_{k+1} = \mathbf{x}_k + h \\ v, & \text{if } \mathbf{x}_{k+1} = \mathbf{x}_k - h \end{cases} \tag{2.16}$$

or in terms of a realization

$$\mathbf{x}_{k+1} = \begin{cases} \mathbf{x}_k + h & \text{with probability } u \\ \mathbf{x}_k - h & \text{with probability } v. \end{cases} \tag{2.17}$$

Using the fact that $v = 1 - u$, (2.15) can be rewritten as

$$p(\mathbf{x}_{k+1}) = \left(u - \frac{1}{2}\right) p(\mathbf{x}_k + h) + \frac{1}{2} p(\mathbf{x}_k + h) - \left(u - \frac{1}{2}\right) p(\mathbf{x}_k - h) + \frac{1}{2} p(\mathbf{x}_k - h) \tag{2.18}$$

Introducing $p(\mathbf{x}_k)$, τ, and h yields as above

$$\begin{aligned} \frac{p(\mathbf{x}_{k+1}) - p(\mathbf{x}_k)}{\tau} &= \frac{h^2}{2\tau} \frac{p(\mathbf{x}_k + h) - 2p(\mathbf{x}_k) + p(\mathbf{x}_k - h)}{h^2} \\ &\quad + \frac{u - \frac{1}{2}}{\tau} p(\mathbf{x}_k + h) - \frac{u - \frac{1}{2}}{\tau} p(\mathbf{x}_k - h) \end{aligned} \tag{2.19}$$

$$\begin{aligned} &= \frac{h^2}{2\tau} \frac{p(\mathbf{x}_k + h) - 2p(\mathbf{x}_k) + p(\mathbf{x}_k - h)}{h^2} \\ &\quad - \frac{2h(\frac{1}{2} - u)}{\tau} \frac{p(\mathbf{x}_k + h) - p(\mathbf{x}_k - h)}{2h} \end{aligned} \tag{2.20}$$

For the transition to the continuous case obtained by taking the limit $h \to 0$ and $\tau \to 0$ while keeping the diffusion constant $D = \frac{h^2}{2\tau}$ fixed, we also fix the *mean drift velocity* $\nu = \frac{2h}{\tau}(\frac{1}{2} - u)$.

As a result, we obtain the *diffusion equation with static mean drift* in one dimension:

$$\partial_t p(\mathbf{x}_t) = \left(D\partial_x^2 - \nu\partial_x\right) p(\mathbf{x}_t). \tag{2.21}$$

As an example, the evolution of diffusion processes with a non-zero drift is shown in Figure 2.4. It can be seen that the drift causes the whole probability mass to be shifted by the process whereas the diffusion increases the variance linearly over time.

In the next step, the above term "mean drift velocity" is motivated. This can easily be seen by considering the mean of the process at time t_k and using the formulation of a realization in (2.17):

$$\mathrm{E}\left[\mathbf{x}_k\right] = u\,\mathrm{E}\left[\mathbf{x}_{k-1} - h\right] + (1-u)\mathrm{E}\left[\mathbf{x}_{k-1} + h\right] \tag{2.22}$$

Due to the linearity of the expectation functional, one directly obtains

$$\begin{aligned}\mathrm{E}\left[\mathbf{x}_k\right] &= u\left(\mathrm{E}\left[\mathbf{x}_{k-1}\right] - h\right) + (1-u)\left(\mathrm{E}\left[\mathbf{x}_{k-1}\right] + h\right) && (2.23)\\ &= \mathrm{E}\left[\mathbf{x}_{k-1}\right] - uh + (1-u)h && (2.24)\\ &= \mathrm{E}\left[\mathbf{x}_{k-1}\right] + 2h\left(\frac{1}{2} - u\right). && (2.25)\end{aligned}$$

Therefore, the increment in the mean for a single time step of duration τ is $2h\left(\frac{1}{2} - u\right)$, therefore, $\nu = \frac{2h}{\tau}\left(\frac{1}{2} - u\right)$ is the mean velocity. In particular, one can see that

$$\mathrm{E}\left[\mathbf{x}_k\right] = k\tau\nu\mathrm{E}\left[\mathbf{x}_0\right]. \tag{2.26}$$

The canonical next step is the generalization to arbitrary dimensions. Therefore, we again consider the multivariate walk from (2.12) in $\mathbb{R}^n$. However, the direction now is not uniformly distributed over all $2n$ options, but probabilities $u_1, \ldots, u_n$ for each dimension are given with $u_i \in (0, 1)$ such that

$$p(\mathbf{x}_{k+1}) = \frac{1}{n}\left(\sum_{i=1}^{n} u_i\, p(\mathbf{x}_k + h\mathbf{e}_i) + (1-u_i)p(\mathbf{x}_k - h\mathbf{e}_i)\right). \tag{2.27}$$

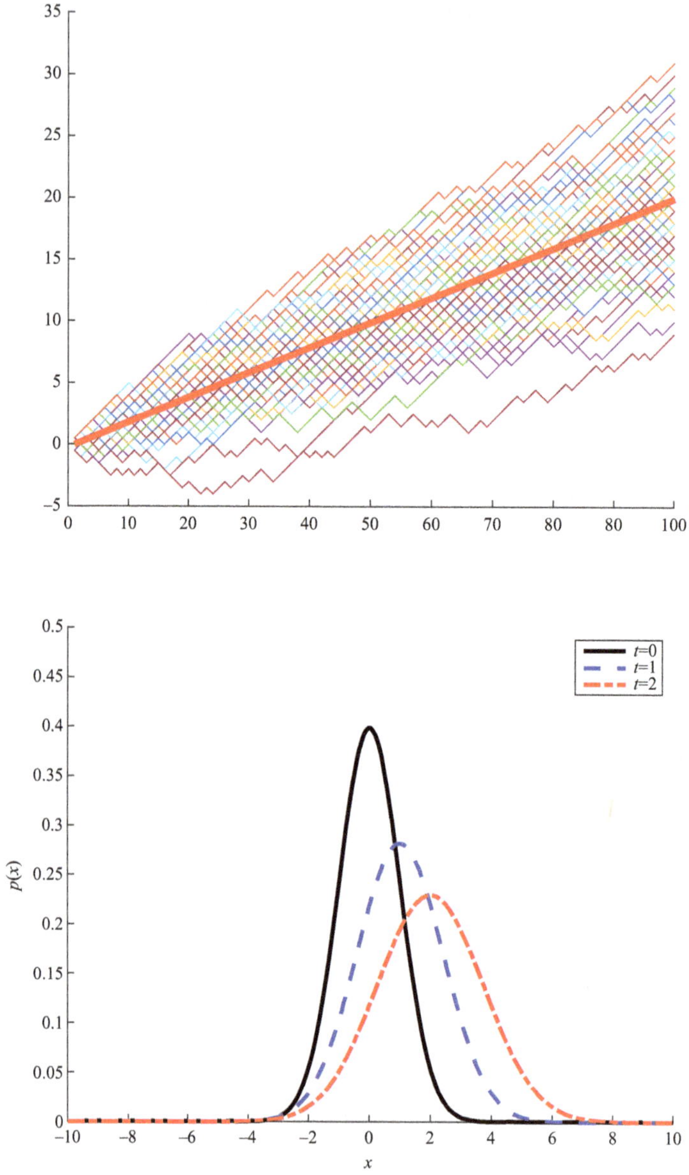

Figure 2.4 *Top: multiple realizations of the Bernoulli model with* $h = 1/2$, $\tau = 1$, *and* $u = 0.3$. *This results in a mean drift of* $v = \frac{2h}{\tau}(\frac{1}{2} - u) = 0.2$, *which is shown as the slope of the red line. Bottom: example evolution of the probability density function of a diffusion with drift* $v = 1$ *for three different time steps.*

For the transition to the continuous formulation, we observe that

$$\begin{aligned}
p(\mathbf{x}_{k+1}) &= \frac{1}{n}\left(\sum_{i=1}^{n} u_i\, p(\mathbf{x}_k + h\mathbf{e}_i) - u_i\, p(\mathbf{x}_k - h\mathbf{e}_i) + u_i\, p(\mathbf{x}_k - h\mathbf{e}_i) + (1-u_i)p(\mathbf{x}_k - h\mathbf{e}_i)\right)\\
&= \frac{1}{n}\left(\sum_{i=1}^{n} u_i\, p(\mathbf{x}_k + h\mathbf{e}_i) - u_i\, p(\mathbf{x}_k - h\mathbf{e}_i) + p(\mathbf{x}_k - h\mathbf{e}_i)\right)\\
&= \frac{1}{n}\left(\sum_{i=1}^{n} u_i\, p(\mathbf{x}_k + h\mathbf{e}_i) + \frac{1}{2}p(\mathbf{x}_k + h\mathbf{e}_i) - \frac{1}{2}p(\mathbf{x}_k + h\mathbf{e}_i) - u_i\, p(\mathbf{x}_k - h\mathbf{e}_i)\right.\\
&\quad \left. + \frac{1}{2}p(\mathbf{x}_k - h\mathbf{e}_i) + \frac{1}{2}p(\mathbf{x}_k - h\mathbf{e}_i)\right)\\
&= \frac{1}{n}\left(\sum_{i=1}^{n}\left(u_i - \frac{1}{2}\right) p(\mathbf{x}_k + h\mathbf{e}_i) + \frac{1}{2}p(\mathbf{x}_k + h\mathbf{e}_i)\right.\\
&\quad \left. - \left(u_i - \frac{1}{2}\right)p(\mathbf{x}_k - h\mathbf{e}_i) + \frac{1}{2}p(\mathbf{x}_k - h\mathbf{e}_i)\right)\\
&= \frac{1}{2n}\left(\sum_{i=1}^{n} p(\mathbf{x}_k + h\mathbf{e}_i) + p(\mathbf{x}_k - h\mathbf{e}_i)\right)\\
&\quad - \frac{1}{n}\left(\sum_{i=1}^{n}\left(\frac{1}{2} - u_i\right)[p(\mathbf{x}_k + h\mathbf{e}_i) - p(\mathbf{x}_k - h\mathbf{e}_i)]\right)
\end{aligned} \tag{2.28}$$

Inserting the terms $p(\mathbf{x}_k)$, τ, and h as above yields

$$\begin{aligned}
\frac{p(\mathbf{x}_{k+1}) - p(\mathbf{x}_k)}{\tau} &= \frac{h^2}{2n\tau}\sum_{i=1}^{n}\frac{p(\mathbf{x}_k + h\mathbf{e}_i) - 2p(\mathbf{x}_k) + p(\mathbf{x}_k - h\mathbf{e}_i)}{h^2}\\
&\quad - \frac{2h}{n\tau}\sum_{i=1}^{n}\left(\frac{1}{2} - u_i\right)\frac{p(\mathbf{x}_k + h\mathbf{e}_i) - p(\mathbf{x}_k - h\mathbf{e}_i)}{2h}
\end{aligned} \tag{2.29}$$

Now it becomes obvious that we have obtained the finite difference formulation for partial derivatives.

> Taking the limit as above yields the continuous case:
>
> $$\partial_t p(\mathbf{x}_t) = \left(-\sum_{i=1}^{n} \nu_i \partial_{x_i} + D\Delta_x\right) p(\mathbf{x}_t), \tag{2.30}$$
>
> where $\nu_1, \ldots, \nu_n$ are the components of the mean velocity vector $\boldsymbol{\nu}$ with $\nu_i = \frac{2h}{n\tau}\left(\frac{1}{2} - u_i\right)$.

It will be shown later on that the diffusion part of the stochastic time evolution

$$D\Delta_x = D\sum_{i=1}^{n} \partial_{x_i}\partial_{x_i} \tag{2.31}$$

can be generalized such that the individual dimensions become correlated:

$$D\Delta_x \longrightarrow \sum_{i,j=1}^{n} b_{i,j}\partial_{x_i}\partial_{x_j}, \tag{2.32}$$

where $\mathbf{B} = (b_{i,j})$ is a square matrix which contains all diffusion coefficients.

In the next section, the constructed stochastic process is made time dependent and combined with deterministic parts.

2.2 Wiener process

The stochastic motion from the previous section is a general approach to model uncertain and noisy behaviour. Next, the Wiener process is introduced as a multivariate Brownian motion with defined initial state $\mathbf{x}_0 = 0$. To this end, some properties are shown first.

Consider first the 1D Brownian motion with

$$p(\mathbf{x}_k|\mathbf{x}_{k-1}) = \begin{cases} \frac{1}{2}, & \text{if } |\mathbf{x}_k - \mathbf{x}_{k-1}| = h \\ 0, & \text{else.} \end{cases} \tag{2.33}$$

To show properties of the Brownian motion, we normalize the scale such that $D = \frac{h^2}{2\tau} = 1/2$, and therefore $h^2 = \tau$. First of all, one directly can see that

$$\mathrm{E}[\mathbf{x}_k] = 1/2\mathrm{E}[\mathbf{x}_{k-1} + h] + 1/2\mathrm{E}[\mathbf{x}_{k-1} - h] \tag{2.34}$$

$$= \mathrm{E}[\mathbf{x}_{k-1}] = \mathrm{E}[\mathbf{x}_0]. \tag{2.35}$$

The second moment is given by

$$\mathrm{var}[\mathbf{x}_k] = \mathrm{E}\left[\mathbf{x}_k^2\right] - \mathrm{E}[\mathbf{x}_k]^2 \tag{2.36}$$

$$= 1/2\mathrm{E}\left[(\mathbf{x}_{k-1} + h)^2\right] + 1/2\mathrm{E}\left[(\mathbf{x}_{k-1} - h)^2\right] - \mathrm{E}[\mathbf{x}_{k-1}]^2 \tag{2.37}$$

$$= \mathrm{E}\left[\mathbf{x}_{k-1}^2\right] + h\mathrm{E}[\mathbf{x}_{k-1}] - h\mathrm{E}[\mathbf{x}_{k-1}] + h^2 - \mathrm{E}[\mathbf{x}_{k-1}]^2 \tag{2.38}$$

$$= \mathrm{var}[\mathbf{x}_{k-1}] + \tau, \tag{2.39}$$

which confirms the normalized fundamental solution given in (2.11). This motivates the conception of the stochastic process $\mathbf{x}_k$ as the sum of its increments [21]:

$$\mathbf{x}_k = \mathbf{x}_0 + \sum_{i=0}^{k-1} \mathrm{d}\mathbf{x}_i, \tag{2.40}$$

where $\mathrm{d}\mathbf{x}_i$ is the ith increment $\mathbf{x}_{i+1} - \mathbf{x}_i$. The increments as a set of random variables are i.i.d.¶ and zero-mean with variance τ. Because of the *Central Limit Theorem* (CLT) in probability theory [22], it holds that

$$z = \sum_{i=0}^{k-1} \frac{\mathrm{d}\mathbf{x}_i}{\sqrt{\tau k}} \tag{2.41}$$

is a random variable whose distribution converges to the standard normal distribution $\mathcal{N}(0, 1)$. In particular, it holds that the distribution of the rescaled process $\sqrt{\tau k} \cdot z$ converges to $\mathcal{N}(0, t_k)$ since $\tau k = t_k$.

When taking the limit $\tau \downarrow 0$ and $k \uparrow \infty$, the increments $\mathrm{d}\mathbf{x}_i$ of the Brownian motion constitute the *Wiener Process* $\mathbf{w}_k$

$$\mathbf{w}_0 := 0, \tag{2.42}$$

$$\mathbf{w}_t := \lim_{\substack{\tau \downarrow 0 \\ k \uparrow \infty \\ \tau k = t}} \frac{1}{\sqrt{k}} \sum_{i=0}^{k-1} \mathrm{d}\mathbf{x}_i. \tag{2.43}$$

Considering the continuous Wiener process $\mathbf{w}_t$ at discrete instants of time t_k, $k = 0, 1, \ldots$, one obtains increments $\mathrm{d}\mathbf{w}_i := \mathbf{w}_{i+1} - \mathbf{w}_i$ which have the following properties:

- Zero-mean: The mean $\mathrm{E}[\mathrm{d}\mathbf{w}_i]$ of some given time interval $t_{i+1} - t_i$ is zero.
- Linear increasing variance: The variance equals the length of its time interval $\Delta_i = t_{i+1} - t_i$.
- Normally distributed: The distribution of $\mathrm{d}\mathbf{w}_i$ is given by $\mathcal{N}(0, t_i)$.
- Time shift invariance: For any $l > 0$, the time shifted process $\mathbf{y}_k = \mathbf{w}_{t_{l+k}} - \mathbf{w}_{t_l}$ is also a Wiener process itself.
- Independent increments: The increment $\mathrm{d}\mathbf{w}_i$ is independent of the history $\mathbf{w}_{t_l}$ for $t_l < t_i$.
- Covariance for overlapping increments: If the time interval of two increments $\mathrm{d}\mathbf{w}_i$ and $\mathrm{d}\mathbf{w}_j$ are overlapping by some interval of length ι, then the covariance is given by $\mathrm{var}\left[\mathrm{d}\mathbf{w}_i, \mathrm{d}\mathbf{w}_j\right] = \iota$.

The continuous Wiener process can be generalized to $\mathbf{w}_t \in \mathbb{R}^n$ such that the increments are distributed according to the multivariate Gaussian $\mathcal{N}(\mathrm{d}\mathbf{w}_i; 0, \Delta_i\mathbf{I})$,

¶Independent and identically distributed.

where $\mathbf{I}$ is the identity matrix. This can be seen by considering the cross-covariance before taking the limit $h \downarrow 0$:

$$\begin{aligned}
\operatorname{var}\left[\mathrm{d}\mathbf{w}_k^{(i)}, \mathrm{d}\mathbf{w}_k^{(j)}\right] &= \mathrm{E}\left[\left(\mathrm{d}\mathbf{w}_k^{(i)} - \mathrm{E}\left[\mathrm{d}\mathbf{w}_k^{(i)}\right]\right)\left(\mathrm{d}\mathbf{w}_k^{(j)} - \mathrm{E}\left[\mathrm{d}\mathbf{w}_k^{(j)}\right]\right)\right] \\
&= \mathrm{E}\left[\left(\mathrm{d}\mathbf{w}_k^{(i)} - \mathrm{E}\left[\mathrm{d}\mathbf{w}_{k-1}^{(i)}\right]\right)\left(\mathrm{d}\mathbf{w}_k^{(j)} - \mathrm{E}\left[\mathrm{d}\mathbf{w}_{k-1}^{(j)}\right]\right)\right] \\
&= \frac{n-1}{n}\mathrm{E}\left[\left(\mathrm{d}\mathbf{w}_{k-1}^{(i)} - \mathrm{E}\left[\mathrm{d}\mathbf{w}_{k-1}^{(i)}\right]\right)\left(\mathrm{d}\mathbf{w}_k^{(j)} - \mathrm{E}\left[\mathrm{d}\mathbf{w}_{k-1}^{(j)}\right]\right)\right] \\
&\quad + \frac{1}{n}\mathrm{E}\left[\left(\mathrm{d}\mathbf{w}_{k-1}^{(i)} \pm h - \mathrm{E}\left[\mathrm{d}\mathbf{w}_{k-1}^{(i)}\right]\right)\left(\mathrm{d}\mathbf{w}_k^{(j)} - \mathrm{E}\left[\mathrm{d}\mathbf{w}_{k-1}^{(j)}\right]\right)\right]
\end{aligned} \tag{2.44}$$

Since in the construction of the multivariate Brownian motion in Section 2.1.1 allowed the choice of only one direction out of $2n$ possibilities, we only enumerate the left (i)-term. We now have two cases: if neither i nor j change the direction, then $\mathrm{d}\mathbf{w}_k^{(l)} = \mathrm{d}\mathbf{w}_{k-1}^{(l)}$ for both $l = i,j$ and therefore $\operatorname{var}\left[\mathrm{d}\mathbf{w}_k^{(i)}, \mathrm{d}\mathbf{w}_k^{(j)}\right] = \operatorname{var}\left[\mathrm{d}\mathbf{w}_{k-1}^{(i)}, \mathrm{d}\mathbf{w}_{k-1}^{(j)}\right] = 0$. If otherwise, w.l.o.g. we may assume the i changes the direction. Then, if $i \neq j$ the cross-terms $h\mathrm{E}\left[\left(\mathrm{d}\mathbf{w}_{k-1}^{(i)} - \mathrm{E}\left[\mathrm{d}\mathbf{w}_{k-1}^{(i)}\right]\right)\right]$ vanish, since the total process is zero-mean. If $i = j$, we obtain $\operatorname{var}\left[\mathrm{d}\mathbf{w}_{k-1}^{(i)}\right] = h^2/n$, which equals τ with the proper choice** of the diffusion constant $D = 1/2$. Together with the arguments from above, we can see that the multivariate Wiener process is distributed according to

$$\mathbf{w}_t \sim \mathcal{N}(\mathbf{w}_t;\, 0,\, t\mathbf{I}). \tag{2.45}$$

Eventually, it should be noted that (2.45) solves the partial differential equation of the diffusion equation with given boundary condition $p(\mathbf{w}_0) = \delta_0$, where δ_0 is the Kronecker–Delta function,

$$\delta_y(x) = \begin{cases} \infty, & x = y \\ 0, & \text{else} \end{cases} \tag{2.46}$$

with

$$\int \mathrm{d}\mathbf{x}\; \delta_y(\mathbf{x}) = 1. \tag{2.47}$$

Therefore $p(\mathbf{w}_0) = \delta_0$ represents the sure knowledge $\mathbf{w}_0 = 0$. Moreover, it holds that the solution of a partial differential equation with initial values is unique, despite the stochastic nature of the process it describes.

**Alternatively, the multivariate Wiener process can also be defined via n independent univariate Wiener processes.

2.3 Stochastic differential equation

By construction, the Wiener process is comprised of infinitesimal increments $\mathbf{dw}_t$, which can be interpreted as the limit of letting the time interval go to zero and the number of increments to infinity such that $k\tau = t$:

$$\mathbf{w}_t = \lim_{\substack{\Delta_t \downarrow 0 \\ k \uparrow \infty \\ k\tau = t}} \sum_{i=0}^{k-1} \mathrm{d}\mathbf{w}_i. \tag{2.48}$$

We define the *Itô integral* as this limit value [23], where the following notation is used:

$$\mathbf{w}_k = \int_{t_0}^{t_k} \mathrm{d}\mathbf{w}_t. \tag{2.49}$$

Though the construction at this point may seem straight forward, a mathematical machinery in the background is necessary in order to ensure that this notation is well-defined in the sense that it exists and is independent of the choice of the time intervals. We will make the distinction between a Riemann and an Îto integral implicit, that is, whenever a stochastic variable is integrated, it refers to the Itô formulation and for deterministic variables the standard Riemann integral is used. We will make use of this distinction for an important class of evolution models, which are a concurrent superposition of a deterministic and a stochastic part given by the integral formulation

$$\mathbf{x}_k = \mathbf{x}_0 + \int_{t_0}^{t_k} \mathrm{d}t\, \mathbf{f}(\mathbf{x}_t) + \int_{t_0}^{t_k} \mathrm{d}\mathbf{w}_t\, \mathbf{B}(\mathbf{x}_t). \tag{2.50}$$

Here, $\mathbf{f} : \mathbb{R}^n \rightarrow \mathbb{R}^n$ is a differentiable function, which models the deterministic part of the time evolution. The stochastic part is the second integral, which is driven by the increments $\mathrm{d}\mathbf{w}_t$ of a Wiener process in $\mathbb{R}^m$. Its effect on the state in $\mathbb{R}^n$ is described by the differentiable mapping $\mathbf{B} : \mathbb{R}^n \rightarrow \mathbb{R}^{n\times m}$. Since the stochastic noise affects the state and the deterministic evolution is state dependent, we may not integrate both elements mutually independent as indicated in (2.50). However, the equation becomes mathematically sound, if we consider the infinitesimal increments [24]. This is encoded in the following notation:

$$\mathrm{d}\mathbf{x}_k = \mathbf{f}(\mathbf{x}_k)\,\mathrm{d}t + \mathbf{B}(\mathbf{x}_k)\,\mathrm{d}\mathbf{w}_k. \tag{2.51}$$

This is called a *Stochastic Difference Equation* (SDE) of an Itô process.

Example 1:
Let us consider the state space $\mathbf{x} = (x, y, \dot{x}, \dot{y})^\top$. It contains the coordinates and the corresponding velocities in two dimensions x and y, respectively. The motion described by $\mathbf{f}(\mathbf{x}) = \mathbf{A}\mathbf{x} = (\dot{x}, \dot{y}, 0, 0)^\top$ with

$$\mathbf{A} = \begin{pmatrix} 0 & 0 & 1 & 0 \\ 0 & 0 & 0 & 1 \\ 0 & 0 & 0 & 0 \\ 0 & 0 & 0 & 0 \end{pmatrix} \tag{2.52}$$

clearly corresponds to a straight motion according to Newton's laws with additional white noise. ■

Example 2:
Based on the same state space as in Example 1, we may also describe a curve-linear motion by the function $\mathbf{f}(\mathbf{x}) = \mathbf{A}\mathbf{x}$ with

$$\mathbf{A} = \begin{pmatrix} 0 & 0 & 1 & 0 \\ 0 & 0 & 0 & 1 \\ 0 & 0 & 0 & -\omega \\ 0 & 0 & \omega & 0 \end{pmatrix} \tag{2.53}$$

for some angular speed $\omega \in \left(-\frac{\pi}{2}, \frac{\pi}{2}\right)$. Sample trajectories for the circular motion are shown in Figure 2.5. ■

The SDE in (2.51) is a generalization of the diffusion equation with a fixed mean velocity $\boldsymbol{\nu} = [\nu_i]_i$ in (2.30).[††] Considering only the deterministic part of the dynamics model directly yields

$$\frac{\mathrm{d}\mathbf{x}_k}{\mathrm{d}t} = \mathbf{f}(\mathbf{x}_k). \tag{2.54}$$

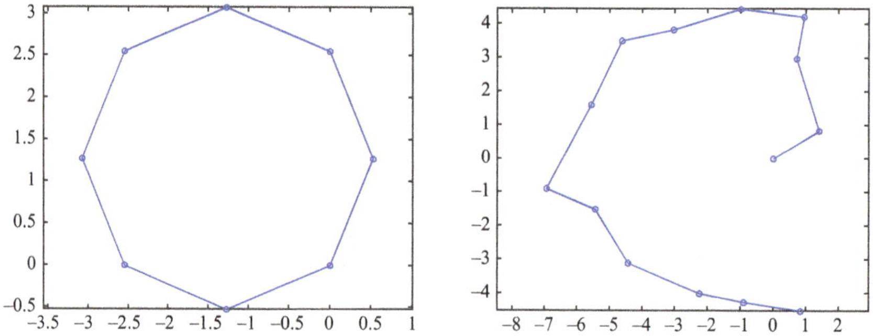

Figure 2.5 Circular motion trajectory for $\omega = \pi/4$ with (right) and without (left) additional white noise

[††]We use this notation to denote the vector comprised of all elements ν_i, $i = 1, \ldots, n$.

Thus, the mapping $\mathbf{f}$ is a state-dependent temporal derivative of the zero-noise process, in other words the mean velocity components at time t_k are given by the components $v_i = [\mathbf{f}(\mathbf{x}_k)]_i$, since $\mathrm{E}[\mathbf{B}\,\mathrm{d}\mathbf{w}_k] = \mathbf{B}\,\mathrm{E}[\mathrm{d}\mathbf{w}_k] = 0$.

On the other hand, considering the stochastic part only we can see that for $\mathbf{f} = 0$, that $\mathrm{d}\mathbf{x}_k = \mathbf{B}(\mathbf{x}_k)\mathrm{d}\mathbf{w}_k$, which describes a multivariate Wiener process in $\mathbb{R}^m$ acting on the state in $\mathbb{R}^n$ as the state-dependent mapping $\mathbf{B}$. Thus, the increment $\mathrm{d}\mathbf{x}_k^{(j)}$ in dimension $j \in \{1, \ldots, n\}$ would be an infinitesimal step of size $h \sum_{i=1}^{n} [\mathbf{B}(\mathbf{x}_k)]_{i,j}$. As a vivid example, one might think of a univariate process with a static function $\mathbf{B} = B \in \mathbb{R}$. This leads to a diffusion constant $D_B = \frac{B^2 h^2}{2\tau}$. In the multivariate case, the diffusion coefficients are given by $\mathbf{B}\mathbf{B}^\top$ [25]. As a consequence, the probability density of an Îto process follows the PDE given by:

$$\partial_t p(\mathbf{x}_k) = \left(-\sum_{i=1}^{n} \partial_{x_i} [\mathbf{f}(\mathbf{x}_k)]_i + \frac{1}{2} \sum_{i,j=1}^{n} \partial_{x_i} \partial_{x_j} [\mathbf{B}(\mathbf{x}_k)\mathbf{B}(\mathbf{x}_k)^\top]_{ij} \right) p(\mathbf{x}_k). \quad (2.55)$$

Equation (2.55) is known as the *Fokker–Planck Equation* (FPE).

In the following section, we consider the linear SDE with time independent drift and diffusion coefficients:

$$\mathrm{d}\mathbf{x}_k = \mathbf{A}\mathbf{x}_k\,\mathrm{d}t + \mathbf{B}\,\mathrm{d}\mathbf{w}_k, \quad (2.56)$$

that is, the drift function $\mathbf{f}(\mathbf{x}) = \mathbf{A}\mathbf{x}$ is linear and time independent and the diffusion coefficient matrix $\mathbf{B}(\mathbf{x}) = \mathbf{B}$ is static. If the sensor information is a time series of data at discrete instants of time $\{t_k\}_{k=0,1,\ldots}$, it is convenient to consider also the discrete version $\{\mathbf{x}_k\}_{k=0,1,\ldots}$ of the Îto process. Let us focus on the deterministic and stochastic part separately as we did above. The first part obviously is governed by the ordinary PDE

$$\frac{\mathrm{d}}{\mathrm{d}t}\mathbf{x}_t = \mathbf{A}\mathbf{x}_t, \quad (2.57)$$

which has the unique solution

$$\mathbf{x}_k = \exp\{\Delta_{k|0}\mathbf{A}\}\mathbf{x}_0, \quad (2.58)$$

where $\Delta_{k|0}$ is the time difference $t_k - t_0$ and the matrix exponential is used with

$$\exp\{\Delta_{k|0}\mathbf{A}\} = \sum_{i=0}^{\infty} (\Delta_{k|0}\mathbf{A})^i, \quad (2.59)$$

$$=: \mathbf{F}_{k|0} \quad (2.60)$$

with $\mathbf{A}^0 := \mathbf{I}$. It is worth noting that the transition matrix $\mathbf{F}_{\cdot|\cdot}$ inherits the semigroup property from the underlying process:

$$\mathbf{F}_{k|k} = \mathbf{I}, \quad (2.61)$$

$$\mathbf{F}_{k|k-2} = \mathbf{F}_{k|k-1}\mathbf{F}_{k-1|k-2}. \quad (2.62)$$

For the stochastic part, we define the accumulated "noise" acting in the state space by

$$\mathbf{w}_{k+1|k} = \int_{t_k}^{t_{k+1}} d\mathbf{w}_t \, \mathbf{F}_{k+1|t}\mathbf{B}, \tag{2.63}$$

where the transition matrix $\mathbf{F}_{k+1|t}$ in the integral is required to map the noise at time t to t_{k+1} and the stochastic integral is defined as

$$\int_{t_k}^{t_{k+1}} \mathbf{F}_{k+1|t}\mathbf{B}d\mathbf{w}_t = \lim_{m\to\infty} \sum_{j=1}^{m} \mathbf{F}_{k+1|\tau^j}\mathbf{B}(\mathbf{w}_{\tau^j} - \mathbf{w}_{\tau^{j-1}}). \tag{2.64}$$

The intermediate time instants τ^j are an equidistant partition of the interval $[k, k+1]$ of length $\frac{1}{m}$. The time discrete noise vectors $\mathbf{w}_{k+1|k}$ themselves are a Gaussian distributed stochastic process. The mean and covariance can be computed by using the definition of the stochastic integral. For the expectation vector one obtains

$$\mathrm{E}\left[\mathbf{w}_{k+1|k}\right] = \mathrm{E}\left[\int_{t_k}^{t_{k+1}} \mathbf{F}_{k+1|t}\mathbf{B}d\mathbf{w}_t\right] \tag{2.65}$$

$$= \mathrm{E}\left[\lim_{m\to\infty} \sum_{j=1}^{m} \mathbf{F}_{k+1|\tau^j}\mathbf{B}(\mathbf{w}_{\tau^j} - \mathbf{w}_{\tau^{j-1}})\right] \tag{2.66}$$

$$= \lim_{m\to\infty} \sum_{j=1}^{m} \mathbf{F}_{k+1|\tau^j}\mathbf{B}\mathrm{E}\left[(\mathbf{w}_{\tau^j} - \mathbf{w}_{\tau^{j-1}})\right] \tag{2.67}$$

$$= \mathbf{0} \tag{2.68}$$

since the expectation of a Wiener increment is zero. For the covariance, we have

$$\mathrm{E}\left[\mathbf{w}_{k+1|k}\mathbf{w}_{k+1|k}^\top\right] = \mathrm{E}\left[\left(\int_{t_k}^{t_{k+1}} \mathbf{F}_{k+1|t}\mathbf{B}d\mathbf{w}_t\right)\left(\int_{t_k}^{t_{k+1}} \mathbf{F}_{k+1|t}\mathbf{B}d\mathbf{w}_t\right)^\top\right] \tag{2.69}$$

$$= \lim_{m\to\infty} \sum_{j=1}^{m} \mathbf{F}_{k+1|\tau^j}\mathbf{B}\,\mathrm{E}\left[(\mathbf{w}_{\tau^j} - \mathbf{w}_{\tau^{j-1}})(\mathbf{w}_{\tau^j} - \mathbf{w}_{\tau^{j-1}})^\top\right]\mathbf{B}^\top\mathbf{F}_{k+1|\tau^j}^\top \tag{2.70}$$

$$= \lim_{m\to\infty} \sum_{j=1}^{m} \mathbf{F}_{k+1|\tau^j}\mathbf{B}\mathbf{B}^\top\mathbf{F}_{k+1|\tau^j}^\top \cdot \frac{1}{m} \tag{2.71}$$

$$= \int_{t_k}^{t_{k+1}} d\tau \, \mathbf{F}_{l+1|\tau}\mathbf{B}\mathbf{B}^\top\mathbf{F}_{k+1|\tau}^\top =: \mathbf{Q}_{k+1|k} \tag{2.72}$$

Together, one obtains a recursive formulation of the time discrete linear Îto process $\{\mathbf{x}_k\}_{k=0,1,...}$ given by

$$\begin{aligned}
\mathbf{x}_{k+1} &= \mathbf{F}_{k+1|k}\mathbf{x}_k + \mathbf{w}_{k+1|k},\\
\mathbf{F}_{k+1|k} &= \exp\{(t_{k+1} - t_k)\mathbf{A}\}\\
\mathbf{w}_{k+1|k} &\sim \mathcal{N}(\mathbf{w}_{k+1|k};\, 0,\, \mathbf{Q}_{k+1|k})\\
\mathbf{Q}_{k+1|k} &= \int_{t_k}^{t_{k+1}} \mathrm{d}\tau\ \mathbf{F}_{l+1|\tau}\mathbf{B}\mathbf{B}^\top\mathbf{F}_{k+1|\tau}^\top.
\end{aligned}$$

Due to the fact that the Wiener noise process has no memory, that is, the current evolution is independent of the past, this also holds for the Îto process. In particular, one obtains by marginalization:

$$p(\mathbf{x}_{k+1}) = \int \mathrm{d}\mathbf{x}_k\ p(\mathbf{x}_{k+1}|\mathbf{x}_k)p(\mathbf{x}_k). \tag{2.73}$$

This can be interpreted as the summation over all possible starting points $\mathbf{x}_k$, which are weighted by a transition kernel given by

$$p(\mathbf{x}_{k+1}|\mathbf{x}_k) = \mathcal{N}(\mathbf{x}_{k+1};\, \mathbf{F}_{k+1|k}\mathbf{x}_k,\, \mathbf{Q}_{k+1|k}). \tag{2.74}$$

Therefore, points far away from the predicted position $\mathbf{F}_{k+1|k}\mathbf{x}_k$ when assuming $\mathbf{x}_k$ as the starting point are punished by exponentially low weights.

2.3.1 Evolution of the mean and covariance

Let us denote

$$\hat{\mathbf{x}}_k := \mathrm{E}[\mathbf{x}_k], \tag{2.75}$$

$$\hat{\mathbf{P}}_k := \mathrm{cov}[\mathbf{x}_k] \tag{2.76}$$

the mean and covariance, respectively, of the stochastic process $\mathbf{x}_k$ at time t_k. Based on the derivations from above, a recursive formulation of the first and second moments of the stochastic process $\mathbf{x}_k$ may now be obtained quite easily. By means of the transition matrix $\mathbf{F}_{k+1|k}$ and the combined noise covariance matrix $\mathbf{Q}_{k+1|k}$, we obtain

$$\hat{\mathbf{x}}_{k+1} := \mathrm{E}\left[\mathbf{F}_{k+1|k}\mathbf{x}_k + \mathbf{w}_{k+1|k}\right] \tag{2.77}$$

$$= \mathbf{F}_{k+1|k}\hat{\mathbf{x}}_k, \tag{2.78}$$

$$\hat{\mathbf{P}}_{k+1} := \mathrm{cov}\left[\mathbf{F}_{k+1|k}\mathbf{x}_k + \mathbf{w}_{k+1|k}\right] \tag{2.79}$$

$$= \mathbf{F}_{k+1|k}\mathrm{cov}[\mathbf{x}_k]\,\mathbf{F}_{k+1|k}^\top + 2\,\mathbf{F}_{k+1|k}\mathrm{E}\left[\mathbf{x}_k\mathbf{w}_{k+1|k}^\top\right] + \mathrm{cov}\left[\mathbf{w}_{k+1|k}\right] \tag{2.80}$$

$$= \mathbf{F}_{k+1|k}\hat{\mathbf{P}}_k\mathbf{F}_{k+1|k}^\top + \mathbf{Q}_{k+1|k}, \tag{2.81}$$

where the last equality holds due to the fact that the noise $\mathbf{w}_{k+1|k}$ is statistically independent from the mean $\hat{\mathbf{x}}_k$ at time t_k.

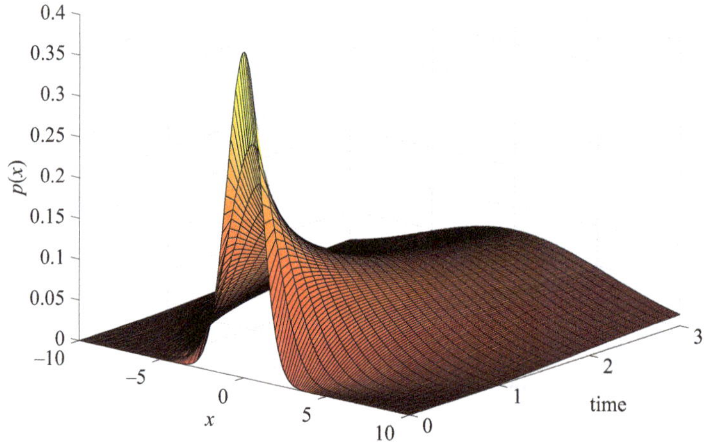

Figure 2.6 Diffusion process of the mass of probability over time of a 1D distribution

Since $\{\mathbf{x}_k\}_k$ is a Gaussian process, its distribution $p(\mathbf{x}_k)$ at a given time index k is fully determined by its mean and covariance:

$$p(\mathbf{x}_k) = \mathcal{N}(\mathbf{x}_k; \hat{\mathbf{x}}_k, \hat{\mathbf{P}}_k). \tag{2.82}$$

Example 3

Assume an initial prior of a 1D linear Îto process $\mathbf{x}_k \in \mathbb{R}$ is given by $p(\mathbf{x}_0) = \mathcal{N}(\mathbf{x}_0; 0, 1)$ and the time evolution is given by $\mathbf{F}_{k|k-1} = 1$ and $\mathbf{Q}_{k|k-1} = 1$ with $\Delta_t = 1$. Then, the mass of probability diffuses along the real axis over time. This process is shown in Figure 2.6.

■

2.4 Path integral formulation

This section is on the connection between the theory of stochastic systems and basic quantum physics which is based on the path integral formulation for dynamical state evolution. The theory of quantum physics reveals that the character of property states for objects with respect to their position and kinematics is inherently of probabilistic nature. According to Heisenberg's principle of uncertainty, the state of particles on a sub-atomic scale can only be described in terms of probabilities. The temporal evolution of these probabilities is given by the Schrödinger equation which is a mathematical analog on to the diffusion equation from above with a complex time component. This analogy already indicates a "principle of uncertainty" on a macroscopic level, which nowadays is known as the laws of statistical mechanics. This research field considers for instance the temporal evolution of a particle distribution influenced by the random collisions with surrounding atoms or molecules as initially considered by Robert Brown and is nowadays known as the Brownian motion.

The connection between target tracking and the laws of quantum physics becomes evident by means of the *path integral*, a methodology introduced by the Nobel prize winner *Richard Feynman* [26] to provide a general and elegant way to calculate the solution of the Schrödinger equation [13]. Feynman's work soon was extended by others and the connection to the Brownian motion was shown by Marc Kac [27]. Back in those days, the theory of target tracking still was in the very early stages of development. The Kalman filter was found in 1960 [6]; however, one can say that data fusion as a research topic as see it today in science was found by the *Joint Directors of Laboratories* (JDL) group only in 1982.

This section considers the path integral formulation of stochastic processes for two reasons: on the one hand, it is shown that a path integral formulation for transition kernels of stochastic processes is possible. On the other hand, analogous terms to classical and quantum physics are established along the derivation of the formulae. The first part revisits the very essence of the path integral in quantum physics. Then, the analog computation is presented for macroscopic objects in terms of their probability distribution.

2.4.1 Path integral in quantum physics

In quantum physics, the state of microscopic systems is fully described by a wave function $\psi : (\mathbf{x}, t) \longrightarrow \mathbb{C}$. Its evolution is governed by the Schrödinger equation

$$\partial_t \psi(\mathbf{x}, t) = -\frac{i}{\hbar}\mathcal{H}\psi(\mathbf{x}, t), \tag{2.83}$$

where $\hbar$ is the reduced Planck constant and $\mathcal{H} = T + V$ is the Hamilton operator which yields the total energy of the system, that in general is comprised of the kinetic energy T and the potential V. Since the kinetic energy T for a particle with mass m in n dimensions is given by

$$T = -\frac{\hbar^2}{2m}\sum_{i=1} \partial_{x_i}^2 \tag{2.84}$$

one can see that there is a close relation to the Fokker–Planck equation in (2.55) based on the *Wick-rotation* [28]

$$it \longrightarrow t \tag{2.85}$$

as illustrated in Figure 2.7.

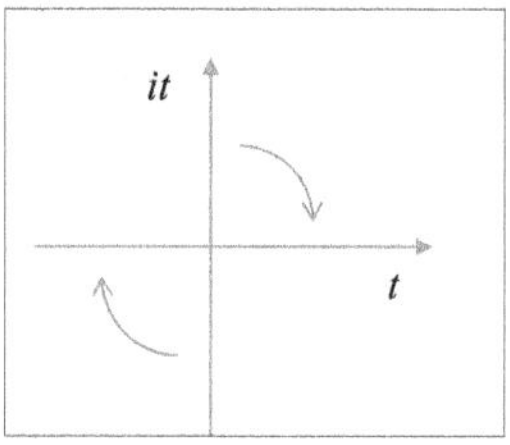

Figure 2.7 The Wick rotation it ⟶ t is the key to transform the Schrödinger equation into a classical stochastic equation

If the system is in an eigenstate $|\alpha\rangle$‡‡ of some time-constant $\mathcal{H}$ with energy E, such that

$$\mathcal{H}|\alpha\rangle = E|\alpha\rangle \tag{2.86}$$

(2.83) reduces to

$$\partial_t \psi(\mathbf{x}, t) = -\frac{i}{\hbar} E \psi(\mathbf{x}, t), \tag{2.87}$$

which is solved by

$$\psi(\mathbf{x}, t) = \frac{1}{c} \psi(\mathbf{x}, 0)\, e^{-\frac{i}{\hbar} tE}, \tag{2.88}$$

where c is a normalization constant such that

$$\int \mathrm{d}\mathbf{x}\; |\psi(\mathbf{x}, t)|^2 = 1. \tag{2.89}$$

The transition to the general case, where $\mathcal{H}$ is time-dependent is obtained by a time discretization $0 = t_0 < t_1 < \cdots < t_k = t$. One can then use (2.88) under the assumption that $\mathcal{H}$ is constant within each fragment. It was the idea of the great physicist Paul Dirac in [29] to iteratively apply short-term kernels. The solution is then obtained by considering the limit of letting the number of discretization points go to infinity. The details are omitted here, but one obtains eventually

$$\psi(\mathbf{x}_k, t) = \int \mathrm{d}\mathbf{x}_0\; \mathbf{K}(\mathbf{x}_k, \mathbf{x}_0)\; \psi(\mathbf{x}_0, t), \tag{2.90}$$

with

$$\mathbf{K}(\mathbf{x}_k, \mathbf{x}_0) = \int \mathrm{d}\mathbf{x}_{k-1} \int \mathrm{d}\mathbf{x}_{k-2} \cdots \int \mathrm{d}\mathbf{x}_1 \prod_{l=0}^{k-1} \mathbf{K}(\mathbf{x}_{l+1}, \mathbf{x}_l). \tag{2.91}$$

‡‡Here, the bra-ket notation for quantum states has been used. This, however, will not be of any importance for the further derivations.

The short-term kernel is according to (2.88) given by

$$\mathbf{K}(\mathbf{x}_{l+1}, \mathbf{x}_l) = e^{-\frac{i}{\hbar}(t_{l+1}-t_l)\left(\frac{p^2(\mathbf{x}_{l+1},\mathbf{x}_l)}{2m}+V(\mathbf{x}_l)\right)} \tag{2.92}$$

$$= e^{\frac{i}{\hbar}(t_{l+1}-t_l)\left(\frac{m}{2}\frac{(\mathbf{x}_{l+1}-\mathbf{x}_l)^2}{(t_{l+1}-t_l)^2}-V(\mathbf{x}_l)\right)} \tag{2.93}$$

$$= e^{\frac{i}{\hbar}(t_{l+1}-t_l)\left(\frac{m}{2}\dot{\mathbf{x}}_l^2-V(\mathbf{x}_l)\right)} \tag{2.94}$$

$$= e^{\frac{i}{\hbar}(t_{l+1}-t_l)\mathcal{L}(\mathbf{x}_l)}, \tag{2.95}$$

where $\mathcal{L} = T - V$ is the Lagrangian, which describes the laws of classical mechanics. The change of sign of the exponent in (2.93) comes from the fact that the momentum operator p has phase which implies that its squared is negative. Taking the limit yields that for the computation of the kernel $\mathbf{K}(\mathbf{x}_k, \mathbf{x}_0)$ the number of integrals goes to infinity and we are literally integrating over the full space of every state at every time with the time window (t_0, t_k). Moreover, it should be noted that the end points $\mathbf{x}_k$ and $\mathbf{x}_0$ are fixed. Therefore, the integral is essentially over all possible trajectories in the state space which connect the two end points. For obvious reasons, very unlikely instants of trajectories are hardly contributing while reasonable instants have higher weight. The contribution is given by the transition kernel. Thus, by introducing a notation for the path integral one obtains

$$\mathbf{K}(\mathbf{x}_k, \mathbf{x}_0) = \int_{\mathbf{x}_0 \rightsquigarrow \mathbf{x}_k} \mathcal{D}\mathbf{x}_t \; e^{\frac{i}{\hbar}\int_{t_0}^{t_k} \mathrm{d}t \, \mathcal{L}(\mathbf{x}_t)} \tag{2.96}$$

$$= \int_{\mathbf{x}_0 \rightsquigarrow \mathbf{x}_k} \mathcal{D}\mathbf{x}_t \; e^{\frac{i}{\hbar}\mathcal{S}(\mathbf{x}_t)}, \tag{2.97}$$

where $\mathcal{S} = \int \mathrm{d}t \, \mathcal{L}(\mathbf{x}_t)$ is known as the classical action. The result (2.97) is known as the path integral formulation of quantum physics.

2.4.2 Path integral for stochastic processes

Let us have a closer look at the path of the stochastic process $\{\mathbf{x}_k\}_k$ in a small scale discretized form with $t_0 < t_1 < \cdots < t_k$ and $t_{l+1} - t_l = \epsilon$ for all l. We have seen in (2.73) that the transition kernel $\mathbf{K}(\mathbf{x}_{l+1}; \mathbf{x}_l)$ is given by

$$p(\mathbf{x}_{l+1}) = \int \mathrm{d}\mathbf{x}_l \; \mathbf{K}(\mathbf{x}_{l+1}; \mathbf{x}_l) \, p(\mathbf{x}_l), \tag{2.98}$$

$$\mathbf{K}(\mathbf{x}_{l+1}; \mathbf{x}_l) = \mathcal{N}(\mathbf{x}_{l+1}; \, \mathbf{F}_{l+1|l}\mathbf{x}_l, \, \mathbf{Q}_{l+1|l}). \tag{2.99}$$

Analogously as above, one may use short-term kernels iteratively in order to formulate the propagator over a large time scale $t_k - t_0$ by recursively applying the kernel on (2.98):

$$p(\mathbf{x}_k) = \int \mathrm{d}\mathbf{x}_{k-1} \int \mathrm{d}\mathbf{x}_{k-2} \cdots \int \mathrm{d}\mathbf{x}_0 \prod_{l=0}^{k-1} \mathbf{K}(\mathbf{x}_{l+1}; \mathbf{x}_l) \, p(\mathbf{x}_0), \tag{2.100}$$

thus we have $\mathbf{K}(\mathbf{x}_k;\mathbf{x}_0) = \int \mathrm{d}\mathbf{x}_{k-1} \int \mathrm{d}\mathbf{x}_{k-2} \cdots \int \mathrm{d}\mathbf{x}_1 \prod_{l=0}^{k-1} \mathbf{K}(\mathbf{x}_{l+1};\mathbf{x}_l)$. This formulation becomes particularly interesting for the limit $\epsilon \to 0$, which yields an infinite dimensional integration along all possible continuous paths starting in $\mathbf{x}_0$ and going to $\mathbf{x}_k$, since these are the only fixed points. As in the quantum physics section, we will use the short notation $\int_{\mathbf{x}_0 \rightsquigarrow \mathbf{x}_k} \mathcal{D}\mathbf{x}$ for this (heuristic) construction. One has that

$$\prod_{l=0}^{k-1} \mathbf{K}(\mathbf{x}_{l+1};\mathbf{x}_l) \propto e^{-\frac{1}{2}\sum_{l=0}^{k-1}(\mathbf{x}_{l+1}-\mathbf{F}_{l+1|l}\mathbf{x}_l)^\top \mathbf{Q}_{l+1|l}^{-1}(\mathbf{x}_{l+1}-\mathbf{F}_{l+1|l}\mathbf{x}_l)}, \tag{2.101}$$

where

$$\sum_{l=0}^{k-1}(\mathbf{x}_{l+1} - \mathbf{F}_{l+1|l}\mathbf{x}_l)^\top \mathbf{Q}_{l+1|l}^{-1}(\mathbf{x}_{l+1} - \mathbf{F}_{l+1|l}\mathbf{x}_l) \tag{2.102}$$

$$= \sum_{l=0}^{k-1} \mathbf{w}_{l+1|l}^\top \mathbf{Q}_{l+1|l}^{-1}\mathbf{w}_{l+1|l} \underset{\epsilon\downarrow 0}{\longrightarrow} \int_{t_0}^{t_k} \mathrm{d}t\ \partial_t(\mathbf{w}_{l+1|l}^\top \mathbf{Q}_{l+1|l}^{-1}\mathbf{w}_{l+1|l}). \tag{2.103}$$

By introducing the free particle energy

$$T = \partial_t(\mathbf{w}_{l+1|l}^\top \mathbf{Q}_{l+1|l}^{-1}\mathbf{w}_{l+1|l}), \tag{2.104}$$

one obtains that the Lagrangian at the absence of a potential $\mathcal{L} = T$ is integrated such that

$$\mathbf{K}(\mathbf{x}_k;\mathbf{x}_0) \propto \int_{\mathbf{x}_0 \rightsquigarrow \mathbf{x}_k} \mathcal{D}\mathbf{x}\ e^{-\frac{1}{2}\mathcal{S}(\mathbf{x})}. \tag{2.105}$$

Here, again the action $\mathcal{S}$ is the integral of the Lagrangian:

$$\mathcal{S}(\mathbf{x}) = \int_{t_0}^{t_k} \mathrm{d}t\ \mathcal{L}(\mathbf{x}_t). \tag{2.106}$$

It should be noted that the proportionality constant of the path integral can easily be obtained by the fact that $\mathbf{K}(\mathbf{x}_k;\mathbf{x}_0)$ is a normalized density function in $\mathbf{x}_k$. This formulation of the transition kernel yields more insights, when sensor data is introduced into the equation. We will see at the end of the next chapter that measurements are actually acting as potentials which attract the trajectory.

Chapter 3
Bayesian state estimation and target tracking

This chapter introduces the theoretical means required for model-based data fusion in applications of target tracking and state estimation. The state is assumed to evolve according to a stochastic process as described in the previous chapter. We have seen that the resulting probability density function has an increasing variance or covariance matrix, respectively, depending on the dimension of the state variable. In terms of information, this reflects the fact that information on the actual state is lost over time due to the stochastic nature of the process. Sensor measurements in contrast provide information on the state, thus, the problem addressed by state estimation is to integrate sensor data into the probability densities in order to derive statistics such as mean and covariance matrix. In applications, one often is interested in extracting information of sensor data produced at discrete instants of time $\{t_l\}_{l=1,\dots,k}$, where k typically represents the current time index. This commonly results in a two-step filter, where at each time stamp the density is propagated based on an assumed dynamical model, referred to as *prediction*, before processing new data in a subsequent *update*.

The famous Bayes theorem is the core of many prediction-update filters, which adds information to an existing *prior* density. The latter density is the given or computed probability density *before* new data from the current time is processed. It refers to the current instant of time t_k, but it is based on the data $\mathcal{Z}^{k-1} = \{Z_1, \dots, Z_{k-1}\}$, which is the time series of sensor observations up to time t_{k-1}. This can be expressed in terms of conditional probability density functions:

$$p(\mathbf{x}_k|\mathcal{Z}^{k-1}) \quad \underset{\text{Prediction}}{\longleftarrow} \quad p(\mathbf{x}_{k-1}|\mathcal{Z}^{k-1}). \tag{3.1}$$

The *posterior*, the density after incorporating the current data $\mathbf{z}_k$ into the prior is given by the conditional pdf $p(\mathbf{x}_k|\mathcal{Z}^k)$, where $\mathcal{Z}^k$ is the time series of measurements up to this time step:

$$p(\mathbf{x}_k|\mathcal{Z}^k) \quad \underset{\text{Update}}{\longleftarrow} \quad p(\mathbf{x}_k|\mathcal{Z}^{k-1}). \tag{3.2}$$

The Bayes theorem which is required for the update step directly follows from the definition of the conditional density. Consider two random variables x and y. Then it holds by definition that

$$p(x|y) = \frac{p(x,y)}{p(y)}. \tag{3.3}$$

Therefore, one directly obtains that

$$p(x|y) = \frac{p(x,y)}{p(y)} \tag{3.4}$$

$$= \frac{p(y,x)}{p(y)} \tag{3.5}$$

$$= \frac{p(y|x)\,p(x)}{p(y)}, \tag{3.6}$$

which is known as the Bayes formula.

Here, we also call the left-hand side of the equation "*posterior*" opposed to the prior on x given by $p(x)$, which does not include the information of an observation y. The term $p(y|x)$ denotes the *"sensor model"* since it is conditioned on the true state and therefore describes the deterministic relation and stochastic behavior of the observation process. Obviously, the sensor model is a probability density function in the sensor observation y as a random variable. If y is given empirically, the density $p(y|x)$ can be considered a likelihood function in x. Thus, the same function is used with x as the variable argument. As a consequence, the likelihood of the state is not a normalized function.

An example of a Bayes update in shown in Figure 3.1. One can see that the prior $p(x)$ in this example is multimodal, that is, there are multiple local maxima. The likelihood $p(z|x)$ in this example is given by a Gamma distribution with shape parameter $a = 4$ and scale parameter $b = 2$. The resulting posterior $p(x|z)$ is not part of a parametric family.* One can observe the effects of the pointwise multiplication of the prior and the likelihood function.

The denominator of the Bayes formula (3.6) is known as the "*normalization constant*," since the shape of the posterior is not affected by it and it may be interpreted as a constant which ensures that the pdf integrates to one. It should be noted that the normalization constant can be written as

$$p(y) = \int \mathrm{d}x\, p(y|x)\, p(x), \tag{3.7}$$

thus it equals the integration of the numerator. Since it is independent of the random variable x of the density on the left-hand side, it does not affect the *shape* of the density but it ensures that it integrates to one. As a consequence, it is sufficient to know the sensor model up to a constant factor c_y with $\ell(y;x) = c_y p(y|x)$ since it cancels out in the Bayes formula. This simplification is sometimes useful for the computation of complex sensor models, for instance, when it can only be evaluated pointwise for some x. If it is known that the posterior $p(x|y)$ is a member of a *parametrized* density family (such as for instance a Gaussian, Bernoulli, Gamma, or Wishart), then normalization is inherently given and a computation of the normalization constant

*It could be approximated by a Gaussian or a Gamma distribution for instance.

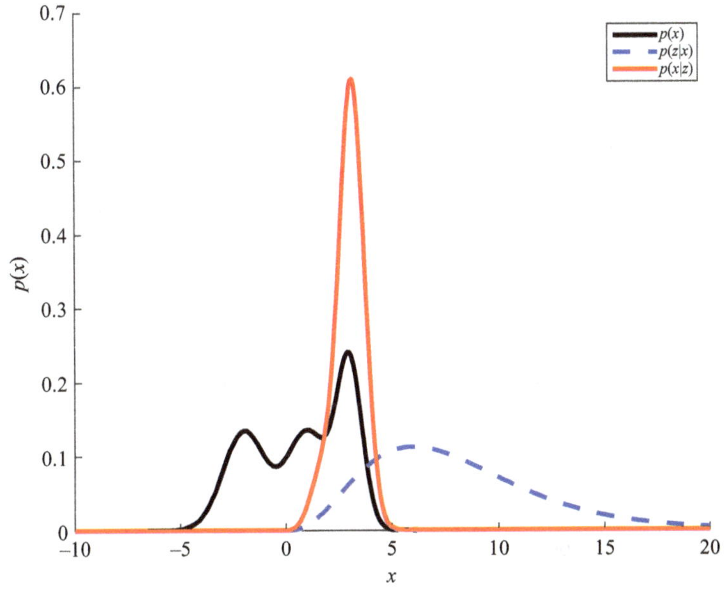

Figure 3.1 *Example of a Bayes update where the likelihood function* $p(z|x)$ *is incorporated into the prior* $p(x)$

$p(y)$ may be omitted since all density members of those families are normalized for all parameters. Moreover, it emphasizes the fact that the pdf $p(y|x)$, which is actually a density in y for a given x is considered a function in x for a given y, since the latter is the empirical observation and therefore fixed.† In other words, one considers the likelihood of the given observation value conditioned on the hypothesis that the true state is given by the variable x. As a function in x this obviously is not normalized.

For a recursive computation of the posterior in state estimation, it is convenient to split the timeseries of measurements into the observations Z_k at time t_k and $\mathcal{Z}^{k-1} = \{Z_1, \ldots, Z_{k-1}\}$.

An application of the Bayes rule yields

$$p(\mathbf{x}_k|\mathcal{Z}^k) = \frac{p(Z_k|\mathbf{x}_k)\, p(\mathbf{x}_k|\mathcal{Z}^{k-1})}{p(Z_k|\mathcal{Z}^{k-1})}. \tag{3.8}$$

This formula is in general intractable, however, for some families of distributions a closed form solution can be computed. The most famous one is for (multivariate)

†The quantum physicist would say that its wave function collapsed.

normal distributions, which yields the update step in the Kalman filter. The Kalman filter will be presented in the next section.

3.1 Fundamental equations of linear estimation

This section will consider linear Gaussian models for the involved densities in the Bayes update for state estimation. Since the resulting equations represent the core of many extensions,[‡] they are called the *Fundamental Equations of Linear Estimation* [30].

It is assumed that a measurement $\mathbf{z}_k$ at time t_k is given, which is corrupted by additive zero-mean normal distributed noise $\mathbf{v}_k$ with covariance matrix $\mathbf{R}_k$:

$$\mathbf{v}_k \sim \mathcal{N}(\mathbf{v}_k;\, \mathbf{O},\, \mathbf{R}_k). \tag{3.9}$$

The additive Gaussian random variable $\mathbf{v}_k$ reflects for instance thermal noise and other errors during the measurement process of the sensor. Clearly, it is stochastically independent from the process noise $\mathbf{w}_{k|k-1}$. Since both have mean zero, it holds that

$$\operatorname{cov}\left[\mathbf{v}_k, \mathbf{w}_{k|k-1}\right] = \mathrm{E}\left[\mathbf{v}_k \mathbf{w}_{k|k-1}^{\mathsf{T}}\right] = 0. \tag{3.10}$$

Furthermore, the sensor measurement $\mathbf{z}_k$ at time t_k is assumed to be in a linear relation to the state space such that one has a *measurement matrix* $\mathbf{H}_k$ with

$$\mathbf{z}_k = \mathbf{H}_k \mathbf{x}_k + \mathbf{v}_k. \tag{3.11}$$

The term $\mathbf{H}_k\mathbf{x}_k$ is deterministic, if the density of $\mathbf{z}_k$ is conditioned on a true state $\mathbf{x}_k$, which implies $\mathrm{E}\left[\mathbf{z}_k - \mathbf{H}_k\mathbf{x}_k\right] = \mathrm{E}\left[\mathbf{v}_k\right] = 0$. The sensor model is therefore given by the Gaussian

$$p(\mathbf{z}_k|\mathbf{x}_k) = \mathcal{N}(\mathbf{z}_k;\, \mathbf{H}_k\mathbf{x}_k,\, \mathbf{R}_k). \tag{3.12}$$

The conjugate prior of a Gaussian is a Gaussian density.[§] Therefore it is assumed that the prior is given as a Gaussian with mean vector $\mathrm{E}\left[\mathbf{x}_k|\mathcal{Z}^{k-1}\right] = \mathbf{x}_{k|k-1}$ and error covariance matrix $\operatorname{cov}\left[\mathbf{x}_k|\mathcal{Z}^{k-1}\right] = \mathbf{P}_{k|k-1}$:

$$p(\mathbf{x}_k|\mathcal{Z}^{k-1}) = \mathcal{N}(\mathbf{x}_k;\, \mathbf{x}_{k|k-1},\, \mathbf{P}_{k|k-1}) \tag{3.13}$$

It has become common in the tracking literature to use the subscript $k|k-1$ or $k|k$ to indicate parameters which refer to a prior or a posterior density, respectively. This, however, is just a notational convention. Using the measurement model, it is easily possible to calculate the parameters of the prior measurement density $p(\mathbf{z}_k|\mathcal{Z}^{k-1}) =$

[‡]The extensions relax the assumptions such that for instance non-linear models, multiple parallel motion models, or imperfect data association may be present.

[§]A conjugate prior is defined as a parametrized probability density family such that the Bayes update step remains in the same family.

$\mathcal{N}(\mathbf{z}_k;\, \mathbf{z}_{k|k-1},\, \mathbf{S}_{k|k-1})$, which is the normalization constant for the Bayes update. One has that

$$\begin{aligned} \mathbf{z}_{k|k-1} &= \mathrm{E}\left[\mathbf{z}_k | \mathcal{Z}^{k-1}\right] && (3.14)\\ &= \mathrm{E}\left[\mathbf{H}_k \mathbf{x}_k + \mathbf{v}_k | \mathcal{Z}^{k-1}\right] && (3.15)\\ &= \mathbf{H}_k\, \mathrm{E}\left[\mathbf{x}_k | \mathcal{Z}^{k-1}\right] && (3.16)\\ &= \mathbf{H}_k \mathbf{x}_{k|k-1}. && (3.17) \end{aligned}$$

In the equations above, (3.16) holds because $\mathbf{z}_k$ is not in $\mathcal{Z}^{k-1}$ and therefore $\mathrm{E}\left[\mathbf{v}_k | \mathcal{Z}^{k-1}\right] = \mathrm{E}\left[\mathbf{v}_k\right] = 0$. For the covariance matrix, one obtains

$$\begin{aligned} \mathbf{S}_{k|k-1} &= \mathrm{cov}\left[\mathbf{z}_k | \mathcal{Z}^{k-1}\right] && (3.18)\\ &= \mathrm{cov}\left[\mathbf{H}_k \mathbf{x}_k + \mathbf{v}_k | \mathcal{Z}^{k-1}\right] && (3.19)\\ &= \mathbf{H}_k\, \mathrm{cov}\left[\mathbf{x}_k | \mathcal{Z}^{k-1}\right] \mathbf{H}_k^\top + \mathrm{cov}\left[\mathbf{v}_k\right] && (3.20)\\ &= \mathbf{H}_k \mathbf{P}_{k|k-1} \mathbf{H}_k^\top + \mathbf{R}_k. && (3.21) \end{aligned}$$

In (3.20), we assumed that the measurement noise of the sensor at time t_k is stochastically independent of the prior estimate error $\mathbf{x}_k - \mathrm{E}\left[\mathbf{x}_k | \mathcal{Z}^{k-1}\right]$. Therefore, it holds that

$$\mathrm{cov}\left[\mathbf{x}_k, \mathbf{v}_k | \mathcal{Z}^{k-1}\right] = \mathrm{E}\left[\left(\mathbf{x}_k - \mathrm{E}\left[\mathbf{x}_k | \mathcal{Z}^{k-1}\right]\right) \mathbf{v}_k^\top\right] = 0. \quad (3.22)$$

> The prior density on the current measurement $\mathbf{z}_k$ conditioned on previous knowledge $\mathcal{Z}^{k-1}$ is given by the normal density
>
> $$p(\mathbf{z}_k | \mathcal{Z}^{k-1}) = \mathcal{N}(\mathbf{z}_k;\, \mathbf{z}_{k|k-1},\, \mathbf{S}_{k|k-1}) \quad (3.23)$$
>
> with
>
> $$\begin{aligned} \mathbf{z}_{k|k-1} &= \mathbf{H}_k \mathbf{x}_{k|k-1} && (3.24)\\ \mathbf{S}_{k|k-1} &= \mathbf{H}_k \mathbf{P}_{k|k-1} \mathbf{H}_k^\top + \mathbf{R}_k. && (3.25) \end{aligned}$$

In order to calculate the posterior density $p(\mathbf{x}_k | \mathcal{Z}^k) = p(\mathbf{x}_k | \mathcal{Z}^{k-1}, \mathbf{z}_k)$, we define the following auxiliary vector:

$$\mathbf{y} = \begin{pmatrix} \mathbf{x}_k \\ \mathbf{z}_k \end{pmatrix}. \quad (3.26)$$

The prior density of $\mathbf{y}$ is also a Gaussian with mean

$$\bar{\mathbf{y}} = \begin{pmatrix} \mathbf{x}_{k|k-1} \\ \mathbf{z}_{k|k-1} \end{pmatrix} \quad (3.27)$$

and covariance matrix

$$\mathbf{P}_{yy} = \begin{pmatrix} \mathbf{P}_{k|k-1} & \mathbf{P}_{xz} \\ \mathbf{P}_{zx} & \mathbf{S}_{k|k-1} \end{pmatrix}, \quad (3.28)$$

where $\mathbf{P}_{xz} = \text{cov}\left[\mathbf{x}_k, \mathbf{z}_k | \mathcal{Z}^{k-1}\right]$ denote the cross-covariance elements. It should be noted that in general the cross-covariance conditioned on the time series of measurements up to time t_{k-1} does not vanish. The posterior density can now be expressed in the following way:

$$p(\mathbf{x}_k \quad | \mathcal{Z}^{k-1}, \mathbf{z}_k) = \frac{p(\mathbf{x}_k, \mathbf{z}_k | \mathcal{Z}^{k-1})}{p(\mathbf{z}_k | \mathcal{Z}^{k-1})} \tag{3.29}$$

$$= \frac{\mathcal{N}(\mathbf{y}; \bar{\mathbf{y}}, \mathbf{P}_{yy})}{\mathcal{N}(\mathbf{z}_k; \mathbf{z}_{k|k-1}, \mathbf{S}_{k|k-1})} \tag{3.30}$$

$$\propto \mathrm{e}^{-\frac{1}{2}\left\{(\mathbf{y}-\bar{\mathbf{y}})^\top \mathbf{P}_{yy}^{-1}(\mathbf{y}-\bar{\mathbf{y}})-(\mathbf{z}_k-\mathbf{z}_{k|k-1})^\top \mathbf{S}_{k|k-1}^{-1}(\mathbf{z}_k-\mathbf{z}_{k|k-1})\right\}} \tag{3.31}$$

$$=: \mathrm{e}^{-\frac{1}{2}q}. \tag{3.32}$$

The argument q of the exponential function above can be rewritten by introducing random variables $\tilde{\mathbf{x}}$ and $\tilde{\mathbf{z}}$ which describe the prior estimation error:

$$\tilde{\mathbf{x}} := \mathbf{x}_k - \mathbf{x}_{k|k-1}, \tag{3.33}$$

$$\tilde{\mathbf{z}} := \mathbf{z}_k - \mathbf{z}_{k|k-1}. \tag{3.34}$$

One has that

$$q = \begin{pmatrix} \tilde{\mathbf{x}} \\ \tilde{\mathbf{z}} \end{pmatrix}^\top \begin{pmatrix} \mathbf{P}_{k|k-1} & \mathbf{P}_{xz} \\ \mathbf{P}_{zx} & \mathbf{S}_{k|k-1} \end{pmatrix}^{-1} \begin{pmatrix} \tilde{\mathbf{x}} \\ \tilde{\mathbf{z}} \end{pmatrix} - \tilde{\mathbf{z}}^\top \mathbf{S}_{k|k-1}^{-1} \tilde{\mathbf{z}} \tag{3.35}$$

$$=: \begin{pmatrix} \tilde{\mathbf{x}} \\ \tilde{\mathbf{z}} \end{pmatrix}^\top \begin{pmatrix} \mathbf{T}_{xx} & \mathbf{T}_{xz} \\ \mathbf{T}_{zx} & \mathbf{T}_{zz} \end{pmatrix} \begin{pmatrix} \tilde{\mathbf{x}} \\ \tilde{\mathbf{z}} \end{pmatrix} - \tilde{\mathbf{z}}^\top \mathbf{S}_{k|k-1}^{-1} \tilde{\mathbf{z}}. \tag{3.36}$$

The matrix inversion lemma (see Section A.1 in the Appendix) yields the following relationship:

$$\mathbf{T}_{xx}^{-1} = \mathbf{P}_{k|k-1} - \mathbf{P}_{xz} \mathbf{S}_{k|k-1}^{-1} \mathbf{P}_{zx}, \tag{3.37}$$

$$\mathbf{P}_{zz}^{-1} = \mathbf{T}_{zz} - \mathbf{T}_{zx} \mathbf{T}_{xx}^{-1} \mathbf{T}_{xz}, \tag{3.38}$$

$$\mathbf{T}_{xx}^{-1} \mathbf{T}_{xz} = -\mathbf{P}_{xz} \mathbf{S}_{k|k-1}^{-1}. \tag{3.39}$$

Therefore, we have

$$\begin{aligned} q = {} & \tilde{\mathbf{x}}^\top \mathbf{T}_{xx} \tilde{\mathbf{x}} + \tilde{\mathbf{x}}^\top \mathbf{T}_{xz} \tilde{\mathbf{z}} \\ & + \tilde{\mathbf{z}}^\top \mathbf{T}_{zx} \tilde{\mathbf{x}} + \tilde{\mathbf{z}}^\top \mathbf{T}_{zz} \tilde{\mathbf{z}} - \tilde{\mathbf{z}}^\top \mathbf{S}_{k|k-1}^{-1} \tilde{\mathbf{z}} \end{aligned} \tag{3.40}$$

$$\begin{aligned} = {} & \left(\tilde{\mathbf{x}} + \mathbf{T}_{xx}^{-1} \mathbf{T}_{xz} \tilde{\mathbf{z}}\right)^\top \mathbf{T}_{xx} \left(\tilde{\mathbf{x}} + \mathbf{T}_{xx}^{-1} \mathbf{T}_{xz} \tilde{\mathbf{z}}\right) \\ & + \tilde{\mathbf{z}}^\top \left(\mathbf{T}_{zz} - \mathbf{T}_{zx} \mathbf{T}_{xx}^{-1} \mathbf{T}_{xz}\right) \tilde{\mathbf{z}} - \tilde{\mathbf{z}}^\top \mathbf{S}_{k|k-1}^{-1} \tilde{\mathbf{z}} \end{aligned} \tag{3.41}$$

$$= \left(\tilde{\mathbf{x}} + \mathbf{T}_{xx}^{-1} \mathbf{T}_{xz} \tilde{\mathbf{z}}\right)^\top \mathbf{T}_{xx} \left(\tilde{\mathbf{x}} + \mathbf{T}_{xx}^{-1} \mathbf{T}_{xz} \tilde{\mathbf{z}}\right). \tag{3.42}$$

One can see that

$$\tilde{\mathbf{x}} + \mathbf{T}_{xx}^{-1} \mathbf{T}_{xz} \tilde{\mathbf{z}} = \mathbf{x}_k - (\mathbf{x}_{k|k-1} + \mathbf{P}_{xz} \mathbf{S}_{k|k-1}^{-1} \tilde{\mathbf{z}}), \tag{3.43}$$

Therefore the posterior density in (3.31) is a Gaussian

$$p(\mathbf{x}_k|\mathcal{Z}^k) = \mathcal{N}(\mathbf{x}_k;\, \mathbf{x}_{k|k}, \mathbf{P}_{k|k}) \tag{3.44}$$

with mean

$$\mathbf{x}_{k|k} := \mathbf{x}_{k|k-1} + \mathbf{P}_{xz}\mathbf{S}_{k|k-1}^{-1}\tilde{\mathbf{z}} \tag{3.45}$$

and covariance matrix

$$\mathbf{P}_{k|k} := \mathbf{P}_{k|k-1} - \mathbf{P}_{xz}\mathbf{S}_{k|k-1}^{-1}\mathbf{P}_{zx}. \tag{3.46}$$

Equations (3.45) and (3.46) are called *fundamental equations of linear estimation*.

3.2 Kalman filter

By means of the above equations, the optimal recursive linear estimator for Gaussian models can easily be derived. The process is started by an initial normal distribution with parameters $\mathbf{x}_{0|0}$ and $\mathbf{P}_{0|0}$. Then, for $k = 1, 2, \ldots$, the latest prior and posterior density functions are calculated in the *"prediction"* and *"filtering"* steps, respectively. This scheme is illustrated in Figure 3.2.

Prediction

In the prediction step, the parameters $\mathbf{x}_{k-1|k-1}$ and $\mathbf{P}_{k-1|k-1}$ from the previous posterior density are used and the time evolution of the density on current state variable from time t_{k-1} to t_k is incorporated into the actual knowledge on the state $\mathbf{x}_k$. The deterministic and stochastic part of the underlying evolution model reflects the dynamics and the process noise, respectively, and, for the Kalman filter, this model is assumed to be linear-Gaussian such that (2.74) holds. It was shown in (2.78) and (2.81) that the prior density is given by a Gaussian with the following parameters:

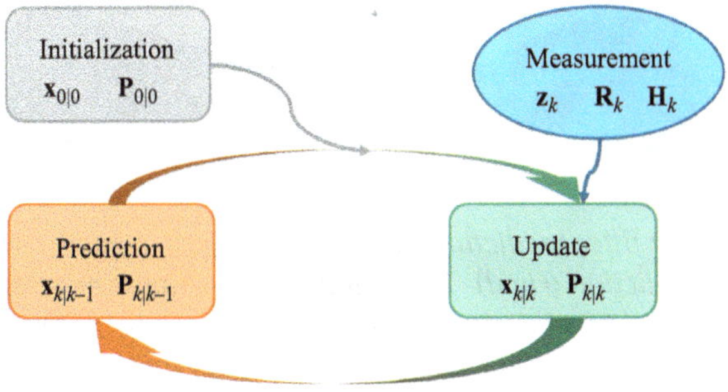

Figure 3.2 Kalman filter recursion

$$p(\mathbf{x}_k|\mathcal{Z}^{k-1}) = \mathcal{N}(\mathbf{x}_k;\ \mathbf{x}_{k|k-1},\ \mathbf{P}_{k|k-1}) \tag{3.47}$$

$$\mathbf{x}_{k|k-1} = \mathbf{F}_{k|k-1}\mathbf{x}_{k-1|k-1} \tag{3.48}$$

$$\mathbf{P}_{k|k-1} = \mathbf{F}_{k|k-1}\mathbf{P}_{k-1|k-1}\mathbf{F}_{k|k-1}^{\top} + \mathbf{Q}_{k|k-1} \tag{3.49}$$

An example for the prediction of a 2D normal distribution based on the formulae above is shown in Figure 3.3. In the diagram on the left side, the initial density is shown. The diagram on the right presents the prediction result, which is given by the lower density. One can see that the drift was applied such that the center has moved forward along one dimension and the covariance matrix was increased, which implies that the Gaussian hat is wider and its top is lower.

Filtering

The filtering step processes the current measurement $\mathbf{z}_k$ and yields the posterior for time t_k. Since the prior and the likelihood are Gaussian and the measurement function $\mathbf{H}_k$ is linear, the fundamental equations of linear estimation may be applied. To this end, the cross-covariance $\mathbf{P}_{xz}$ needs to be computed. One has

$$\mathbf{P}_{xz} = \operatorname{cov}\left[\mathbf{x}_k, \mathbf{z}_k|\mathcal{Z}^{k-1}\right] \tag{3.50}$$

$$= \mathrm{E}\left[(\mathbf{x}_k - \mathbf{x}_{k|k-1})(\mathbf{z}_k - \mathbf{z}_{k|k-1})^{\top}\right] \tag{3.51}$$

$$= \mathrm{E}\left[(\mathbf{x}_k - \mathbf{x}_{k|k-1})(\mathbf{H}_k\mathbf{x}_k + \mathbf{v}_k - \mathbf{H}_k\mathbf{x}_{k|k-1})^{\top}\right] \tag{3.52}$$

$$= \mathrm{E}\left[(\mathbf{x}_k - \mathbf{x}_{k|k-1})(\mathbf{x}_k - \mathbf{x}_{k|k-1})^{\top}\right]\mathbf{H}_k^{\top} \tag{3.53}$$

$$= \operatorname{cov}\left[\mathbf{x}_k|\mathcal{Z}^{k-1}\right]\mathbf{H}_k^{\top} \tag{3.54}$$

$$= \mathbf{P}_{k|k-1}\mathbf{H}_k^{\top}, \tag{3.55}$$

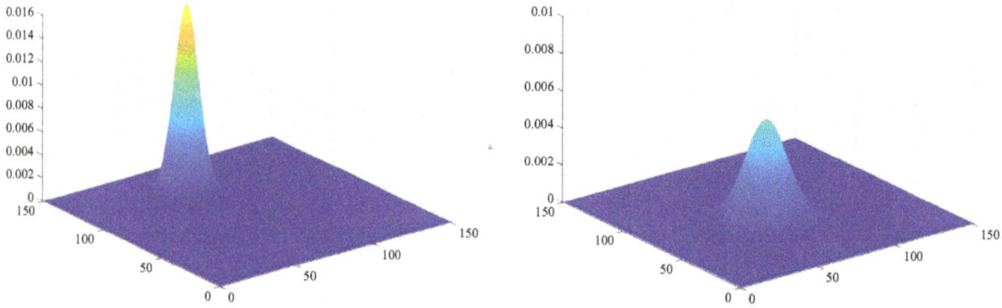

Figure 3.3 Kalman filter prediction: initial density (left) together with the result of the prediction (right). The drift due to the transition matrix leads to a displacement of the center and the process noise covariance increases the width of the Gaussian. This reflects the fact that information is lost, if time passes and new measurements are not yet incorporated into the estimation process.

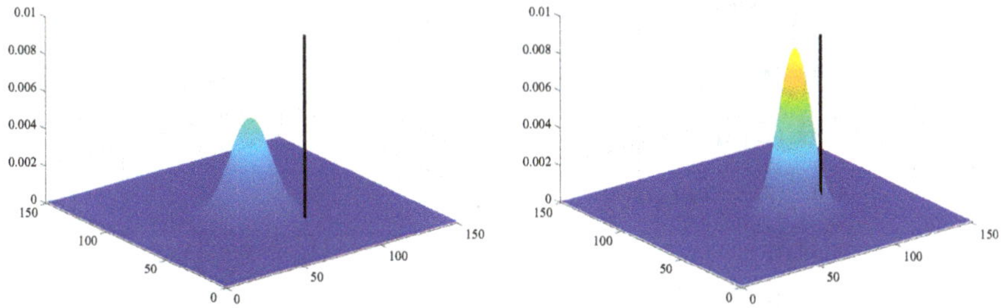

Figure 3.4 *Kalman update: a measurement $\mathbf{z}_k$ (black line) for time t_k is processed. Based on the Kalman filter formulae, the prior density (left) is shifted towards the measurement along the innovation vector $\boldsymbol{\nu}_k$ depending on the gain matrix $\mathbf{W}_{k|k-1}$ to obtain the posterior (right).*

where (3.53) holds due to the fact that $\text{cov}\left[\mathbf{x}_k, \mathbf{v}_k | \mathcal{Z}^{k-1}\right] = 0$ since the measurement $\mathbf{z}_k$ is not included in $\mathcal{Z}^{k-1}$ and therefore its noise is uncorrelated with the prior estimate.

Using the fundamental equations (3.45) and (3.46) yield the famous Kalman filter update:

$$\mathbf{x}_{k|k} := \mathbf{x}_{k|k-1} + \mathbf{W}_{k|k-1}\boldsymbol{\nu}_k \tag{3.56}$$

$$\mathbf{P}_{k|k} := \mathbf{P}_{k|k-1} + \mathbf{W}_{k|k-1}\mathbf{S}_{k|k-1}\mathbf{W}_{k|k-1}^{\top} \tag{3.57}$$

$$\mathbf{W}_{k|k-1} = \mathbf{P}_{k|k-1}\mathbf{H}_k^{\top}\mathbf{S}_{k|k-1}^{-1} \tag{3.58}$$

$$\mathbf{S}_{k|k-1} = \mathbf{H}_k\mathbf{P}_{k|k-1}\mathbf{H}_k^{\top} + \mathbf{R}_k \tag{3.59}$$

$$\boldsymbol{\nu}_k = \mathbf{z}_k - \mathbf{z}_{k|k-1} \tag{3.60}$$

$$\mathbf{z}_{k|k-1} = \mathbf{H}_k\mathbf{x}_{k|k-1}. \tag{3.61}$$

An illustrative example is shown in Figure 3.4, where a measurement (black line) is integrated into a prior by means of the Kalman filter. The gain matrix $\mathbf{W}_{k|k-1}$ steers the correction of the prior based on a ratio between the prior knowledge and the information of the measurement. Therefore, a large gain matrix can have two reasons, either a prior with a large covariance matrix or a likelihood with a very small covariance matrix.

The vector $\boldsymbol{\nu}_k$ is called *"innovation"* and the matrix $\mathbf{W}_{k|k-1}$ is the *"Kalman gain"* since it steers the effect of the correction term on the state estimate. It becomes obvious that the posterior covariance is "smaller" than the prior covariance, which reflects the fact that information has been gained and included into the estimation process. Moreover, it can easily be seen that the innovation is zero-mean and $\mathbf{S}_{k|k-1}$ is the *"innovation error covariance"*:

$$\mathrm{E}\left[\boldsymbol{\nu}_k | \mathcal{Z}^{k-1}\right] = \mathrm{E}\left[\mathbf{H}_k \mathbf{x}_k + \mathbf{v}_k - \mathbf{H}_k \mathbf{x}_{k|k-1}\right] \tag{3.62}$$

$$= \mathbf{H}_k(\mathrm{E}\left[\mathbf{x}_k | \mathcal{Z}^{k-1}\right] - \mathbf{x}_{k|k-1}) + \mathrm{E}\left[\mathbf{v}_k\right] = 0 \tag{3.63}$$

$$\mathrm{cov}\left[\boldsymbol{\nu}_k\right] = \mathrm{E}\left[(\mathbf{z}_k - \mathbf{z}_{k|k-1})(\mathbf{z}_k - \mathbf{z}_{k|k-1})^\top\right] \tag{3.64}$$

$$= \mathbf{H}_k \mathrm{E}\left[(\mathbf{x}_k - \mathbf{x}_{k|k-1})(\mathbf{x}_k - \mathbf{x}_{k|k-1})^\top | \mathcal{Z}^{k-1}\right] \mathbf{H}_k^\top + \mathrm{cov}\left[\mathbf{v}_k \mathbf{v}_k^\top\right] \tag{3.65}$$

$$= \mathbf{H}_k \mathbf{P}_{k|k-1} \mathbf{H}_k^\top + \mathbf{R}_k = \mathbf{S}_{k|k-1} \tag{3.66}$$

The innovation of a sensor reading $\mathbf{z}_k$ is particularly important for instance in the presence of false alarms. There, one may test for the obtained value of $\boldsymbol{\nu}_k = \mathbf{z}_k - \mathbf{H}_k \mathbf{x}_{k|k-1}$ whether it obeys its expected density function

$$\boldsymbol{\nu}_k \sim \mathcal{N}(\boldsymbol{\nu}_k;\, 0,\, \mathbf{S}_{k|k-1}). \tag{3.67}$$

If it does not fit with respect to a given level of significance, chances are high that chances are high that it is clutter. From a theoretical perspective, we now have achieved the proof of the first formulation of the "*product formula*."

The product formula is a useful tool to transform Gaussian densities. Using the Gaussian models from above in the Bayes formula (3.6) yields for normal densities $\mathcal{N}(\mathbf{z};\, \mathbf{Hx},\, \mathbf{R})$ and $\mathcal{N}(\mathbf{x};\, \mathbf{y},\, \mathbf{P})$ that

$$p(x|y)\, p(y) = p(y|x)\, p(x) \tag{3.68}$$

$$\mathcal{N}(\mathbf{x};\, \hat{\mathbf{y}},\, \hat{\mathbf{P}})\, \mathcal{N}(\mathbf{z};\, \mathbf{Hy},\, \mathbf{S}) = \mathcal{N}(\mathbf{z};\, \mathbf{Hx},\, \mathbf{R})\, \mathcal{N}(\mathbf{x};\, \mathbf{y},\, \mathbf{P}), \tag{3.69}$$

where

$$\hat{\mathbf{y}} = \mathbf{y} + \mathbf{W}(\mathbf{z} - \mathbf{Hy}) \tag{3.70}$$

$$\hat{\mathbf{P}} = \mathbf{P} - \mathbf{WSW}^\top \tag{3.71}$$

$$\mathbf{W} = \mathbf{PH}^\top \mathbf{S}^{-1} \tag{3.72}$$

$$\mathbf{S} = \mathbf{HPH}^\top + \mathbf{R}. \tag{3.73}$$

3.3 Least squares

In this section, we are going on a little excursion to a slightly different setting. Assume that we are given only a data vector $\mathbf{z}$ and want to infer an estimate $\hat{\mathbf{x}}$ of the state variable $\mathbf{x}$. As in the Kalman filter scenario, a linear dependence and additive normal distributed noise is assumed so that the conditional pdf of $\mathbf{z}$ is given by

$$p(\mathbf{z}|\mathbf{x}) = \mathcal{N}(\mathbf{z};\, \mathbf{Hx},\, \mathbf{R}) \tag{3.74}$$

where $\mathbf{R} = \text{cov}[\mathbf{z}]$ and $\mathbf{H}$ is the measurement matrix. Without prior information, an estimate of the state can be computed by the *"maximum likelihood estimator"* (MLE):

$$\hat{\mathbf{x}} = \arg\max_{\mathbf{x}} p(\mathbf{z}|\mathbf{x}) \tag{3.75}$$

which in our case is of course equivalent to minimizing the negative argument of the exponential function:

$$\hat{\mathbf{x}} = \arg\min_{\mathbf{x}} \left\{ {}^1\!/_2 \, (\mathbf{z} - \mathbf{H}\mathbf{x})^\top \mathbf{R}^{-1} (\mathbf{z} - \mathbf{H}\mathbf{x}) \right\}. \tag{3.76}$$

Since the minimization objective function is convex, having a zero-gradient is a sufficient condition for the solution. One may therefore write

$$0 = \nabla_{\hat{\mathbf{x}}} \left[{}^1\!/_2 \, (\mathbf{z} - \mathbf{H}\hat{\mathbf{x}})^\top \mathbf{R}^{-1} (\mathbf{z} - \mathbf{H}\hat{\mathbf{x}}) \right] \tag{3.77}$$

$$= (\mathbf{z} - \mathbf{H}\hat{\mathbf{x}})^\top \mathbf{R}^{-1} \mathbf{H}. \tag{3.78}$$

By transposing both sides of the equation one obtains

$$0 = \mathbf{H}^\top \mathbf{R}^{-1} \mathbf{z} - \mathbf{H}^\top \mathbf{R}^{-1} \mathbf{H} \hat{\mathbf{x}}. \tag{3.79}$$

Solving for $\hat{\mathbf{x}}$ yields

$$\hat{\mathbf{x}} = (\mathbf{H}^\top \mathbf{R}^{-1} \mathbf{H})^{-1} \mathbf{H}^\top \mathbf{R}^{-1} \mathbf{z}. \tag{3.80}$$

In particular, the state is called *"observable"* based on the information contained in $\mathbf{z}$, if the inverse of the term $\mathbf{H}^\top \mathbf{R}^{-1} \mathbf{H}$ exists. In the context of data fusion, it is important to compute the covariance matrix of an estimate. For the Least Squares (LS) method, one has that

$$\begin{aligned}
\text{cov}[\hat{\mathbf{x}}] &= \mathrm{E}\left[(\mathbf{x} - \hat{\mathbf{x}})(\mathbf{x} - \hat{\mathbf{x}})^\top\right] \\
&= \mathrm{E}\left[(\mathbf{x} - (\mathbf{H}^\top \mathbf{R}^{-1} \mathbf{H})^{-1} \mathbf{H}^\top \mathbf{R}^{-1} \mathbf{z})(\mathbf{x} - (\mathbf{H}^\top \mathbf{R}^{-1} \mathbf{H})^{-1} \mathbf{H}^\top \mathbf{R}^{-1} \mathbf{z})^\top\right] \\
&= \mathrm{E}\left[(\mathbf{H}^\top \mathbf{R}^{-1} \mathbf{H})^{-1}((\mathbf{H}^\top \mathbf{R}^{-1} \mathbf{H})\mathbf{x} - \mathbf{H}^\top \mathbf{R}^{-1} \mathbf{z})((\mathbf{H}^\top \mathbf{R}^{-1} \mathbf{H})\mathbf{x} - \mathbf{H}^\top \mathbf{R}^{-1} \mathbf{z})^\top (\mathbf{H}^\top \mathbf{R}^{-1} \mathbf{H})^{-1}\right] \\
&= \mathrm{E}\left[(\mathbf{H}^\top \mathbf{R}^{-1} \mathbf{H})^{-1} \mathbf{H}^\top \mathbf{R}^{-1} (\mathbf{H}\mathbf{x} - \mathbf{z})(\mathbf{H}\mathbf{x} - \mathbf{z})^\top (\mathbf{H}^\top \mathbf{R}^{-1})^\top (\mathbf{H}^\top \mathbf{R}^{-1} \mathbf{H})^{-1}\right] \\
&= (\mathbf{H}^\top \mathbf{R}^{-1} \mathbf{H})^{-1} \mathbf{H}^\top \mathbf{R}^{-1} \mathrm{E}\left[(\mathbf{H}\mathbf{x} - \mathbf{z})(\mathbf{H}\mathbf{x} - \mathbf{z})^\top\right] \mathbf{R}^{-1} \mathbf{H} (\mathbf{H}^\top \mathbf{R}^{-1} \mathbf{H})^{-1} \\
&= (\mathbf{H}^\top \mathbf{R}^{-1} \mathbf{H})^{-1}.
\end{aligned} \tag{3.81}$$

Here it was used that $\mathrm{E}\left[(\mathbf{H}\mathbf{x} - \mathbf{z})(\mathbf{H}\mathbf{x} - \mathbf{z})^\top\right]$ is definition of the measurement covariance $\mathbf{R}_k$.

As a conclusion, the LS estimate and its covariance are given by the formulae

$$\mathbf{P} = (\mathbf{H}^\top \mathbf{R}^{-1} \mathbf{H})^{-1} \tag{3.82}$$

$$\hat{\mathbf{x}} = \mathbf{P}\mathbf{H}^\top \mathbf{R}^{-1} \mathbf{z}. \tag{3.83}$$

3.4 Kalman filter as a least squares solution

As stated above, the LS method may be applied, if the data $\mathbf{z}$ has a linear dependency on the state $\mathbf{x}$ with additive normal distributed noise:

$$\mathbf{z} = \mathbf{H}\mathbf{x} + \mathbf{v} \tag{3.84}$$

$$\mathbf{v} \sim \mathcal{N}(0, \mathbf{R}). \tag{3.85}$$

This implies, in particular, that the state and the data refer to the same instant of time. Since the prior estimate $\mathbf{x}_{k|k-1}$ in the Kalman filter recursion and the measurement $\mathbf{z}_k$ fulfill the requirements, one may construct the pseudo data measurement $\mathbf{z}^\star$ comprised of both:

$$\mathbf{z}^\star := \begin{pmatrix} \mathbf{x}_{k|k-1} \\ \mathbf{z}_k \end{pmatrix}. \tag{3.86}$$

Now the prior estimate is considered part of the input data for the fusion process, thus it is treated as random variable with mean $\mathbf{x}_k$ and covariance matrix $\mathbf{P}_{k|k-1}$. Therefore, one has for the mean of the constructed pseudo measurement

$$\mathrm{E}\left[\mathbf{z}^\star|\mathbf{x}_k\right] = \begin{pmatrix} \mathbf{x}_k \\ \mathbf{H}_k\mathbf{x}_k \end{pmatrix} \tag{3.87}$$

$$= \mathbf{H}^\star\mathbf{x}_k \tag{3.88}$$

$$\mathbf{H}^\star := \begin{pmatrix} \mathbf{I} \\ \mathbf{H}_k \end{pmatrix} \tag{3.89}$$

and its covariance

$$\mathrm{cov}\left[\mathbf{z}^\star|\mathbf{x}_k\right] = \mathbf{R}^\star = \begin{pmatrix} \mathbf{P}_{k|k-1} & 0 \\ 0 & \mathbf{R}_k \end{pmatrix}. \tag{3.90}$$

Filling in the LS equations from (3.82) yields the posterior covariance given by

$$\mathbf{P}_{k|k} = (\mathbf{H}^{\star\top}\mathbf{R}^{\star-1}\mathbf{H}^\star)^{-1} \tag{3.91}$$

$$= \left((\mathbf{I}\ \mathbf{H}_k^\top) \begin{pmatrix} \mathbf{P}_{k|k-1} & \\ & \mathbf{R}_k \end{pmatrix}^{-1} \begin{pmatrix} \mathbf{I} \\ \mathbf{H}_k \end{pmatrix} \right)^{-1} \tag{3.92}$$

$$= \left((\mathbf{I}\ \mathbf{H}_k^\top) \begin{pmatrix} \mathbf{P}_{k|k-1}^{-1} & \\ & \mathbf{R}_k^{-1} \end{pmatrix} \begin{pmatrix} \mathbf{I} \\ \mathbf{H}_k \end{pmatrix} \right)^{-1} \tag{3.93}$$

$$= \left((\mathbf{P}_{k|k-1}^{-1}\ \mathbf{H}_k^\top\mathbf{R}_k^{-1}) \begin{pmatrix} \mathbf{I} \\ \mathbf{H}_k \end{pmatrix} \right)^{-1} \tag{3.94}$$

$$= (\mathbf{P}_{k|k-1}^{-1} + \mathbf{H}_k^\top\mathbf{R}_k^{-1}\mathbf{H}_k)^{-1}. \tag{3.95}$$

Analogously one is able to derive the posterior mean from (3.83) as follows:

$$\begin{aligned}
\mathbf{x}_{k|k} &= \mathbf{P}_{k|k}(\mathbf{H}^{\star\top}\mathbf{R}^{\star-1}\mathbf{z}^{\star}) && (3.96)\\
&= \mathbf{P}_{k|k}\left(\begin{pmatrix}\mathbf{I} & \mathbf{H}_k^{\top}\end{pmatrix}\begin{pmatrix}\mathbf{P}_{k|k-1}^{-1} & \\ & \mathbf{R}_k^{-1}\end{pmatrix}\begin{pmatrix}\mathbf{x}_{k|k-1}\\ \mathbf{z}_k\end{pmatrix}\right) && (3.97)\\
&= \mathbf{P}_{k|k}\left(\begin{pmatrix}\mathbf{P}_{k|k-1}^{-1} & \mathbf{H}_k^{\top}\mathbf{R}_k^{-1}\end{pmatrix}\begin{pmatrix}\mathbf{x}_{k|k-1}\\ \mathbf{z}_k\end{pmatrix}\right) && (3.98)\\
&= \mathbf{P}_{k|k}(\mathbf{P}_{k|k-1}^{-1}\mathbf{x}_{k|k-1} + \mathbf{H}_k^{\top}\mathbf{R}_k^{-1}\mathbf{z}_k). && (3.99)
\end{aligned}$$

Consequently, both can be written in terms of inverted covariance matrices:

$$\begin{aligned}
\mathbf{P}_{k|k}^{-1} &= \mathbf{P}_{k|k-1}^{-1} + \mathbf{H}_k^{\top}\mathbf{R}_k^{-1}\mathbf{H}_k && (3.100)\\
\mathbf{P}_{k|k}^{-1}\mathbf{x}_{k|k} &= \mathbf{P}_{k|k-1}^{-1}\mathbf{x}_{k|k-1} + \mathbf{H}_k^{\top}\mathbf{R}_k^{-1}\mathbf{z}_k. && (3.101)
\end{aligned}$$

This motivates the introduction of "*information parameters*," since an inverse covariance matrix reflects the amount and quality of the information on the state variable.

We define

$$\begin{aligned}
\mathbf{i}_{k|k} &:= \mathbf{P}_{k|k}^{-1}\mathbf{x}_{k|k} && \text{“posterior information state”} && (3.102)\\
\mathbf{I}_{k|k} &:= \mathbf{P}_{k|k}^{-1} && \text{“posterior information matrix.”} && (3.103)
\end{aligned}$$

Likewise the prior information parameters are defined analogously as $\mathbf{i}_{k|k-1}$ and $\mathbf{I}_{k|k-1}$.

In particular in case of a multivariate normal distributed variable, it holds that the inverted covariance matrix equals the *"Fisher Information Matrix"* (FIM) [31].

Using this notation, one obtains the update equations in terms of information parameters:

$$\begin{aligned}
\mathbf{I}_{k|k} &= \mathbf{I}_{k|k-1} + \mathbf{I}_k && (3.104)\\
\mathbf{i}_{k|k} &= \mathbf{i}_{k|k-1} + \mathbf{i}_k && (3.105)\\
\mathbf{I}_k &:= \mathbf{H}_k^{\top}\mathbf{R}_k^{\top\,-1}\mathbf{H}_k && (3.106)\\
\mathbf{i}_k &:= \mathbf{H}_k^{\top}\mathbf{R}_k^{\top\,-1}\mathbf{z}_k. && (3.107)
\end{aligned}$$

This formulation of the Kalman update is called "information filter."

Thus, we can infer that uncorrelated information "adds together." The information filter form is particular useful for instance in scenarios with "zero knowledge initialization," where a track is formed without any prior. This can be expressed by setting $\mathbf{I}_{0|0} = 0$.

By means of the *Inversion Lemma* (IL, see Section A.1 in the Appendix), one can recognize that the information filter update is equivalent to the standard Kalman filter equations. One has for the covariance

$$\mathbf{P}_{k|k} = (\mathbf{P}_{k|k-1}^{-1} + \mathbf{H}_k^\top \mathbf{R}_k^{-1} \mathbf{H}_k)^{-1} \tag{3.108}$$

$$= \mathbf{P}_{k|k-1} - \mathbf{P}_{k|k-1} \mathbf{H}_k^\top (\mathbf{H}_k \mathbf{P}_{k|k-1} \mathbf{H}_k^\top + \mathbf{R}_k)^{-1} \mathbf{H}_k \mathbf{P}_{k|k-1} \tag{3.109}$$

$$= \mathbf{P}_{k|k-1} - \mathbf{P}_{k|k-1} \mathbf{H}_k^\top \mathbf{S}_k^{-1} \mathbf{H}_k \mathbf{P}_{k|k-1} \tag{3.110}$$

$$= \mathbf{P}_{k|k-1} - \mathbf{P}_{k|k-1} \mathbf{H}_k^\top \mathbf{S}_k^{-1} \mathbf{S}_k \mathbf{S}_k^{-1} \mathbf{H}_k \mathbf{P}_{k|k-1} \tag{3.111}$$

$$\mathbf{W}_{k|k-1} := \mathbf{P}_{k|k-1} \mathbf{H}_k^\top \mathbf{S}_k^{-1} \tag{3.112}$$

$$\mathbf{P}_{k|k} = \mathbf{P}_{k|k-1} - \mathbf{W}_{k|k-1} \mathbf{S}_k \mathbf{W}_{k|k-1}^\top . \tag{3.113}$$

Analogously, one obtains for the mean

$$\mathbf{x}_{k|k} = \mathbf{P}_{k|k} (\mathbf{P}_{k|k-1}^{-1} \mathbf{x}_{k|k-1} + \mathbf{H}_k^\top \mathbf{R}_k^{-1} \mathbf{z}_k) \tag{3.114}$$

$$= \mathbf{P}_{k|k} \left((\mathbf{P}_{k|k-1}^{-1} + \mathbf{H}_k^\top \mathbf{R}_k^{-1} \mathbf{H}_k - \mathbf{H}_k^\top \mathbf{R}_k^{-1} \mathbf{H}_k) \mathbf{x}_{k|k-1} + \mathbf{H}_k^\top \mathbf{R}_k^{-1} \mathbf{z}_k \right) \tag{3.115}$$

$$= \underbrace{\mathbf{P}_{k|k} \underbrace{(\mathbf{P}_{k|k-1}^{-1} + \mathbf{H}_k^\top \mathbf{R}_k^{-1} \mathbf{H}_k)}_{=\mathbf{P}_{k|k}^{-1}}}_{=\mathbf{I}} \mathbf{x}_{k|k-1} - \mathbf{P}_{k|k} \mathbf{H}_k^\top \mathbf{R}_k^{-1} \mathbf{H}_k \mathbf{x}_{k|k-1} + \mathbf{P}_{k|k} \mathbf{H}_k^\top \mathbf{R}_k^{-1} \mathbf{z}_k \tag{3.116}$$

$$= \mathbf{x}_{k|k-1} + \mathbf{P}_{k|k} \mathbf{H}_k^\top \mathbf{R}_k^{-1} (\mathbf{z}_k - \mathbf{H}_k \mathbf{x}_{k|k-1}) \tag{3.117}$$

$$= \mathbf{x}_{k|k-1} + (\mathbf{P}_{k|k-1}^{-1} + \mathbf{H}_k^\top \mathbf{R}_k^{-1} \mathbf{H}_k)^{-1} \mathbf{H}_k^\top \mathbf{R}_k^{-1} (\mathbf{z}_k - \mathbf{H}_k \mathbf{x}_{k|k-1}) \tag{3.118}$$

$$\underset{\text{IL}}{=} \mathbf{x}_{k|k-1} - \mathbf{P}_{k|k-1} \mathbf{H}_k^\top (- \mathbf{H}_k \mathbf{P}_{k|k-1} \mathbf{H}_k^\top - \mathbf{R}_k)^{-1} (\mathbf{z}_k - \mathbf{H}_k \mathbf{x}_{k|k-1}) \tag{3.119}$$

$$= \mathbf{x}_{k|k-1} + \mathbf{P}_{k|k-1} \mathbf{H}_k^\top (\mathbf{H}_k \mathbf{P}_{k|k-1} \mathbf{H}_k^\top + \mathbf{R}_k)^{-1} (\mathbf{z}_k - \mathbf{H}_k \mathbf{x}_{k|k-1}) \tag{3.120}$$

$$= \mathbf{x}_{k|k-1} + \mathbf{P}_{k|k-1} \mathbf{H}_k^\top \mathbf{S}_k^{-1} (\mathbf{z}_k - \mathbf{H}_k \mathbf{x}_{k|k-1}) \tag{3.121}$$

$$= \mathbf{x}_{k|k-1} + \mathbf{W}_{k|k-1} (\mathbf{z}_k - \mathbf{H}_k \mathbf{x}_{k|k-1}), \tag{3.122}$$

which proves the equivalence.

As a consequence, the product formula for Gaussians may be extended such that two equivalent versions exist:

$$\mathcal{N}(\mathbf{x};\, \bar{\mathbf{y}},\, \bar{\mathbf{P}}) \cdot \mathcal{N}(\mathbf{z};\, \mathbf{H}\mathbf{y},\, \mathbf{S}) = \mathcal{N}(\mathbf{x};\, \mathbf{y},\, \mathbf{P}) \cdot \mathcal{N}(\mathbf{z};\, \mathbf{H}\mathbf{x},\, \mathbf{R}) \tag{3.123}$$

$$\bar{\mathbf{P}} = \begin{cases} \mathbf{P} - \mathbf{W}\mathbf{S}\mathbf{W}^\top \\ (\mathbf{P}^{-1} + \mathbf{H}^\top\mathbf{R}^{-1}\mathbf{H})^{-1} \end{cases}$$

$$\bar{\mathbf{y}} = \begin{cases} \mathbf{y} + \mathbf{W}\boldsymbol{\nu} \\ \bar{\mathbf{P}}\left(\mathbf{P}^{-1}\mathbf{y} + \mathbf{H}^\top\mathbf{R}^{-1}\mathbf{z}\right) \end{cases} \tag{3.124}$$

$$\mathbf{S} = \mathbf{H}\mathbf{P}\mathbf{H}^\top + \mathbf{R} \tag{3.125}$$

$$\mathbf{W} = \mathbf{P}\mathbf{H}^\top\mathbf{S}^{-1} \tag{3.126}$$

$$\boldsymbol{\nu} = \mathbf{z} - \mathbf{H}\mathbf{y} \tag{3.127}$$

This product formula will play an important role in the upcoming derivations on distributed state estimation, since it provides the means to manipulate the product of two Gaussians in a way such that one of the Gaussians becomes independent of the state variable $\mathbf{x}$.

3.5 Kalman filter as an orthogonal projection

We have already seen two equivalent derivations of the Kalman update formulae. Thus, why consider a third one? In fact, the Kalman filter derived as an orthogonal projection is the most complex but also the most geometrical insightful way. Its derivations may seem a little abstract for a moment, but eventually, the theory will provide beautiful insight. In particular, the geometric interpretation enables a visual imagination, which allows for a deep understanding of the method.

One may begin with the definition of a symmetric bilinear form on the vector space of multivariate random variables:

$$\langle \mathbf{x}, \mathbf{y} \rangle := \operatorname{cov}[\mathbf{x}, \mathbf{y}] \tag{3.128}$$

$$= \mathrm{E}\left[(\mathbf{x} - \mathrm{E}[\mathbf{x}])(\mathbf{y} - \mathrm{E}[\mathbf{y}])^\top\right]. \tag{3.129}$$

It should be noted that covariance *matrices* here play the role of *scalars*. Since $\langle \mathbf{x}, \mathbf{x} \rangle$ is positive,[‖] $\langle \cdot, \cdot \rangle$ defines a matrix-valued inner product. This allows us to define a matrix-valued "norm" of a multivariate random variable $\mathbf{x}$ by

$$\|\mathbf{x}\| := \sqrt{\langle \mathbf{x}, \mathbf{x} \rangle}, \tag{3.130}$$

where the square root is some mapping such that $\sqrt{\mathbf{C}} = \mathbf{A}$, if $\mathbf{A}\mathbf{A}^\top = \mathbf{C}$. The *Cholesky* decomposition is an example. We will also use the notation $\|\mathbf{A}\|^2 = \mathbf{C}$.

[‖] In the sense that the resulting matrix is positive definite.

In addition, we define the *"linear span"* of a set of vectors $\mathcal{W} = \{\mathbf{w}_1, \ldots, \mathbf{w}_l\}$ with $\mathbf{w}_i \in \mathbb{R}^q$ as

$$\mathcal{Y}(\mathbf{w}_1, \ldots, \mathbf{w}_l) := \left\{ \mathbf{y} \,\middle|\, \exists \mathbf{P}_i \in \mathbb{R}^{n \times q} : \quad \mathbf{y} = \sum_{i=1}^{l} \mathbf{P}_i \mathbf{w}_i \right\}. \tag{3.131}$$

This definition uses matrix-valued coefficients $\mathbf{P}_i$ for the basis $\mathcal{W}$.

Let $\mathbf{x}$ be the state variable which is to be estimated. The following theorem provides a first glimpse on the expressiveness of the above definitions.

> *Theorem of the orthogonal estimate*
> *Let* $\mathcal{Y} = \mathcal{Y}(\mathbf{w}_1, \ldots, \mathbf{w}_l)$ *for* $\mathbf{w}_i \in \mathbb{R}^q$ *for* $i = 1, \ldots, l$ *and* $\mathbf{x} \in \mathbb{R}^n$. *It holds that*
>
> $$\hat{\mathbf{y}} = \arg\min_{\mathbf{y} \in \mathcal{Y}} \operatorname{tr}\left(\|\mathbf{y} - \mathbf{x}\|^2\right) \quad \Leftrightarrow \quad \langle \mathbf{x} - \hat{\mathbf{y}}, \mathbf{w}_i \rangle = 0 \quad \forall i. \tag{3.132}$$

Since $\|\mathbf{y} - \mathbf{x}\|^2$ is actually the estimation error covariance for some estimate $\mathbf{y}$ of the state $\mathbf{x}$, the statement means that the "minimum variance estimate" is equivalent to an orthogonal projection of $\mathbf{x}$ onto $\mathcal{Y}$, where the estimation error $\mathbf{x} - \hat{\mathbf{y}}$ is orthogonal to the vectors in $\mathcal{Y}$. In this sense, the minimum variance estimate is the closest point to the span of $\mathcal{W}$. A visualization of the orthogonal estimate can be found in Figure 3.5.

This can be seen as follows:

Proof of "$\Rightarrow$":
Assume $\langle \mathbf{x} - \hat{\mathbf{y}}, \mathbf{w}_i \rangle := \mathbf{C} > 0$ for a fixed i. Denote $\mathbf{A} = \operatorname{cov}[\mathbf{w}_i]$. Since $\mathbf{A}$ is symmetric and positive definite it holds that $\mathbf{C}\mathbf{A}^{-1}\mathbf{C}^\top$ is so as well. Therefore, $\operatorname{tr}\left(\mathbf{C}\mathbf{A}^{-1}\mathbf{C}\right) > 0$.
Set $\bar{\mathbf{y}} := \hat{\mathbf{y}} - \mathbf{C}\mathbf{A}^{-1}\mathbf{w}_i \in \mathcal{Y}$. Then

$$\begin{aligned}
\operatorname{tr}\left(\|\bar{\mathbf{y}} - \mathbf{x}\|^2\right) &= \operatorname{tr}\left(\operatorname{cov}[\bar{\mathbf{y}} - \mathbf{x}]\right) && (3.133)\\
&= \operatorname{tr}\left(\operatorname{cov}\left[\hat{\mathbf{y}} - \mathbf{C}\mathbf{A}^{-1}\mathbf{w}_i - \mathbf{x}\right]\right) && (3.134)\\
&= \operatorname{tr}\left(\operatorname{cov}\left[\hat{\mathbf{y}} - \mathbf{x}\right]\right) - \operatorname{tr}\left(\mathbf{C}\mathbf{A}^{-1}\operatorname{cov}\left[\mathbf{w}_i, (\hat{\mathbf{y}} - \mathbf{x})\right]\right)\\
&\quad - \operatorname{tr}\left(\operatorname{cov}\left[(\hat{\mathbf{y}} - \mathbf{x}), \mathbf{w}_i\right]\mathbf{A}^{-1}\mathbf{C}\right) + \operatorname{tr}\left(\mathbf{C}\mathbf{A}^{-1}\operatorname{cov}[\mathbf{w}_i]\,\mathbf{A}^{-1}\mathbf{C}\right) && (3.135)\\
&= \operatorname{tr}\left(\operatorname{cov}\left[\hat{\mathbf{y}} - \mathbf{x}\right] - \mathbf{C}\mathbf{A}^{-1}\mathbf{C} - \mathbf{C}\mathbf{A}^{-1}\mathbf{C} + \mathbf{C}\mathbf{A}^{-1}\mathbf{A}\mathbf{A}^{-1}\mathbf{C}\right) && (3.136)\\
&= \operatorname{tr}\left(\left\|\hat{\mathbf{y}} - \mathbf{x}\right\|^2\right) - \operatorname{tr}\left(\mathbf{C}\mathbf{A}^{-1}\mathbf{C}\right). && (3.137)
\end{aligned}$$

This contradicts the definition of $\hat{\mathbf{y}}$.

Proof of "$\Leftarrow$":
Let $\langle \mathbf{x} - \hat{\mathbf{y}}, \mathbf{w}_i \rangle = 0$ for all $i = 1, \ldots, l$. For a given $\mathbf{y} \in \mathcal{Y}$ it holds that $\boldsymbol{\nu} = \mathbf{y} - \hat{\mathbf{y}} \in \mathcal{Y}$. Therefore, $\boldsymbol{\nu} = \sum_i \mathbf{P}_i \mathbf{w}_i$ for some $\mathbf{P}_i$. One obtains

$$\operatorname{tr}\left(\|\mathbf{y} - \mathbf{x}\|^2\right) = \operatorname{tr}\left(\left\|\hat{\mathbf{y}} + \boldsymbol{\nu} - \mathbf{x}\right\|^2\right) \tag{3.138}$$

$$= \operatorname{tr}\left(\operatorname{cov}\left[\hat{\mathbf{y}} - \mathbf{x}\right] + \operatorname{cov}\left[(\hat{\mathbf{y}} - \mathbf{x}), \boldsymbol{\nu}\right] + \operatorname{cov}\left[\boldsymbol{\nu}, (\hat{\mathbf{y}} - \mathbf{x})\right] + \operatorname{cov}\left[\boldsymbol{\nu}\right]\right). \tag{3.139}$$

Due to the construction of $\boldsymbol{\nu}$ and the linearity of the covariance, it holds that

$$\operatorname{cov}\left[(\hat{\mathbf{y}} - \mathbf{x}), \boldsymbol{\nu}\right] = \sum_{i=1}^{l} \operatorname{cov}\left[(\hat{\mathbf{y}} - \mathbf{x}), \mathbf{w}_i\right] \mathbf{P}_i^\top = 0. \tag{3.140}$$

Using (3.140) in (3.139) yields

$$\operatorname{tr}\left(\|\mathbf{y} - \mathbf{x}\|^2\right) = \operatorname{tr}\left(\left\|\hat{\mathbf{y}} - \mathbf{x}\right\|^2\right) + \operatorname{tr}\left(\|\boldsymbol{\nu}\|^2\right) > \operatorname{tr}\left(\left\|\hat{\mathbf{y}} - \mathbf{x}\right\|^2\right). \tag{3.141}$$

This concludes the proof.

The concept of the Kalman filter update in this geometric theory is now, to construct the orthogonal projection such that it obeys both sides of the Orthogonal Estimate Theorem (3.132). As a consequence, one will see that Kalman estimate is the minimum variance estimate.

To this end, we define the *"measurement span"* $\mathcal{Y}^k$

$$\mathcal{Y}^k := \mathcal{Y}(\mathbf{z}_1, \ldots, \mathbf{z}_k), \tag{3.142}$$

the estimate $\hat{\mathbf{y}}_k$ from the measurement span

$$\hat{\mathbf{y}}_k := \arg\min_{\mathbf{y} \in \mathcal{Y}^k} \operatorname{tr}\left(\|(\mathbf{x}_k - \mathbf{y})\|^2\right), \tag{3.143}$$

and the innovation $\boldsymbol{\nu}_k$

$$\boldsymbol{\nu}_k := \mathbf{z}_k - \mathbf{H}_k \bar{\mathbf{y}}_{k-1} \tag{3.144}$$

with

$$\bar{\mathbf{y}}_{k-1} := \arg\min_{\mathbf{y} \in \mathcal{Y}^{k-1}} \operatorname{tr}\left(\|(\mathbf{x}_k - \mathbf{y})\|^2\right). \tag{3.145}$$

It holds for the current measurement noise $\mathbf{v}_k$ that $\operatorname{cov}[(\mathbf{x}_k - \bar{\mathbf{y}}_{k-1}), \mathbf{v}_k] = 0$ since $\mathbf{z}_k$ is not included in $\mathcal{Y}^{k-1}$. Therefore, it follows from $\mathbf{z}_k = \mathbf{H}_k \mathbf{x}_k + \mathbf{v}_k$ and $\operatorname{cov}[\mathbf{v}_k] = \mathbf{R}_k$ that

$$\langle \boldsymbol{\nu}_k, \boldsymbol{\nu}_k \rangle = \|\boldsymbol{\nu}_k\|^2 = \mathbf{H}_k \operatorname{cov}\left[(\mathbf{x}_k - \bar{\mathbf{y}}_{k-1})\right] \mathbf{H}_k^\top + \mathbf{R}_k. \tag{3.146}$$

For two time indices l and k with $l \neq k$, we may assume w.l.o.g.[¶] that $k > l$. One has due to statistical independence that

$$\langle \mathbf{v}_k, \mathbf{v}_l \rangle = 0, \tag{3.147}$$

$$\langle \mathbf{v}_k, (\mathbf{x}_l - \bar{\mathbf{y}}_{l-1}) \rangle = 0, \tag{3.148}$$

where $\mathbf{x}_l - \bar{\mathbf{y}}_{l-1}$ is the prior estimation error at time l. Moreover, we denote $\bar{\mathbf{y}}_{l-1} =: \sum_{i=1}^{l-1} \bar{\mathbf{P}}_i \mathbf{z}_i$. Using these equations, one obtains that

$$\begin{aligned}
\langle \boldsymbol{\nu}_k, \boldsymbol{\nu}_l \rangle &= \langle \mathbf{z}_k - \mathbf{H}_k \bar{\mathbf{y}}_{k-1}, \mathbf{z}_l - \mathbf{H}_l \bar{\mathbf{y}}_{l-1} \rangle && (3.149)\\
&= \langle \mathbf{H}_k(\mathbf{x}_k - \bar{\mathbf{y}}_{k-1}) + \mathbf{v}_k, \mathbf{H}_l(\mathbf{x}_l - \bar{\mathbf{y}}_{l-1}) + \mathbf{v}_l \rangle && (3.150)\\
&= \mathbf{H}_k \langle (\mathbf{x}_k - \bar{\mathbf{y}}_{k-1}), (\mathbf{x}_l - \bar{\mathbf{y}}_{l-1}) \rangle \mathbf{H}_l^\top + \langle \mathbf{v}_k, (\mathbf{x}_l - \bar{\mathbf{y}}_{l-1}) \rangle \mathbf{H}_l^\top \\
&\quad + \mathbf{H}_k \langle (\mathbf{x}_k - \bar{\mathbf{y}}_{k-1}), \mathbf{v}_l \rangle + \langle \mathbf{v}_k, \mathbf{v}_l \rangle && (3.151)\\
&= \mathbf{H}_k \langle (\mathbf{x}_k - \bar{\mathbf{y}}_{k-1}), (\mathbf{x}_l - \bar{\mathbf{y}}_{l-1}) \rangle \mathbf{H}_l^\top + \mathbf{H}_k \langle (\mathbf{x}_k - \bar{\mathbf{y}}_{k-1}), \mathbf{v}_l \rangle && (3.152)\\
&= \mathbf{H}_k \langle (\mathbf{x}_k - \bar{\mathbf{y}}_{k-1}), \mathbf{H}_l(\mathbf{x}_l - \bar{\mathbf{y}}_{l-1}) + \mathbf{v}_l \rangle && (3.153)\\
&= \mathbf{H}_k \langle (\mathbf{x}_k - \bar{\mathbf{y}}_{k-1}), \mathbf{z}_l - \mathbf{H}_l \bar{\mathbf{y}}_{l-1} \rangle && (3.154)\\
&= \mathbf{H}_k \langle (\mathbf{x}_k - \bar{\mathbf{y}}_{k-1}), \mathbf{z}_l \rangle - \mathbf{H}_k \langle (\mathbf{x}_k - \bar{\mathbf{y}}_{k-1}), \bar{\mathbf{y}}_{l-1} \rangle \mathbf{H}_l^\top && (3.155)\\
&= \mathbf{H}_k \langle (\mathbf{x}_k - \bar{\mathbf{y}}_{k-1}), \mathbf{z}_l \rangle - \sum_{i=1}^{l-1} \mathbf{H}_k \langle (\mathbf{x}_k - \bar{\mathbf{y}}_{k-1}), \mathbf{z}_i \rangle \bar{\mathbf{P}}_i^\top \mathbf{H}_l^\top && (3.156)\\
&= 0. && (3.157)
\end{aligned}$$

The last equation holds due to the Theorem of the Orthogonal Estimate.

As a consequence, one may conclude that the innovations $\{\boldsymbol{\nu}_l\}_l$ form an orthogonal series:

$$\langle \boldsymbol{\nu}_k, \boldsymbol{\nu}_l \rangle = \begin{cases} \mathbf{H}_k \mathrm{cov}\left[(\mathbf{x}_k - \bar{\mathbf{y}}_{k-1})\right] \mathbf{H}_k^\top + \mathbf{R}_k, & \text{if } l = k \\ 0 & \text{else.} \end{cases} \tag{3.158}$$

Therefore, the vectors

$$\mathbf{e}_j := \left(\sqrt{\mathbf{S}_j}\right)^{-1} \boldsymbol{\nu}_j \tag{3.159}$$

with $\mathbf{S}_j := \mathrm{cov}\left[\boldsymbol{\nu}_j\right]$ define an orthonormal series with

$$\langle \mathbf{e}_i, \mathbf{e}_j \rangle = \begin{cases} \mathbf{I}, & \text{if } i = j \\ 0, & \text{else.} \end{cases} \tag{3.160}$$

This enables us to define the orthogonal projection of the true state $\mathbf{x}_k$:

$$\hat{\mathbf{x}}_k = \sum_{i=1}^{k} \langle \mathbf{x}_k, \mathbf{e}_i \rangle \mathbf{e}_i \tag{3.161}$$

¶Without loss of generality.

with $\hat{\mathbf{x}}_k \in \mathcal{Y}^k$. This motivates the geographical interpretation of the minimum variance estimate as a geometrical projection onto the measurement span as illustrated in Figure 3.5.

We observe that

$$\langle \hat{\mathbf{x}}_k, \mathbf{e}_j \rangle = \sum_{i=1}^{k} \langle \mathbf{x}_k, \mathbf{e}_i \rangle \langle \mathbf{e}_i, \mathbf{e}_j \rangle \tag{3.162}$$

$$= \langle \mathbf{x}_k, \mathbf{e}_j \rangle, \tag{3.163}$$

and therefore

$$\langle \hat{\mathbf{x}}_k - \mathbf{x}_k, \mathbf{e}_j \rangle = 0. \tag{3.164}$$

Since $\bar{\mathbf{y}}_i \in \mathcal{Y}^k$ for all i, it can easily be seen that $\mathcal{Y}(\mathbf{e}_1, \ldots, \mathbf{e}_k) = \mathcal{Y}^k$. Therefore, one obtains that

$$\langle \hat{\mathbf{x}}_k - \mathbf{x}_k, \mathbf{z}_j \rangle = 0 \tag{3.165}$$

for all j. As a consequence of this construction, we may apply the Theorem of Orthogonal Projection which says that $\hat{\mathbf{x}}_k$ is the minimum variance estimate:

$$\hat{\mathbf{x}}_k = \arg\min_{\mathbf{y} \in \mathcal{Y}^k} \{\mathrm{tr}\,(\mathrm{cov}\,[(\mathbf{x} - \mathbf{y})])\}. \tag{3.166}$$

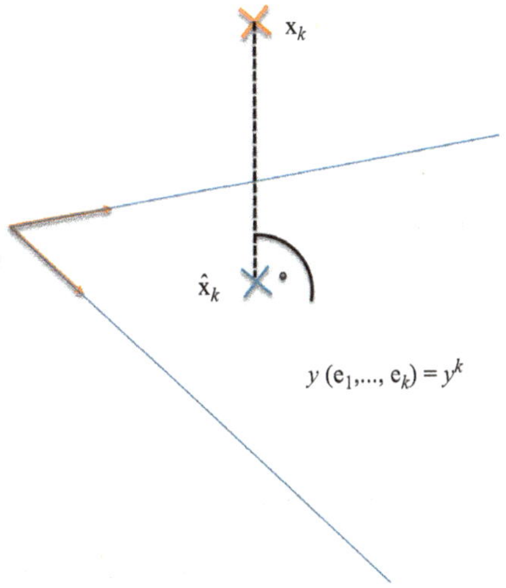

Figure 3.5 The construction of the estimate $\hat{\mathbf{x}}_k$ is the orthogonal projection of the true state $\mathbf{x}_k$ onto the linear span $\mathcal{Y}^k$. Based on the Theorem of the Orthogonal Estimate we know that this is the Minimum Error Variance mean.

In order to obtain the Kalman filter equations, a recursive formulation is required. Using the definition from above and the known state transition model $\mathbf{x}_k = \mathbf{F}_{k|k-1}\mathbf{x}_{k-1} + \mathbf{w}_{k|k-1}$ as derived in the previous chapter, one can infer

$$\hat{\mathbf{x}}_k = \sum_{i=1}^{k} \langle \mathbf{x}_k, \mathbf{e}_i \rangle \mathbf{e}_i \tag{3.167}$$

$$= \sum_{i=1}^{k-1} \langle \mathbf{x}_k, \mathbf{e}_i \rangle \mathbf{e}_i + \langle \mathbf{x}_k, \mathbf{e}_k \rangle \mathbf{e}_k \tag{3.168}$$

$$= \sum_{i=1}^{k-1} \langle \mathbf{F}_{k|k-1}\mathbf{x}_{k-1} + \mathbf{w}_{k|k-1}, \mathbf{e}_i \rangle \mathbf{e}_i + \langle \mathbf{x}_k, \mathbf{e}_k \rangle \mathbf{e}_k \tag{3.169}$$

$$= \mathbf{F}_{k|k-1} \sum_{i=1}^{k-1} \langle \mathbf{x}_{k-1}, \mathbf{e}_i \rangle \mathbf{e}_i + \sum_{i=1}^{k-1} \langle \mathbf{w}_{k|k-1}, \mathbf{e}_i \rangle \mathbf{e}_i + \langle \mathbf{x}_k, \mathbf{e}_k \rangle \mathbf{e}_k \tag{3.170}$$

$$= \mathbf{F}_{k|k-1} \sum_{i=1}^{k-1} \langle \mathbf{x}_{k-1}, \mathbf{e}_i \rangle \mathbf{e}_i + \langle \mathbf{x}_k, \mathbf{e}_k \rangle \mathbf{e}_k \tag{3.171}$$

$$= \mathbf{F}_{k|k-1}\hat{\mathbf{x}}_{k-1} + \langle \mathbf{x}_k, \mathbf{e}_k \rangle \mathbf{e}_k. \tag{3.172}$$

Therefore, we define

$$\mathbf{x}_{k|k-1} := \mathbf{F}_{k|k-1}\hat{\mathbf{x}}_{k-1} \tag{3.173}$$

$$\mathbf{x}_{k|k} := \mathbf{x}_{k|k-1} + \langle \mathbf{x}_k, \mathbf{e}_k \rangle \mathbf{e}_k \tag{3.174}$$

Due to the statistical independence of $\mathbf{x}_{k|k-1}$ and $\mathbf{e}_k$ which follows from the fact that the normalized innovation $\mathbf{e}_k$ is not contained in the prior estimate $\mathbf{x}_{k|k-1}$, one has that

$$\operatorname{cov}\left[\mathbf{z}_k - \mathbf{H}_k\mathbf{x}_{k|k-1}, \mathbf{e}_k\right] = \operatorname{cov}\left[\mathbf{z}_k, \mathbf{e}_k\right] \tag{3.175}$$

$$= \operatorname{cov}\left[\boldsymbol{\nu}_k + \mathbf{H}_k\bar{\mathbf{y}}_{k-1} - \mathbf{H}_k\mathbf{x}_{k|k-1}, \mathbf{e}_k\right] \tag{3.176}$$

$$= \operatorname{cov}\left[\boldsymbol{\nu}_k, \mathbf{e}_k\right] \tag{3.177}$$

$$= \operatorname{cov}\left[\boldsymbol{\nu}_k, \sqrt{\mathbf{S}_k}^{-1}\boldsymbol{\nu}_k\right] \tag{3.178}$$

$$= \operatorname{cov}\left[\boldsymbol{\nu}_k, \boldsymbol{\nu}_k\right]\sqrt{\mathbf{S}_k}^{-1} \tag{3.179}$$

$$= \sqrt{\mathbf{S}_k}. \tag{3.180}$$

On the other hand, we find that for $j < k$ when using the measurement model $\mathbf{z}_k = \mathbf{H}_k\mathbf{x}_k + \mathbf{v}_k$:

$$\begin{aligned}
\operatorname{cov}\left[\mathbf{z}_k - \mathbf{H}_k\mathbf{x}_{k|k-1}, \mathbf{e}_j\right] &= \operatorname{cov}\left[\mathbf{H}_k\mathbf{x}_k + \mathbf{v}_k - \mathbf{H}_k\mathbf{x}_{k|k-1}, \mathbf{e}_j\right] && (3.181)\\
&= \operatorname{cov}\left[\mathbf{H}_k\mathbf{x}_k - \mathbf{H}_k\mathbf{x}_{k|k-1}, \mathbf{e}_j\right] && (3.182)\\
&= \mathbf{H}_k\operatorname{cov}\left[\mathbf{x}_k - \mathbf{x}_{k|k-1}, \mathbf{e}_j\right] && (3.183)\\
&= \mathbf{H}_k\operatorname{cov}\left[\mathbf{x}_k - \mathbf{x}_{k|k} + \langle\mathbf{x}_k, \mathbf{e}_k\rangle\mathbf{e}_k, \mathbf{e}_j\right] && (3.184)\\
&= \mathbf{H}_k\operatorname{cov}\left[\mathbf{x}_k - \mathbf{x}_{k|k}, \mathbf{e}_j\right] + \langle\mathbf{x}_k, \mathbf{e}_k\rangle\operatorname{cov}\left[\mathbf{e}_k, \mathbf{e}_j\right] && (3.185)\\
&= 0. && (3.186)
\end{aligned}$$

Since $\mathbf{z}_k - \mathbf{H}_k\mathbf{x}_{k|k-1} \in \mathcal{Y}^k = \mathcal{Y}(\mathbf{e}_1, \ldots, \mathbf{e}_k)$, there is a representation of the form

$$\mathbf{z}_k - \mathbf{H}_k\mathbf{x}_{k|k-1} =: \sum_{i=1}^{k} \mathbf{P}_i^{(\nu)}\mathbf{e}_i. \tag{3.187}$$

However, due to the previous results, we have shown that

$$\operatorname{cov}\left[\sum_{i=1}^{k} \mathbf{P}_i^{(\nu)}\mathbf{e}_i, \mathbf{e}_j\right] = \begin{cases} \sqrt{\mathbf{S}_k}, & \text{if } j = k \\ 0, & \text{else.} \end{cases} \tag{3.188}$$

Therefore, it holds that

$$\mathbf{P}_i^{(\nu)} = \begin{cases} \sqrt{\mathbf{S}_k}, & \text{if } i = k \\ 0, & \text{else.} \end{cases} \tag{3.189}$$

Thus, one may write

$$\mathbf{z}_k - \mathbf{H}_k\mathbf{x}_{k|k-1} = \sqrt{\mathbf{S}_k}\mathbf{e}_k. \tag{3.190}$$

As a consequence, the correction term for the posterior in (3.174) may be rewritten as

$$\begin{aligned}
\langle\mathbf{x}_k, \mathbf{e}_k\rangle\mathbf{e}_k &= \langle\mathbf{x}_k, \mathbf{e}_k\rangle\sqrt{\mathbf{S}_k}^{-1}\sqrt{\mathbf{S}_k}\mathbf{e}_k && (3.191)\\
&= \langle\mathbf{x}_k, \mathbf{e}_k\rangle\sqrt{\mathbf{S}_k}^{-1}(\mathbf{z}_k - \mathbf{H}_k\mathbf{x}_{k|k-1}). && (3.192)
\end{aligned}$$

Thus, one obtains a gain matrix $\mathbf{W}_{k|k-1}$ with

$$\begin{aligned}
\mathbf{x}_{k|k} &= \mathbf{x}_{k|k-1} + \mathbf{W}_{k|k-1}(\mathbf{z}_k - \mathbf{H}_k\mathbf{x}_{k|k-1}) && (3.193)\\
\mathbf{W}_{k|k-1} &:= \langle\mathbf{x}_k, \mathbf{e}_k\rangle\sqrt{\mathbf{S}_k}^{-1}. && (3.194)
\end{aligned}$$

As in the Kalman filter convention, we denote

$$\begin{aligned}
\mathbf{P}_{k|k} &:= \operatorname{cov}\left[\mathbf{x}_k - \mathbf{x}_{k|k}\right] && (3.195)\\
\mathbf{P}_{k|k-1} &:= \operatorname{cov}\left[\mathbf{x}_k - \mathbf{x}_{k|k-1}\right] && (3.196)\\
&= \mathbf{F}_{k|k-1}\mathbf{P}_{k-1|k-1}\mathbf{F}_{k|k-1}^{\top} + \mathbf{Q}_{k|k-1}. && (3.197)
\end{aligned}$$

Now it holds that

$$\mathrm{E}\left[\mathbf{x}_k|\mathcal{Z}^{k-1}\right] = \mathbf{x}_{k|k-1} \tag{3.198}$$

$$\mathrm{E}\left[\mathbf{x}_{k|k-1}|\mathcal{Z}^{k-1}\right] = \mathbf{x}_{k-1}. \tag{3.199}$$

Therefore, one has

$$\mathrm{cov}\left[(\mathbf{x}_{k-1} - \mathbf{x}_{k-1|k-1}), \mathbf{x}_{k-1|k-1}\right] = 0 \tag{3.200}$$

and analogously

$$\mathrm{cov}\left[(\mathbf{x}_k - \mathbf{x}_{k|k}), \mathbf{z}_k\right] = 0. \tag{3.201}$$

The term $\mathbf{x}_k - \mathbf{x}_{k|k}$ can be written as

$$\mathbf{x}_k - \mathbf{x}_{k|k} = \mathbf{F}_{k|k-1}\mathbf{x}_{k-1} + \mathbf{w}_{k|k-1} - \mathbf{F}_{k|k-1}\mathbf{x}_{k-1|k-1} - \mathbf{W}_{k|k-1}(\mathbf{z}_k - \mathbf{H}_k\mathbf{x}_{k|k-1}) \tag{3.202}$$

$$= \mathbf{F}_{k|k-1}(\mathbf{x}_{k-1} - \mathbf{x}_{k-1|k-1}) + \mathbf{w}_{k|k-1} - \mathbf{W}_{k|k-1}(\mathbf{z}_k - \mathbf{H}_k\mathbf{x}_{k|k-1}). \tag{3.203}$$

Using the measurement model $\mathbf{z}_k = \mathbf{H}_k\mathbf{x}_k + \mathbf{v}_k$ yields

$$\mathbf{W}_{k|k-1}(\mathbf{z}_k - \mathbf{H}_k\mathbf{x}_{k|k-1}) = \mathbf{W}_{k|k-1}(\mathbf{H}_k\mathbf{x}_k + \mathbf{v}_k - \mathbf{H}_k\mathbf{x}_{k|k-1}) \tag{3.204}$$

$$= \mathbf{W}_{k|k-1}\mathbf{H}_k(\mathbf{x}_k - \mathbf{x}_{k|k-1}) + \mathbf{W}_{k|k-1}\mathbf{v}_k \tag{3.205}$$

$$= \mathbf{W}_{k|k-1}\mathbf{H}_k(\mathbf{F}_{k|k-1}(\mathbf{x}_{k-1} - \mathbf{x}_{k-1|k-1}) + \mathbf{w}_{k|k-1}) - \mathbf{W}_{k|k-1}\mathbf{v}_k \tag{3.206}$$

One can now apply (3.206) to (3.203) and obtain

$$\begin{aligned}\mathbf{x}_k - \mathbf{x}_{k|k} &= (\mathbf{I} - \mathbf{W}_{k|k-1}\mathbf{H}_k)\mathbf{F}_{k|k-1}(\mathbf{x}_{k-1} - \mathbf{x}_{k-1|k-1}) \\ &\quad + (\mathbf{I} - \mathbf{W}_{k|k-1}\mathbf{H}_k)\mathbf{w}_{k|k-1} + \mathbf{W}_{k|k-1}\mathbf{v}_k.\end{aligned} \tag{3.207}$$

On the other hand, one may use the evolution model, the measurement model and an additive zero to write $\mathbf{z}_k$ as

$$\mathbf{z}_k = \mathbf{H}_k(\mathbf{F}_{k|k-1}(\mathbf{x}_{k-1} - \mathbf{x}_{k-1|k-1} + \mathbf{x}_{k-1|k-1}) + \mathbf{H}_k\mathbf{w}_{k|k-1} + \mathbf{v}_k. \tag{3.208}$$

In the next step, one applies (3.207) and (3.208) in (3.201). This leads to the following derivation:

$$\begin{aligned}0 &= \mathrm{cov}\left[(\mathbf{x}_k - \mathbf{x}_{k|k}), \mathbf{z}_k\right] \\ &= \mathrm{cov}\left[(\mathbf{I} - \mathbf{W}_{k|k-1}\mathbf{H}_k)\mathbf{F}_{k|k-1}(\mathbf{x}_{k-1} - \mathbf{x}_{k-1|k-1}), \mathbf{H}_k\mathbf{F}_{k|k-1}(\mathbf{x}_{k-1} - \mathbf{x}_{k-1|k-1} + \mathbf{x}_{k-1|k-1}) + \mathbf{H}_k\mathbf{w}_{k|k-1} + \mathbf{v}_k\right] \\ &\quad + \mathrm{cov}\left[(\mathbf{I} - \mathbf{W}_{k|k-1}\mathbf{H}_k)\mathbf{w}_{k|k-1}, \mathbf{H}_k\mathbf{F}_{k|k-1}(\mathbf{x}_{k-1} - \mathbf{x}_{k-1|k-1} + \mathbf{x}_{k-1|k-1}) + \mathbf{H}_k\mathbf{w}_{k|k-1} + \mathbf{v}_k\right] \\ &\quad - \mathrm{cov}\left[\mathbf{W}_{k|k-1}\mathbf{v}_k, \mathbf{H}_k\mathbf{F}_{k|k-1}\mathbf{x}_{k-1} + \mathbf{H}_k\mathbf{w}_{k|k-1} + \mathbf{v}_k\right].\end{aligned} \tag{3.209}$$

Further calculation with the stochastic independence of $\mathbf{w}_{k|k-1}$, $\mathbf{v}_k$, and $\mathbf{x}_{k-1} - \mathbf{x}_{k-1|k-1}$ yields

$$\begin{aligned}0 &= (\mathbf{I} - \mathbf{W}_{k|k-1}\mathbf{H}_k)\mathbf{F}_{k|k-1}\mathbf{P}_{k-1|k-1}\mathbf{F}_{k|k-1}^{\top}\mathbf{H}_k^{\top} \\ &\quad + (\mathbf{I} - \mathbf{W}_{k|k-1}\mathbf{H}_k)\mathbf{Q}_{k|k-1}\mathbf{H}_k^{\top} + \mathbf{W}_{k|k-1}\mathbf{R}_k\end{aligned} \tag{3.210}$$

$$= (\mathbf{I} - \mathbf{W}_{k|k-1}\mathbf{H}_k)\mathbf{P}_{k|k-1}\mathbf{H}_k^{\top} + \mathbf{W}_{k|k-1}\mathbf{R}_k. \tag{3.211}$$

One can now solve the above equation for $\mathbf{W}_{k|k-1}$:

$$\mathbf{W}_{k|k-1}\left(\mathbf{H}_k\mathbf{P}_{k|k-1}\mathbf{H}_k^\top + \mathbf{R}_k\right) = \mathbf{P}_{k|k-1}\mathbf{H}_k^\top \tag{3.212}$$

and obtain

$$\mathbf{W}_{k|k-1} = \mathbf{P}_{k|k-1}\mathbf{H}_k^\top\left(\mathbf{H}_k\mathbf{P}_{k|k-1}\mathbf{H}_k^\top + \mathbf{R}_k\right)^{-1}, \tag{3.213}$$

which represents the Kalman gain matrix.

The last missing piece for the full derivation of the Kalman filter update formulae is the posterior covariance matrix $\mathbf{P}_{k|k}$. This is quite straightforward though we will first find the "*Joseph form*." For the derivation, the $(\cdot)^2$ notation for vectors is used for the sake of notational simplicity. The second factor is transposed as usual for the covariance matrices:

$$\begin{aligned}
\mathbf{P}_{k|k} &= \mathrm{E}\left[(\mathbf{x}_k - \mathbf{x}_{k|k})^2\right] && (3.214)\\
&= \mathrm{E}\left[(\mathbf{x}_k - \mathbf{x}_{k|k-1} - \mathbf{W}_{k|k-1}(\mathbf{z}_k - \mathbf{H}_k\mathbf{x}_{k|k-1}))^2\right] && (3.215)\\
&= \mathrm{E}\left[(\mathbf{x}_k - \mathbf{x}_{k|k-1} - \mathbf{W}_{k|k-1}(\mathbf{H}_k\mathbf{x}_k + \mathbf{v}_k - \mathbf{H}_k\mathbf{x}_{k|k-1}))^2\right] && (3.216)\\
&= \mathrm{E}\left[((\mathbf{I} - \mathbf{W}_{k|k-1}\mathbf{H}_k)(\mathbf{x}_k - \mathbf{x}_{k|k-1}) - \mathbf{W}_{k|k-1}\mathbf{v}_k)^2\right] && (3.217)\\
&= (\mathbf{I} - \mathbf{W}_{k|k-1}\mathbf{H}_k)\mathrm{E}\left[(\mathbf{x}_k - \mathbf{x}_{k|k-1})^2\right](\mathbf{I} - \mathbf{W}_{k|k-1}\mathbf{H}_k)^\top \\
&\quad +2(\mathbf{I} - \mathbf{W}_{k|k-1}\mathbf{H}_k)\mathrm{E}\left[(\mathbf{x}_k - \mathbf{x}_{k|k-1})\mathbf{v}_k\right] + \mathbf{W}_{k|k-1}\mathrm{E}\left[\mathbf{v}_k\mathbf{v}_k^\top\right]\mathbf{W}_{k|k-1}^\top && (3.218)\\
&= (\mathbf{I} - \mathbf{W}_{k|k-1}\mathbf{H}_k)\mathbf{P}_{k|k-1}(\mathbf{I} - \mathbf{W}_{k|k-1}\mathbf{H}_k)^\top + \mathbf{W}_{k|k-1}\mathbf{R}_k\mathbf{W}_{k|k-1}^\top. && (3.219)
\end{aligned}$$

Thus, the Joseph form (3.219) provides the error covariance for an arbitrary gain matrix $\mathbf{W}_{k|k-1}$. We will see that this form is equivalent to the previously derived Kalman equations in the case of the *optimal* gain. At first, it should be noted that

$$\begin{aligned}
\mathbf{P}_{k|k-1}\mathbf{H}_k^\top &= \mathbf{P}_{k|k-1}\mathbf{H}_k^\top\mathbf{S}_k^{-1}\mathbf{S}_k && (3.220)\\
&= \mathbf{W}_{k|k-1}\mathbf{S}_k && (3.221)\\
&= \mathbf{W}_{k|k-1}\left(\mathbf{H}_k\mathbf{P}_{k|k-1}\mathbf{H}_k^\top + \mathbf{R}_k\right). && (3.222)
\end{aligned}$$

Using this fact leads us to the result

$$\begin{aligned}
&(\mathbf{I} - \mathbf{W}_{k|k-1}\mathbf{H}_k)\mathbf{P}_{k|k-1}\mathbf{H}_k^\top\mathbf{W}_{k|k-1}^\top = \\
&\qquad \mathbf{P}_{k|k-1}\mathbf{H}_k^\top\mathbf{W}_{k|k-1}^\top - \mathbf{W}_{k|k-1}\mathbf{H}_k\mathbf{P}_{k|k-1}\mathbf{H}_k^\top\mathbf{W}_{k|k-1}^\top && (3.223)\\
&= \mathbf{W}_{k|k-1}\left(\mathbf{H}_k\mathbf{P}_{k|k-1}\mathbf{H}_k^\top + \mathbf{R}_k\right)\mathbf{W}_{k|k-1}^\top - \mathbf{W}_{k|k-1}\mathbf{H}_k\mathbf{P}_{k|k-1}\mathbf{H}_k^\top\mathbf{W}_{k|k-1}^\top && (3.224)\\
&= \mathbf{W}_{k|k-1}\mathbf{R}_k\mathbf{W}_{k|k-1}^\top. && (3.225)
\end{aligned}$$

Therefore, the posterior covariance reduces to

$$\begin{aligned}
\mathbf{P}_{k|k} &= (\mathbf{I} - \mathbf{W}_{k|k-1}\mathbf{H}_k)\mathbf{P}_{k|k-1}(\mathbf{I} - \mathbf{W}_{k|k-1}\mathbf{H}_k)^\top \\
&\quad + \mathbf{W}_{k|k-1}\mathbf{R}_k\mathbf{W}_{k|k-1}^\top && (3.226) \\
&= (\mathbf{I} - \mathbf{W}_{k|k-1}\mathbf{H}_k)\mathbf{P}_{k|k-1}(\mathbf{I} - \mathbf{W}_{k|k-1}\mathbf{H}_k)^\top \\
&\quad + (\mathbf{I} - \mathbf{W}_{k|k-1}\mathbf{H}_k)\mathbf{P}_{k|k-1}(\mathbf{W}_{k|k-1}\mathbf{H}_k)^\top \\
&= (\mathbf{I} - \mathbf{W}_{k|k-1}\mathbf{H}_k)\mathbf{P}_{k|k-1} && (3.227) \\
&= \mathbf{P}_{k|k-1} - \mathbf{W}_{k|k-1}\mathbf{H}_k\mathbf{P}_{k|k-1} && (3.228) \\
&= \mathbf{P}_{k|k-1} - \mathbf{W}_{k|k-1}\left(\mathbf{H}_k\mathbf{P}_{k|k-1}\mathbf{H}_k^\top + \mathbf{R}_k\right)\mathbf{W}_{k|k-1}^\top && (3.229) \\
&= \mathbf{P}_{k|k-1} - \mathbf{W}_{k|k-1}\mathbf{S}_k\mathbf{W}_{k|k-1}^\top. && (3.230)
\end{aligned}$$

This concludes the derivation of the Kalman update equations by means of an orthogonal projection. Though the computation is tedious to some extent, the geometric interpretation merits a unique insight on its construction. We have seen that the Kalman posterior estimate is a minimum variance estimate of a matrix-valued combination of all measurements:

$$\mathbf{x}_{k|k} = \sum_{l=1}^{k} \mathbf{P}^{(l)}\mathbf{z}_l \tag{3.231}$$

for some matrices $\mathbf{P}^{(l)}$. Since this was constructed using an orthogonal projection, we have seen that the estimation error is orthogonal to the measurement span as sketched in Figure 3.5.

3.6 Data association

The state estimation considerations above have assumed that there is no measurement ambiguity, that is, at each time stamp, the sensor produces exactly one measurement, which was stems from an object of interest. The latter is described in its dynamic properties by a stochastic process. In many real-world applications, the assumption of a unique measurement to track association as in the Kalman filer does not hold. Instead, the sensor may produce a random number m_k of observations at each time instant t_k, where $m_k \geq 0$. Since the system produces at most one true measurement per target, it follows that one has at least $m_k - 1$ false alarms. One may even consider the case that the target was not detected at all at time step k. This implies an ambiguity in the measurement interpretation, which depends on the model for false alarms and non-detections for the particular sensor. The false measurements appear for instance due to thermal noise in a receiver circuit or simply stem from objects which are not of interest, the so-called clutter. For the application of the Kalman filter in order to update a track or a list of tracks, solutions to the data association problem are required. In addition, this problem is highly connected to the question of track existence, since it is unclear when and where tracks start and end, if measurement ambiguity is present.

Throughout the past decades of research, various solutions to the data association for Kalman filter tracks have been published [32,33]. Here, we discuss the most important ones in order of increasing complexity. The problem of track existence [34] is the focus of Section 3.10.

3.6.1 Nearest neighbor

The simplest approach to solve this challenge is called *nearest neighbor* (NN). This approach involves associating the particular measurement $\mathbf{z}_k^j \in \{\mathbf{z}_k^1, \ldots, \mathbf{z}_k^{m_k}\}$ with the expected measurement $\bar{\mathbf{z}}_k = \mathbf{H}_k \mathbf{x}_{k|k-1}$, as defined previously in (3.24), such that the distance between them is minimized with respect to a predefined metric. If no measurement was produced at time step t_k, that is ($m_k = 0$), then the prior density is used in the following prediction step of the Bayes cycle. Otherwise, the association for track i with prior mean $\mathbf{x}_{k|k-1}^i$ and covariance matrix $\mathbf{P}_{k|k-1}^i$, the associated measurement $\mathbf{z}_k^\star$ is found by minimizing for instance the *Mahalanobis distance* of the expected measurement, which is given by

$$\mathbf{z}_k^\star = \arg\min_{\mathbf{z}_k^j} \; (\mathbf{z}_k^j - \bar{\mathbf{z}}_k^i)^\top (\mathbf{S}_k^{(i,j)})^{-1} (\mathbf{z}_k^j - \bar{\mathbf{z}}_k^i), \tag{3.232}$$

where

$$\bar{\mathbf{z}}_k^i = \mathbf{H}_k \mathbf{x}_{k|k-1}^i \tag{3.233}$$

$$\mathbf{S}_k^{(i,j)} = \mathbf{H}_k \mathbf{P}_{k|k-1}^i \mathbf{H}_k^\top + \mathbf{R}_k^j. \tag{3.234}$$

This process is schematically shown in Figure 3.6, where three measurements were produced by the sensor. The ellipse shows the error ellipse of the expected measurement. As a consequence, the Mahalanobis distance is minimized by $\mathbf{z}_k^2$, which can then be used for the update step via the Kalman filter equations.

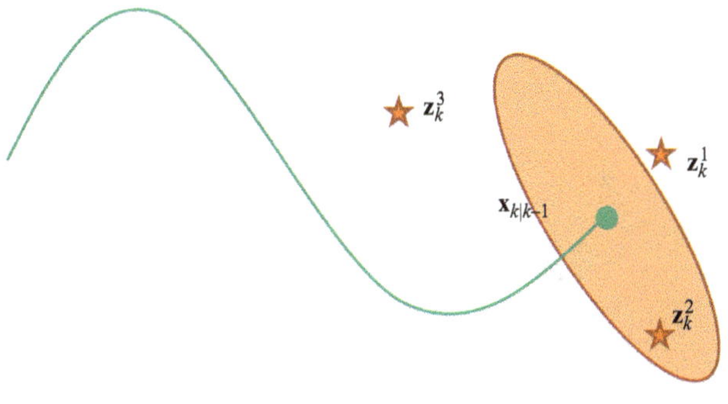

Figure 3.6 Single track scenario for the NN approach. The Mahalanobis distance towards the expected measurement (at $\mathbf{H}_k \mathbf{x}_{k|k-1}$) is minimized by $\mathbf{z}_k^2$.

3.6.2 Global nearest neighbor

The situation becomes more tricky as soon as non-separated tracks come into play. This is demonstrated in the example shown in Figure 3.7, where two tracks have a potential association conflict. One can see that measurement $\mathbf{z}_k^1$ is a good candidate for both track one and two, respectively. Consequently, measurement $\mathbf{z}_k^2$ could be associated with track two or be considered a false alarm. The local (with respect to a single track only) view would yield that measurement $\mathbf{z}_k^1$ is associated to track two. Considering the global (with respect to all tracks) picture shows that track two has another measurement to be associated with $\mathbf{z}_k^2$. Minimizing the global association costs would therefore lead to measurement one being associated with track one.

To solve this algorithmically, a cost matrix $\mathbf{C} = (c_{ij})$ is set up. Here, $i = 1, \ldots, n+1$ is the track index for n tracks and $j = 1, \ldots, m+1$ is the measurement index for m observations. One then computes

$$c_{ij} = \begin{cases} p_D \, \mathcal{N}(\mathbf{z}_k^j; \mathbf{H}_k \mathbf{x}_{k|k-1}^i, \mathbf{S}_k^{i,j}), & \text{if } i \leq n, j \leq m \\ (1 - p_D(\mathbf{x}_{k|k}^i)), & \text{if } i \leq n, j = m+1 \\ \rho_F(\mathbf{z}_k^j), & \text{if } i = n+1, j \leq m \\ \infty & \text{if } i = n+1, j = m+1. \end{cases} \tag{3.235}$$

Here, $p_D(\mathbf{x}_k)$ models the probability for detecting a target with state $\mathbf{x}_k$ and $\rho_F(\mathbf{z}_k)$ represents the density function of false alarms in the measurement space for an observation $\mathbf{z}_k$.

The optimal association $\hat{a} : 1, \ldots, n \longrightarrow 1, \ldots, m+1$ is then given by the minimization of the costs

$$\hat{a} := \arg\min_a \sum_{i=1}^{n} c_{ia(i)} \tag{3.236}$$

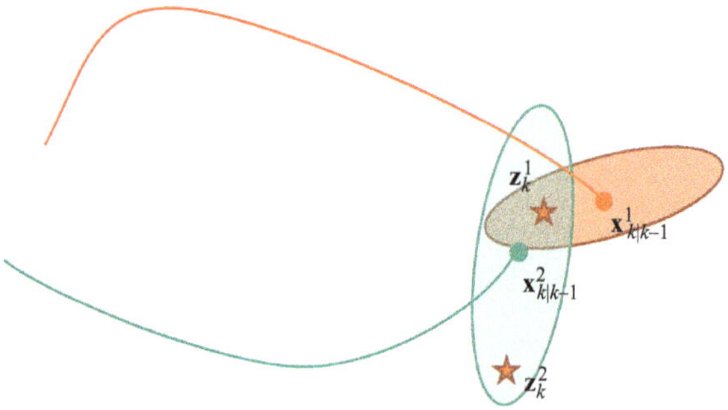

Figure 3.7 Two track scenario where an association conflict occurs for measurement $\mathbf{z}_k^1$. Minimizing the global costs in the global NN *(GNN) solves this problem as it accounts for the second option $\mathbf{z}_k^2$ for track two.*

under the constraint that $a(i) \neq a(j)$ for all $i \neq j$, if $a(i) \neq m+1$. This combinatorial optimization problem can be solved by the *Munkres* algorithm or any of its variants in polynomial time.

Example
In the following numerical setup, four targets are simulated in the *Stone Soup* framework [35]** with crossing trajectories. A linear Gaussian sensor observes the 2D positions with additive white noise ($\mathbf{R}_k = \mathbf{I}$). In addition, clutter measurements are produced uniformly distributed in the field of view. Tracks are initialized at the starting position of each target.

The tracking result of Kalman filters following the GNN association rule is shown in Figure 3.8. In both cases, the number of false alarms is Poisson distributed $n_F \sim p_F(n)$. In the upper image (a), the mean number of false alarms is set to 10, whereas in the lower one (b) the mean is set to 20.

One can see that for the low clutter scenario (a), the GNN allows at least to roughly follow the targets by means of the Kalman tracks, whereas if the high clutter is present as in (b), only one out of four tracks does not diverge (red ground truth, purple track). A similar situation occurs, if the number of false alarms is low but the number of crossing tracks raises. The increasing number of possible associations also significantly affects the run time of the GNN algorithm.

3.6.3 Probabilistic data association

Both, the NN and the GNN approach, have in common that for each track a *hard* decision is made, that is, some possibly relevant data is disregarded completely, since unique decisions are made for each association possibility. This disregards some of the inherent ambiguity in critical scenarios. For instance, if a clutter measurement is close to the true detection of a target, there might be not enough information to decide correctly at this very time step. However, if one might be able to "carry" the ambiguity over time for some time steps, this lack of knowledge may well be resolved.

Since one can model lack of knowledge and ambiguity in terms of probability, it seems natural to do *soft* decisions based on the estimated relevance. To make this more clear, let us consider the following single track scenario given in Figure 3.9.

The idea of all probabilistic approaches for data association is an *enumeration* of possible hypotheses. In the scenario above, one has three optional measurements for the given track, which lead to three hypotheses, h_1, h_2, and h_3. Each of which associates one measurement to the track and declares the other two to be false alarms. However, one has the additional hypothesis h_0, which declares all three measurements to be false alarms and assumes the target itself to be undetected. Under the assumption that a target produces at most one measurement, the given list of interpretations enumerates all possible hypotheses in this example. But how does this process go on? If one ends up with a set of posterior hypotheses, how could the process continue in time step $k+1$? The key word to answer these questions is the "*conjugate prior*," which means, that the prior density is of the same type or density family as the posterior.

**https://github.com/dstl/Stone-Soup

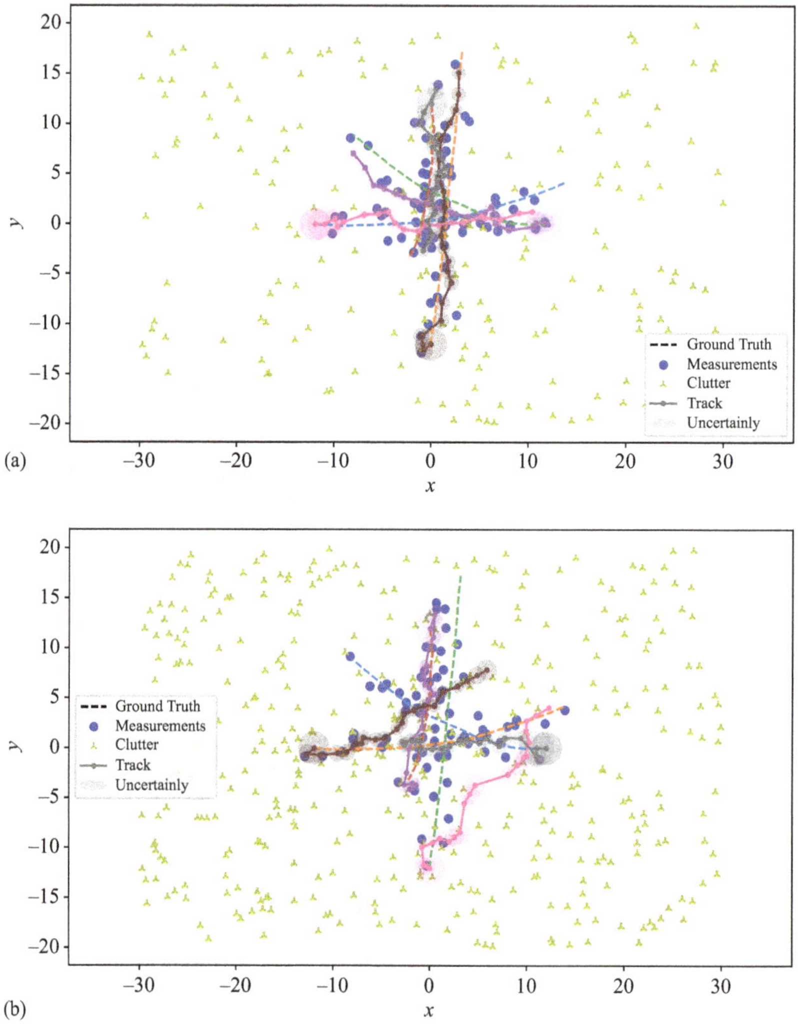

Figure 3.8 Simulation of four crossing targets and the trajectories of the Kalman filter tracks when using the GNN association rule. The mean number of false alarms was set to 10 in (a) and 20 in (b), respectively.

Given a conjugate prior, the Bayesian recursion may go on in an infinite loop. If the track in this example is produced by a Kalman filter, clearly the prior is a Gaussian. Therefore, the posterior must be a Gaussian, too. This is achieved by *merging* all of the obtained posterior hypotheses into a Gaussian density. The details of this process are described in the following.

At first, one may conclude that there are $m_k + 1$ hypotheses for the data association problem, if m_k measurements were observed at time step k. Since they are

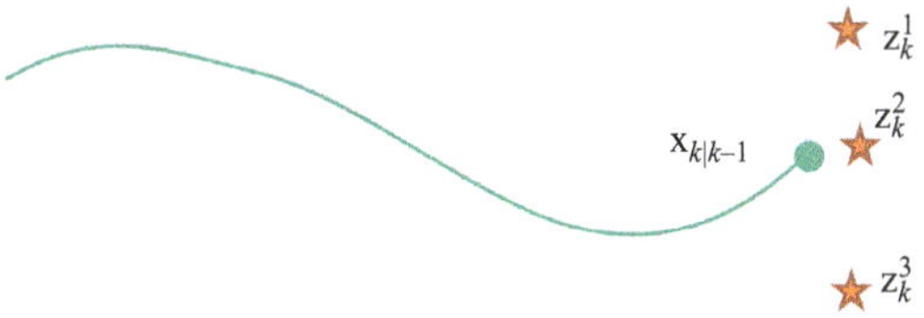

Figure 3.9 Data association problem with one track and three measurements

mutually exclusive, they are "*or*"-connected, which translates into sums of probabilities.[††] The law of marginalization for the random variable D which denotes the event of a detection yields that the likelihood of the set of measurements $Z_k = \{\mathbf{z}_k^1, \ldots, \mathbf{z}_k^{m_k}\}$ can be expressed as

$$p(Z_k|\mathbf{x}_k) = p(Z_k, D|\mathbf{x}_k) + p(Z_k, \neg D|\mathbf{x}_k) \tag{3.237}$$

$$= p(Z_k|\mathbf{x}_k, D)\, p(D) + p(Z_k|\mathbf{x}_k, \neg D)\, p(\neg D). \tag{3.238}$$

Analogously, if the target is detected, one can hypothesize that in the enumeration of all received measurements the actual detection is present:

$$p(Z_k|\mathbf{x}_k, D) = \sum_{j=1}^{m_k} p(Z_k|\mathbf{x}_k, D, h_j)\, p(h_j|\mathbf{x}_k). \tag{3.239}$$

Here, $p(h_j|\mathbf{x}_k) = p(h_j)$ represents the prior probability of hypothesis h_j. Since no prior knowledge on the likelihood of occurrence is available, the uniform distribution

$$p(h_j) = \frac{1}{m_k} \tag{3.240}$$

is chosen. The likelihood conditioned on an interpretation hypothesis h_j is given by

$$p(Z_k|\mathbf{x}_k, D, h_j) = p(\mathbf{z}_k^j|\mathbf{x}_k)\, p_F(m_k - 1) \prod_{l \neq j} p_{FA}(\mathbf{z}_k^l). \tag{3.241}$$

Here, $p_F(n)$ is the probability of observing n false alarms and $p_{FA}(\mathbf{z})$ represents the spatial distribution of a false alarm $\mathbf{z}$. Introducing the notion of a (constant[‡‡]) probability of detection

$$p(D|\mathbf{x}_k) = p_D \tag{3.242}$$

one can directly infer that $p(\neg D|\mathbf{x}_k) = 1 - p_D$. For the sake of simplicity, the following model for the false alarms is used.[§§] Without prior knowledge on the clutter measurements, it is assumed that the spatial distribution p_{FA} is a uniform distribution

[††]And probability densities likewise.
[‡‡]Actually, the probability of detection may well be modeled state dependent as $p_D(\mathbf{x}_k)$. This will not change the upcoming derivations. For the sake of simplicity, it is assumed constant in most formulations.
[§§]Again, various more sophisticated models exist. The models presented here are basic, but have proven useful in countless applications.

in the *Field of View* (FoV) with size |FoV| such that $p_{FA} = 1/|\mathrm{FoV}|$. The number of false measurements m is modeled as a Poisson distribution with mean value λ which is a fixed modeling parameter:

$$p_F(m) = \left(\frac{\lambda^m}{m!}\right) \mathrm{e}^{-\lambda}. \tag{3.243}$$

The parameter λ can be expressed in terms of the so-called *false alarm rate* ρ_F given by

$$\rho_F = \frac{\lambda}{|\mathrm{FoV}|}. \tag{3.244}$$

Using these models in (3.241), one can continue to develop the terms in greater detail:

$$p(Z_k|\mathbf{x}_k, D, h_j) = p_F(m_k - 1)\, |\mathrm{FoV}|^{-(m_k-1)}\, \mathcal{N}(\mathbf{z}_k^j;\, \mathbf{H}_k\mathbf{x}_k,\, \mathbf{R}_k), \tag{3.245}$$

where the linear Gaussian model for $p(\mathbf{z}_k^j|\mathbf{x}_k)$ was applied.

For the non-detection case, one has

$$\begin{aligned} p(Z_k|\mathbf{x}_k, \neg D) &= p_F(m_k) \prod_{l=1}^{m_k} p_{FA}(\mathbf{z}_k^l) && (3.246)\\ &= p_F(m_k)\, |\mathrm{FoV}|^{-m_k}. && (3.247) \end{aligned}$$

Now, for the Poisson distributed number of clutter measurements, the following equation holds:

$$p_F(m) = \frac{\lambda}{m} \cdot p_F(m-1). \tag{3.248}$$

Therefore, by means of the false alarm rate $\rho_F = \frac{\lambda}{|\mathrm{FoV}|}$, we have that

$$\begin{aligned} |\mathrm{FoV}|^{-(m_k-1)} p_F(m_k - 1)\frac{1}{m_k} &= |\mathrm{FoV}|\frac{1}{\lambda}|\mathrm{FoV}|^{-m_k} p_F(m_k) && (3.249)\\ &= \frac{1}{\rho_F}|\mathrm{FoV}|^{-m_k} p_F(m_k). && (3.250) \end{aligned}$$

For the likelihood function, we therefore have the following result:

$$\begin{aligned} p(Z_k|\mathbf{x}_k) = {} & (|\mathrm{FoV}|^{-m_k} p_F(m_k))((1 - p_D) \\ & + \frac{p_D}{\rho_F} \sum_{j=1}^{m_k} \mathcal{N}(\mathbf{z}_j;\, \mathbf{H}_k\mathbf{x}_k,\, \mathbf{R}_k)). \end{aligned} \tag{3.251}$$

Furthermore, since terms of proportionality cancel in the Bayes formula [see paragraph before the Bayes formula in (3.8)], if they are independent of the random variables of the density, one may write

$$p(Z_k|\mathbf{x}_k) \propto (1 - p_D) + \frac{p_D}{\rho_F} \sum_{j=1}^{m_k} \mathcal{N}(\mathbf{z}_k^j;\, \mathbf{H}_k\mathbf{x}_k,\, \mathbf{R}_k). \tag{3.252}$$

3.7 Probabilistic data association filter

In the *Probabilistic Data Association Filter* (PDAF), the track prior and posterior are each represented by a Gaussian density. As a consequence, the prediction is equivalent to the Kalman filter case. For the update step, the PDA likelihood in (3.252) is applied. Let us assume that the prior is given by the Gaussian

$$p(\mathbf{x}_k|\mathcal{Z}^{k-1}) = \mathcal{N}(\mathbf{x}_k;\ \mathbf{x}_{k|k-1},\ \mathbf{P}_{k|k-1}). \tag{3.253}$$

In order to use the Bayes formula,

$$p(\mathbf{x}_k|\mathcal{Z}^k) = \frac{p(\mathbf{x}_k|\mathcal{Z}^{k-1})\, p(Z_k|\mathbf{x}_k)}{\int \mathrm{d}\mathbf{x}_k\, p(\mathbf{x}_k|\mathcal{Z}^{k-1})\, p(Z_k|\mathbf{x}_k)} \tag{3.254}$$

with the prior and the likelihood from above, we are going to compute the multiplication of the terms in the numerator and normalize the result afterwards. By means of the product formula (3.123), one obtains

$$\begin{aligned}\mathcal{N}(\mathbf{x}_k;\ \mathbf{x}_{k|k-1},\ \mathbf{P}_{k|k-1}) \cdot \left((1-p_D) + \frac{p_D}{\rho_F}\sum_{j=1}^{m_k} \mathcal{N}(\mathbf{z}_k^j;\ \mathbf{H}_k\mathbf{x}_k,\ \mathbf{R}_k)\right) \\ = \sum_{j=0}^{m_k} p_k^{j\star}\, \mathcal{N}(\mathbf{x}_k;\ \mathbf{x}_{k|k}^j,\ \mathbf{P}_{k|k}^j),\end{aligned} \tag{3.255}$$

where

$$\mathbf{x}_{k|k}^j = \begin{cases} \mathbf{x}_{k|k-1}, & \text{if } j = 0 \\ \mathbf{x}_{k|k-1} + \mathbf{W}_{k|k-1}^j(\mathbf{z}_k^j - \mathbf{H}_k\mathbf{x}_{k|k-1}), & \text{if } j \geq 1 \end{cases} \tag{3.256}$$

$$\mathbf{P}_{k|k}^j = \begin{cases} \mathbf{P}_{k|k-1}, & \text{if } j = 0 \\ \mathbf{P}_{k|k-1} - \mathbf{W}_{k|k-1}^j\mathbf{S}_k^j\mathbf{W}_{k|k-1}^{j\top}, & \text{if } j \geq 1 \end{cases} \tag{3.257}$$

$$p_k^{j\star} = \begin{cases} (1-p_D) & \text{if } j = 0 \\ \frac{p_D}{\rho_F}\mathcal{N}(\mathbf{z}_k^j;\ \mathbf{H}_k\mathbf{x}_{k|k-1},\ \mathbf{S}_k^j) & \text{if } j \geq 1 \end{cases} \tag{3.258}$$

$$\mathbf{W}_{k|k-1}^j = \mathbf{P}_{k|k-1}\mathbf{H}_k^\top(\mathbf{S}_k^j)^{-1} \tag{3.259}$$

$$\mathbf{S}_k^j = \mathbf{H}_k\mathbf{P}_{k|k-1}\mathbf{H}_k^\top + \mathbf{R}_k^j. \tag{3.260}$$

The denominator in (3.254) is now straight forward to compute since the Gaussians integrate to one:

$$\int \mathrm{d}\mathbf{x}_k \sum_{j=0}^{m_k} p_k^{j\star}\, \mathcal{N}(\mathbf{x}_k;\ \mathbf{x}_{k|k}^j,\ \mathbf{P}_{k|k}^j) = \sum_{j=0}^{m_k} p_k^{j\star}. \tag{3.261}$$

As a consequence the weights are normalized by

$$p_k^j = \frac{p_k^{j\star}}{\sum_{l=0}^{m_k} p_k^{l\star}} \tag{3.262}$$

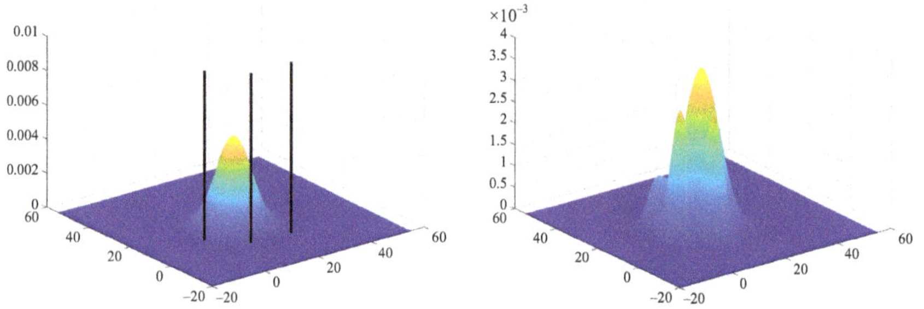

Figure 3.10 Update of a prior (left) with multiple measurements indicated by the black lines. Since the assumption is that a target spawns at most one measurement, the measurements must be interpreted mutually exclusive. This yields the Gaussian mixture on the right.

such that $\sum p_k^j = 1$. Therefore, the posterior is given by the following *Gaussian Mixture Density*

$$p(\mathbf{x}_k | \mathcal{Z}^k) = \sum_{j=0}^{m_k} p_k^j \, \mathcal{N}(\mathbf{x}_k;\, \mathbf{x}_{k|k}^j,\, \mathbf{P}_{k|k}^j). \tag{3.263}$$

A simulated example is shown in Figure 3.10, where three measurements (black lines) are obtained from the sensor and the prior density is on the left side. The enumeration of the four resulting hypotheses and the corresponding update yields a Gaussian mixture on the right side.

Since we want to obey to the concept of a conjugate prior (compare Section 3.7), it is necessary to approximate the Gaussian Mixture by a single Gaussian. This is done via *Moment Matching*, which is presented in the following paragraph.

3.7.1 Moment matching

The idea of moment matching is to approximate an arbitrary density by a Gaussian by matching its first (expectation value) and second (covariance matrix) moments.

In the case of a Gaussian mixture with N components, the given density has the form

$$p(\mathbf{x}) = \sum_{i=1}^{N} p_i \, \mathcal{N}(\mathbf{x};\, \hat{\mathbf{x}}^i,\, \mathbf{P}^i), \tag{3.264}$$

with $\sum p_i = 1$. Computing the mean, one directly obtains

$$\hat{\mathbf{x}} = \mathrm{E}\,[\mathbf{x}] = \sum_{i=1}^{N} p_i \, \hat{\mathbf{x}}^i. \tag{3.265}$$

The covariance matrix can be derived analogously:

$$\operatorname{cov}[\mathbf{x}] = \mathrm{E}\left[(\mathbf{x}-\hat{\mathbf{x}})(\mathbf{x}-\hat{\mathbf{x}})^\top\right] \tag{3.266}$$

$$= \sum_{i=1}^{N} p_i \,\mathrm{E}\left[(\mathbf{x}-\hat{\mathbf{x}})(\mathbf{x}-\hat{\mathbf{x}})^\top\right] \tag{3.267}$$

$$= \sum_{i=1}^{N} p_i \,\mathrm{E}\left[(\mathbf{x}-\mathbf{x}^i+\mathbf{x}^i-\hat{\mathbf{x}})(\mathbf{x}-\mathbf{x}^i+\mathbf{x}^i-\hat{\mathbf{x}})^\top\right] \tag{3.268}$$

$$= \sum_{i=1}^{N} p_i \left(\mathbf{P}^i + (\mathbf{x}^i-\hat{\mathbf{x}})(\mathbf{x}^i-\hat{\mathbf{x}})^\top\right). \tag{3.269}$$

Here the term $(\mathbf{x}^i-\hat{\mathbf{x}})(\mathbf{x}^i-\hat{\mathbf{x}})^\top$ is called the *spread term* and accounts for the dislocation of the center of the ith component.

This leads us to the approximation of a Gaussian mixture via moment matching by a single Gaussian:

$$\sum_{i=1}^{N} p_i \,\mathcal{N}(\mathbf{x};\hat{\mathbf{x}}^i,\mathbf{P}^i) \approx \mathcal{N}(\mathbf{x};\hat{\mathbf{x}},\mathbf{P}), \tag{3.270}$$

where

$$\hat{\mathbf{x}} = \sum_{i=1}^{N} p_i \,\hat{\mathbf{x}}^i \tag{3.271}$$

$$\mathbf{P} = \sum_{i=1}^{N} p_i \left(\mathbf{P}^i + (\mathbf{x}^i-\hat{\mathbf{x}})(\mathbf{x}^i-\hat{\mathbf{x}})^\top\right). \tag{3.272}$$

Applying the moment matching equations to the PDAF posterior in (3.263) yields

$$p(\mathbf{x}_k|\mathcal{Z}^k) \approx \mathcal{N}(\mathbf{x};\mathbf{x}_{k|k},\mathbf{P}_{k|k}), \tag{3.273}$$

where the posterior parameters are given by

$$\mathbf{x}_{k|k} = \sum_{j=0}^{m_k} p_k^j \mathbf{x}_{k|k}^j \tag{3.274}$$

$$\mathbf{P}_{k|k} = \sum_{j=0}^{m_k} p_k^j \left(\mathbf{P}_{k|k}^j + (\mathbf{x}_{k|k}^j-\mathbf{x}_{k|k})(\mathbf{x}_{k|k}^j-\mathbf{x}_{k|k})^\top\right). \tag{3.275}$$

The result of the moment matching for the example in Figure 3.10 is shown in Figure 3.11.

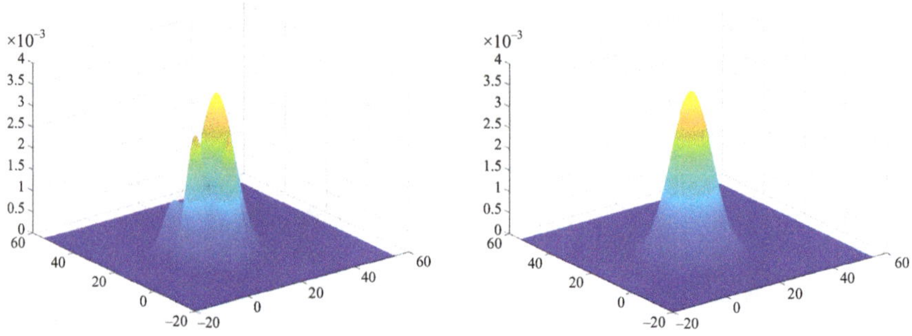

Figure 3.11 In the PDAF, the Gaussian mixture (left) based on the Bayes formula has to be approximated by a single Gaussian density (right). This is obtained via moment matching.

Using the definitions of the parameters for the individual components of the Gaussian mixture in (3.256) and (3.257), the posterior parameters may be rewritten:

$$\mathbf{x}_{k|k} = p_k^0 \mathbf{x}_{k|k-1} + \sum_{j=1}^{m_k} p_k^j \left\{ \mathbf{x}_{k|k-1} + \mathbf{W}_{k|k-1}^j \boldsymbol{\nu}_k^j \right\}, \tag{3.276}$$

where $\boldsymbol{\nu}_k^j = \mathbf{z}_k^j - \mathbf{H}_k \mathbf{x}_{k|k-1}$ denotes the innovation for measurement $\mathbf{z}_k^j$. This leads to

$$\mathbf{x}_{k|k} = \mathbf{x}_{k|k-1} + \sum_{j=1}^{m_k} \left\{ p_k^j \mathbf{W}_{k|k-1}^j \boldsymbol{\nu}_k^j \right\}. \tag{3.277}$$

Given that the all measurement matrices $\mathbf{H}_k^j$ and the measurement error covariance matrices $\mathbf{R}_k^j$ are identical, one has that $\mathbf{W}_{k|k-1} = \mathbf{W}_{k|k-1}^j$ for all j. In this case, one may introduce the *combined innovation*

$$\boldsymbol{\nu}_k = \sum_{j=1}^{m_k} p_k^j \boldsymbol{\nu}_k^j \tag{3.278}$$

such that

$$\mathbf{x}_{k|k} = \mathbf{x}_{k|k-1} + \mathbf{W}_{k|k-1} \boldsymbol{\nu}_k. \tag{3.279}$$

It should be noted that the sum of the weights in (3.278) is not normalized, since the possibility of a non-detection is represented by p_k^0. The update equation in (3.279) reflects, however, that the PDAF combines the actual *observations* to a pseudo measurement based on the probabilistic models.

Analogously, one has for the posterior covariance matrix

$$\mathbf{P}_{k|k} = p_k^0 \mathbf{P}_{k|k-1} + \sum_{j=1}^{m_k} p_k^j \left(\mathbf{P}_{k|k-1} - \mathbf{W}_{k|k-1}^j \mathbf{S}_k^j \mathbf{W}_{k|k-1}^{j\top} + (\mathbf{x}_{k|k}^j - \mathbf{x}_{k|k})(\mathbf{x}_{k|k}^j - \mathbf{x}_{k|k})^\top) \right) \tag{3.280}$$

$$= \mathbf{P}_{k|k-1} - \sum_{j=1}^{m_k} p_k^j \mathbf{W}_{k|k-1}^j \mathbf{S}_k^j \mathbf{W}_{k|k-1}^{j\top} + \sum_{j=1}^{m_k} p_k^j \mathbf{W}_{k|k-1}^j (\boldsymbol{\nu}_k^j - \boldsymbol{\nu})(\boldsymbol{\nu}_k^j - \boldsymbol{\nu})^\top \mathbf{W}_{k|k-1}^{j\top}. \tag{3.281}$$

Also for the covariance, the case of fixed measurement models for all measurements leads to a simplified equation:

$$\begin{aligned} \mathbf{P}_{k|k} &= \mathbf{P}_{k|k-1} - (1 - p_k^0)\mathbf{W}_{k|k-1}\mathbf{S}_k\mathbf{W}_{k|k-1}^\top \\ &+ \mathbf{W}_{k|k-1} \left[\sum_{j=1}^{m_k} p_k^j (\boldsymbol{\nu}_k^j - \boldsymbol{\nu})(\boldsymbol{\nu}_k^j - \boldsymbol{\nu})^\top \right] \mathbf{W}_{k|k-1}^\top. \end{aligned} \tag{3.282}$$

These updated formulae succeed in the intended *soft decision* for the measurement association, since ambiguity is reflected in the association of multiple measurements to a single track.

A further enhancement in the tracking precision is achieved by bookkeeping association hypotheses over time instead of merging them at each time step as in the PDAF above. This will be focus of the next section.

3.8 Multi-hypotheses-tracker

Hypotheses-based target tracking has become the de facto standard for target tracking in the presence of non-detections and false alarms. Nowadays, there exist a multitude of variants based on slightly different models for target existence or based on Poisson Point Processes [36], which estimate the target densities in a joint state space in terms of an intensity function. Those relations are revealed in a particular beautiful manner by considering the corresponding probability (density) generating functions and the implied analytical combinatorics [37], which is not the scope of this book. Therefore, we will focus on a basic derivation of the *multiple-hypotheses-tracker* (MHT) [32,38,39].

As indicated at the end of the previous section, one may hypothesize all possible interpretations of the current set of measurements. Given a time series of unique interpretations for each time step, Kalman conditions apply and therefore the conditional posterior can be computed. Since the hypotheses are or-connected, this yields a normalized Gaussian mixture, whose components grow exponentially over time, because in the update step each prior component is multiplied with all conditional likelihood functions for the latest data hypothesis.

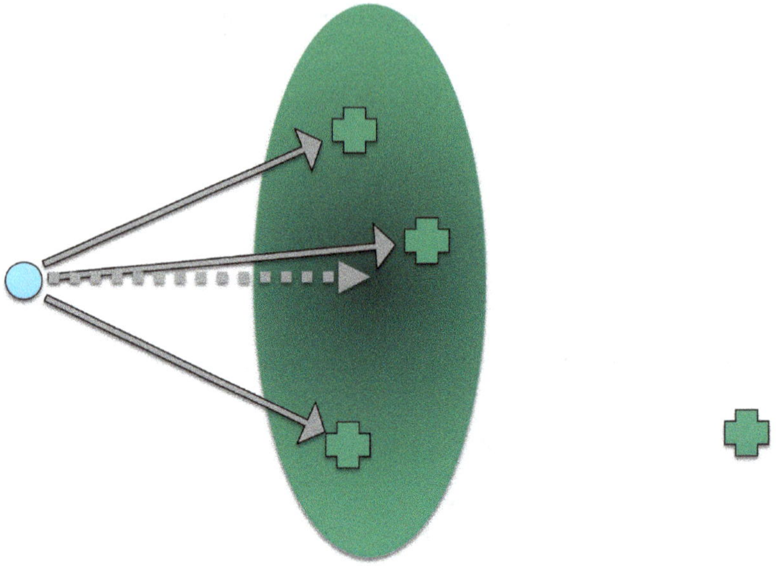

Figure 3.12 A multiple hypotheses continuation due to false measurements and ambiguity (crosses). Each sensor return within a gate (ellipse) produces a new hypothesis.

As a pre-filtering step, *gating* is applied in order to reduce the number of associated measurements, in which all observations outside a confidence region[|||] are discarded. In this scheme, each measurement within a certain gate creates a new hypothesis. If the probability associated with a hypothesis becomes too low it is removed, so that a manageable number of hypotheses are maintained. Figure 3.12 shows a possible scenario in a MHT update step. Only measurements within the gate (ellipse) centered at position of the expected measurement are considered and create new hypotheses.

Let $j_k = 0$ denote the data interpretation hypothesis that the object has not been detected at all by the sensor at time t_k and so all measurements have to be considered as clutter. A hypothesis denoted by $j_k \in \{1, \ldots, m_k\}$ refers to the interpretation that the object has been detected, $\mathbf{z}_k^{j_k} \in Z_k$ being the corresponding measurement, and the remaining sensor data being clutter. Evidently, the set $\{0, \ldots, m_k\}$ describes mutually exclusive and exhaustive data interpretations.

[|||] In the Gaussian case, the *gate* corresponds to an ellipsoid around the expected measurement vector. Direction and length of all half-axes are given by the Eigenvectors and -values of the innovation covariance. This may be seen as a multi-dimensional generalization of the confidence interval in the 1D case. In most applications, a confidence level is chosen between 90% and 99.9%.

Above in (3.251), we have seen that the likelihood function for ambiguous data $Z_k = \{\mathbf{z}_k^j\}_{j=1,\dots,m_k}$ is proportional to a weighted sum of Gaussians plus a constant:

$$p(Z_k|\mathbf{x}_k) = \sum_{j_k=0}^{m_k} p(Z_k|j_k, \mathbf{x}_k)\, p(j_k) \tag{3.283}$$

$$\propto (1 - P_D)\rho_F + P_D \sum_{j_k=1}^{m_k} \mathcal{N}(\mathbf{z}_k^{j_k};\ \mathbf{H}_k\mathbf{x}_k,\ \mathbf{R}_k). \tag{3.284}$$

Data interpretation hypotheses are the basis for all MHT techniques, however, eventually, there is a multitude of different implementations, which mainly depend on the hypotheses handling in order their number limited over time and whether they are track based or measurement based [40]. The latter hypothesis possible assignments from the data to all targets (including potentially new tracks).

In this section, the track-based MHT is presented, that is, it is assumed, that all targets are stochastically independent and therefore the joint probability density factorizes into single tracks. The MHT scheme then handles the hypothesis-based data association for each track individually. In order to maintain the ambiguity reflected by the variety of the hypotheses, the MHT prior and posterior densities are Gaussian mixtures. The associated origin of a time series $\mathcal{Z}^{k-1} = \{Z_{k-1}, \dots, Z_1\}$ of sensor data accumulated up to the time t_{k-1} can be interpreted by interpretation histories,

$$\mathbf{j}_{k-1} = (j_{k-1}, \dots, j_1), \qquad \text{where}\, j_l \in \{0, \dots, m_l\} \quad \forall l \tag{3.285}$$

that assume at each data collection time t_l, $1 \leq l \leq k-1$, a certain data interpretation j_l to be true. Via marginalization, the filtering density $p(\mathbf{x}_k|\mathcal{Z}^k)$ at time t_{k-1} can be written as a mixture over such interpretation histories $\mathbf{j}_k$:

$$p(\mathbf{x}_{k-1}|\mathcal{Z}^{k-1}) = \sum_{\mathbf{j}_{k-1}} p(\mathbf{j}_{k-1}|\mathcal{Z}^k)\, p(\mathbf{x}_{k-1}|\mathbf{j}_{k-1}, \mathcal{Z}^{k-1}) \tag{3.286}$$

$$=: \sum_{\mathbf{j}_{k-1}} p_{k-1|k-1}^{\mathbf{j}_{k-1}}\ \mathcal{N}(\mathbf{x}_{k-1};\ \mathbf{x}_{k-1|k-1}^{\mathbf{j}_{k-1}},\ \mathbf{P}_{k-1|k-1}^{\mathbf{j}_{k-1}}), \tag{3.287}$$

where the parameters $\mathbf{x}_{k-1|k-1}^{\mathbf{j}_{k-1}}$ and $\mathbf{P}_{k-1|k-1}^{\mathbf{j}_{k-1}}$ are calculated by setting assuming that to the hypothesis $\mathbf{j}_{k-1}$ is true, that is therefore without ambiguity and can be calculated according to the Kalman filter equations. In order to see the recursive computation of the mixture weights $p_{k|k}^{\mathbf{j}_k}$, the prediction-filtering cycle from time step $k-1$ to k is reviewed in the following.

Similar to the Kalman filter, the prior density is easily obtained via marginalization:

$$p(\mathbf{x}_k|\mathcal{Z}^{k-1}) = \int \mathrm{d}\mathbf{x}_{k-1}\, p(\mathbf{x}_k|\mathbf{x}_{k-1})\, p(\mathbf{x}_{k-1}|\mathcal{Z}^{k-1}) \tag{3.288}$$

$$= \int \mathrm{d}\mathbf{x}_{k-1}\, \mathcal{N}(\mathbf{x}_k;\, \mathbf{F}_{k|k-1}\mathbf{x}_{k-1},\, \mathbf{Q}_{k|k-1})$$

$$\sum_{\mathbf{j}_{k-1}} p_{k-1|k-1}^{\mathbf{j}_{k-1}}\, \mathcal{N}(\mathbf{x}_{k-1};\, \mathbf{x}_{k-1|k-1}^{\mathbf{j}_{k-1}},\, \mathbf{P}_{k-1|k-1}^{\mathbf{j}_{k-1}}) \tag{3.289}$$

$$= \sum_{\mathbf{j}_{k-1}} p_{k-1|k-1}^{\mathbf{j}_{k-1}} \int \mathrm{d}\mathbf{x}_{k-1}\, \mathcal{N}(\mathbf{x}_k;\, \mathbf{F}_{k|k-1}\mathbf{x}_{k-1},\, \mathbf{Q}_{k|k-1})$$

$$\mathcal{N}(\mathbf{x}_{k-1};\, \mathbf{x}_{k-1|k-1}^{\mathbf{j}_{k-1}},\, \mathbf{P}_{k-1|k-1}^{\mathbf{j}_{k-1}}) \tag{3.290}$$

$$= \sum_{\mathbf{j}_{k-1}} p_{k-1|k-1}^{\mathbf{j}_{k-1}}\, \mathcal{N}(\mathbf{x}_k;\, \mathbf{x}_{k|k-1}^{\mathbf{j}_{k-1}},\, \mathbf{P}_{k|k-1}^{\mathbf{j}_{k-1}}) \tag{3.291}$$

As a consequence, one can see that the all tracks are predicted using the transition model $p(\mathbf{x}_k|\mathbf{x}_{k-1})$ while the weights remain constant during the prediction step. In particular, for a given component $j_{k-1} \in \mathbf{j}_{k-1}$ of the Gaussian mixture of the previous posterior, the prior parameters are given by

$$\mathbf{x}_{k|k-1}^{j_{k-1}} = \mathbf{F}_{k|k-1}\mathbf{x}_{k-1|k-1}^{j_{k-1}} \tag{3.292}$$

$$\mathbf{P}_{k|k-1}^{j_{k-1}} = \mathbf{F}_{k|k-1}\mathbf{P}_{k-1|k-1}^{j_{k-1}}\mathbf{F}_{k|k-1}^{\top} + \mathbf{Q}_{k|k-1}. \tag{3.293}$$

In the filtering step, one has that the numerator of the Bayes formula using the prior from above and the likelihood in (3.284) is given by

$$\sum_{\mathbf{j}_{k-1}} p_{k-1|k-1}^{\mathbf{j}_{k-1}} \quad \mathcal{N}(\mathbf{x}_k;\, \mathbf{x}_{k|k-1}^{\mathbf{j}_{k-1}},\, \mathbf{P}_{k|k-1}^{\mathbf{j}_{k-1}}) \cdot \left((1-p_D) + \frac{p_D}{\rho_F}\sum_{j=1}^{m_k} \mathcal{N}(\mathbf{z}_k^j;\, \mathbf{H}_k\mathbf{x}_k,\, \mathbf{R}_k)\right)$$

$$= \sum_{\mathbf{j}_{k-1}} p_{k-1|k-1}^{\mathbf{j}_{k-1}} \sum_{j=0}^{m_k} p_k^{j,\mathbf{j}_{k-1}\star}\, \mathcal{N}(\mathbf{x}_k;\, \mathbf{x}_{k|k}^{j,\mathbf{j}_{k-1}},\, \mathbf{P}_{k|k}^{j,\mathbf{j}_{k-1}}), \tag{3.294}$$

where

$$\mathbf{x}_{k|k}^{j,\mathbf{j}_{k-1}} = \begin{cases} \mathbf{x}_{k|k-1}^{\mathbf{j}_{k-1}}, & \text{if } j = 0 \\ \mathbf{x}_{k|k}^{\mathbf{j}_{k-1}} + \mathbf{W}_{k|k-1}^{j,\mathbf{j}_{k-1}}(\mathbf{z}_k^j - \mathbf{H}_k\mathbf{x}_{k|k-1}^{\mathbf{j}_{k-1}}), & \text{if } j \geq 1 \end{cases} \tag{3.295}$$

$$\mathbf{P}_{k|k}^{j,\mathbf{j}_{k-1}} = \begin{cases} \mathbf{P}_{k|k}^{\mathbf{j}_{k-1}}, & \text{if } j = 0 \\ \mathbf{P}_{k|k}^{\mathbf{j}_{k-1}} - \mathbf{W}_{k|k-1}^{j,\mathbf{j}_{k-1}}\mathbf{S}_k^{j,\mathbf{j}_{k-1}}\mathbf{W}_{k|k-1}^{j,\mathbf{j}_{k-1}\top}, & \text{if } j \geq 1 \end{cases} \tag{3.296}$$

$$p_k^{j,\mathbf{j}_{k-1}\star} = \begin{cases} (1 - p_D) \cdot p_{k-1|k-1}^{\mathbf{j}_{k-1}} & \text{if } j = 0 \\ \frac{p_D}{\rho_F} \mathcal{N}(\mathbf{z}_k^j;\, \mathbf{H}_k\mathbf{x}_{k|k-1},\, \mathbf{S}_k^{j,\mathbf{j}_{k-1}}) \cdot p_{k-1|k-1}^{\mathbf{j}_{k-1}} & \text{if } j \geq 1 \end{cases} \tag{3.297}$$

$$\mathbf{W}_{k|k-1}^{j,\mathbf{j}_{k-1}} = \mathbf{P}_{k|k-1}^{\mathbf{j}_{k-1}}\mathbf{H}_k^\top(\mathbf{S}_k^{j,\mathbf{j}_{k-1}})^{-1} \tag{3.298}$$

$$\mathbf{S}_k^{j,\mathbf{j}_{k-1}} = \mathbf{H}_k\mathbf{P}_{k|k-1}^{\mathbf{j}_{k-1}}\mathbf{H}_k^\top + \mathbf{R}_k^j. \tag{3.299}$$

The normalization of the posterior density is obtained via a rescaling of the weights:

$$p_{k|k}^{\mathbf{j}_k} = \frac{p_k^{j,\mathbf{j}_{k-1}\star}}{\sum_{j,\mathbf{j}_{k-1}} p_k^{j,\mathbf{j}_{k-1}\star}}. \tag{3.300}$$

One finally arrives at the Gaussian mixture, which represents the posterior density

$$p(\mathbf{x}_k|\mathcal{Z}^k) = \sum_{\mathbf{j}_k} p_{k|k}^{\mathbf{j}_k}\, \mathcal{N}(\mathbf{x}_k;\, \mathbf{x}_{k|k}^{\mathbf{j}_k},\, \mathbf{P}_{k|k}^{\mathbf{j}_k}) \tag{3.301}$$

with $\mathbf{j}_k = \{\{j\}_{j=0,\ldots,m_k} \times \mathbf{j}_{k-1}\}$ such that $|\mathbf{j}_k| = (m_k + 1) \cdot |\mathbf{j}_{k-1}|$. This obviously implies that the number of hypotheses grows exponentially in time as visualized in Figure 3.13.

Therefore, mixture reduction techniques have to be applied in order to make the MHT algorithm feasible:

- *Merging:* Similar hypotheses can be merged together. The test for similarity usually is based on the Mahalanobis distance. If for two hypotheses j_1 and j_2, respectively, with $j_1, j_2 \in \mathbf{j}_k$ it holds that

$$(\mathbf{x}_{k|k}^{j_1} - \mathbf{x}_{k|k}^{j_2})^\top \left(\mathbf{P}_{k|k}^{j_1} + \mathbf{P}_{k|k}^{j_2}\right)^{-1} (\mathbf{x}_{k|k}^{j_1} - \mathbf{x}_{k|k}^{j_2}) < \mu_{\text{merge}} \tag{3.302}$$

  for a fixed threshold μ_{merge}, their respective components in the posterior Gaussian mixture are merged. Actually, a group of multiple similar hypotheses $\mathbf{j}_g \subset \mathbf{j}_k$ can

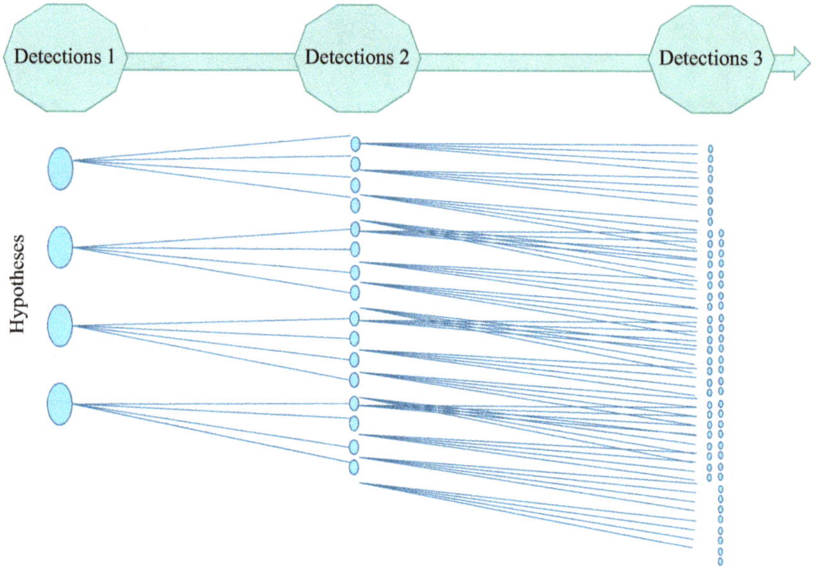

Figure 3.13 Visualization of the growing memory disaster due to the exponential growth in the number of data interpretation hypotheses over time

be searched by their respective Mahalanobis distance and eventually combined to a single hypothesis $j^\star$ via moment matching such that

$$\mathbf{x}_{k|k}^{j^\star} = \sum_{j \in \mathbf{j}_g} \frac{p_{k|k}^j}{\sum_{l \in \mathbf{j}_g} p_{k|k}^l} \mathbf{x}_{k|k}^j \tag{3.303}$$

$$\mathbf{P}_{k|k}^{j^\star} = \sum_{j \in \mathbf{j}_g} \frac{p_{k|k}^j}{\sum_{l \in \mathbf{j}_g} p_{k|k}^l} \left(\mathbf{P}_{k|k}^j + \boldsymbol{\nu}^j \boldsymbol{\nu}^{j\top} \right) \tag{3.304}$$

$$\boldsymbol{\nu}^j = \mathbf{x}_{k|k}^j - \mathbf{x}_{k|k}^{j^\star}. \tag{3.305}$$

- *Pruning:* Hypotheses with very low weights hardly contribute to the final result of the estimate and may therefore be neglected. To this end, a second model parameter μ_{prune} is introduced as a threshold such that a hypothesis $j \in \mathbf{j}_k$ is deleted, if $p_{k|k}^j < \mu_{\text{prune}}$. Finally, the weights of the remaining hypotheses $\tilde{\mathbf{j}}_k$ have to be renormalized such that

$$\sum_{l \in \tilde{\mathbf{j}}_k} p_{k|k}^l = 1. \tag{3.306}$$

This concludes the track-based MHT, a sophisticated multi-target tracking algorithm in scenarios with non-detections and false alarms. Its strength is concurrently its weakness: the large number of model parameters for the sensor data likelihood and the thresholds for managing the number of hypotheses enable a detailed fine tuning and adaptation. However, it can also be tedious to find parameters which are general enough.

The following section is going to introduce another layer of detailed modeling by introducing the possibility of using multiple evolution models in parallel.

3.9 Interacting multiple model filter

We have seen in the previous section that sensor properties can well be modeled with great detail by introducing additional parameters and hypothesizing different interpretation of empirical evidence. A similar scheme can improve the state estimation process by enhancing the evolution model of the system parameters [30,41]. The most intuitive way is to use multiple evolution models in parallel and to run a bank of M Kalman filters such that

$$p(\mathbf{x}_k|\mathcal{Z}^k) = \sum_{i=1}^{M} p(i)\, p(\mathbf{x}_k|\mathcal{Z}^k, i) \tag{3.307}$$

and $p(\mathbf{x}_k|\mathcal{Z}^k, i)$ was computed by exclusively using model i. However, it is quite obvious that in particular in target tracking, there are *phases* of distinct motion such as high-g maneuvers and straight forward movement. This motivates the notion of interaction between the models, such that they may be in effect for a certain time. This may be modeled as a Markov chain where the transition probabilities for switching from model i_{k-1} to i_k at time t_k are given by

$$p(i_k|i_{k-1}) = p_{i_k,i_{k-1}} \tag{3.308}$$

where $(p_{i_k,i_{k-1}})_{i_k,i_{k-1}}$ is a probability matrix. Such a scheme is visualized in Figure 3.14 for $M = 3$, where the arrows indicate possible transitions. The related probabilities must obey the condition that the sum of all outgoing paths of a node equals one, that is

$$\sum_{i_k=1}^{M} p_{i_k,i_{k-1}} = 1 \qquad \forall\, i_{k-1}. \tag{3.309}$$

Under the assumption of linear Gaussian transition densities for all M models, let them be given by

$$p(\mathbf{x}_k|\mathbf{x}_{k-1}, i_k) = \mathcal{N}(\mathbf{x}_k;\, \mathbf{F}^{i_k}_{k|k-1}\mathbf{x}_k,\, \mathbf{Q}^{i_k}_{k|k-1}). \tag{3.310}$$

Furthermore, it is assumed that the previous posterior from time t_{k-1} is given by a Gaussian mixture with M components:

$$p(\mathbf{x}_{k-1}|\mathcal{Z}^{k-1}) = \sum_{i_{k-1}=1}^{M} p^{i_{k-1}}_{k-1|k-1}\, \mathcal{N}(\mathbf{x}_{k-1};\, \mathbf{x}^{i_{k-1}}_{k-1|k-1},\, \mathbf{P}^{i_{k-1}}_{k-1|k-1}). \tag{3.311}$$

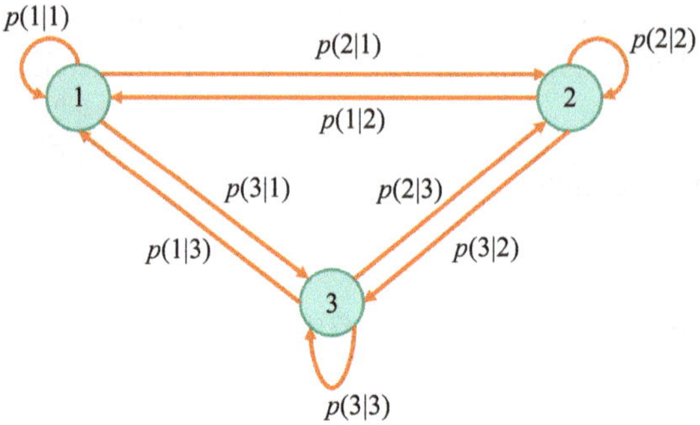

Figure 3.14 Transition model for a Markov chain with three nodes

For the computation of the prior density, the marginalization introduces both, the $\mathbf{x}_{k-1}$ and i_k:

$$p(\mathbf{x}_k|\mathcal{Z}^{k-1}) = \sum_{i_k=1}^{M} \int \mathrm{d}\mathbf{x}_{k-1}\, p(\mathbf{x}_k, i_k, \mathbf{x}_{k-1}|\mathcal{Z}^{k-1}) \tag{3.312}$$

$$= \sum_{i_k, i_{k-1}=1}^{M} \int \mathrm{d}\mathbf{x}_{k-1}\, p(\mathbf{x}_k, i_k, \mathbf{x}_{k-1}, i_{k-1}|\mathcal{Z}^{k-1}) \tag{3.313}$$

$$= \sum_{i_k, i_{k-1}=1}^{M} \int \mathrm{d}\mathbf{x}_{k-1}\, p(\mathbf{x}_k, i_k|\mathbf{x}_{k-1}, i_{k-1})\, p(\mathbf{x}_{k-1}, i_{k-1}|\mathcal{Z}^{k-1}). \tag{3.314}$$

According to the product formula (3.123), the Kalman filter prediction may be applied for each transition mode such that

$$\int \mathrm{d}\mathbf{x}_{k-1}\, p(\mathbf{x}_k, i_k|\mathbf{x}_{k-1}, i_{k-1})\, p(\mathbf{x}_{k-1}, i_{k-1}|\mathcal{Z}^{k-1}) =$$

$$p_{i_k, i_{k-1}} \mathcal{N}(\mathbf{x}_k;\, \mathbf{F}^{i_k}_{k|k-1}\mathbf{x}_k,\, \mathbf{Q}^{i_k}_{k|k-1})\, p^{i_{k-1}}_{k-1|k-1} \mathcal{N}(\mathbf{x}_{k-1};\, \mathbf{x}^{i_{k-1}}_{k-1|k-1},\, \mathbf{P}^{i_{k-1}}_{k-1|k-1}) \tag{3.315}$$

$$= p_{i_k, i_{k-1}} p^{i_{k-1}}_{k-1|k-1} \mathcal{N}(\mathbf{x}_k;\, \mathbf{x}^{i_k, i_{k-1}}_{k|k-1},\, \mathbf{P}^{i_k, i_{k-1}}_{k|k-1}), \tag{3.316}$$

with

$$\mathbf{x}^{i_k, i_{k-1}}_{k|k-1} = \mathbf{F}^{i_k}_{k|k-1}\mathbf{x}^{i_{k-1}}_{k-1|k-1}, \tag{3.317}$$

$$\mathbf{P}^{i_k, i_{k-1}}_{k|k-1} = \mathbf{F}^{i_k}_{k|k-1}\mathbf{P}^{i_{k-1}}_{k-1|k-1}\mathbf{F}^{i_k\,\top}_{k|k-1} + \mathbf{Q}^{i_k}_{k|k-1}. \tag{3.318}$$

It is not surprising that due to the double sum, one obtains M^2 components. In order to reduce the number again to M, all components, which had a transition into the same model i_k, are merged via moment matching:

$$p(\mathbf{x}_k, i_k|\mathcal{Z}^{k-1}) \approx p^{i_k}_{k|k-1}\, \mathcal{N}(\mathbf{x}_k;\, \mathbf{x}^{i_k}_{k|k-1},\, \mathbf{P}^{i_k}_{k|k-1}), \tag{3.319}$$

where

$$p_{k|k-1}^{i_k} = \sum_{i_{k-1}} p_{i_k,i_{k-1}} p_{k-1|k-1}^{i_{k-1}} \tag{3.320}$$

$$\mathbf{x}_{k|k-1}^{i_k} = \frac{1}{p_{k|k-1}^{i_k}} \sum_{i_{k-1}} p_{i_k,i_{k-1}} p_{k-1|k-1}^{i_{k-1}} \mathbf{x}_{k|k-1}^{i_k,i_{k-1}} \tag{3.321}$$

$$\mathbf{P}_{k|k-1}^{i_k} = \frac{1}{p_{k|k-1}^{i_k}} \sum_{i_{k-1}} p_{i_k,i_{k-1}} p_{k-1|k-1}^{i_{k-1}} \left(\mathbf{P}_{k|k-1}^{i_k,i_{k-1}} + \boldsymbol{\nu}^{i_k,i_{k-1}} \boldsymbol{\nu}^{i_k,i_{k-1}\top} \right) \tag{3.322}$$

$$\boldsymbol{\nu}^{i_k,i_{k-1}} = \mathbf{x}_{k|k-1}^{i_k,i_{k-1}} - \mathbf{x}_{k|k-1}^{i_k}. \tag{3.323}$$

This process of enumerating all possible transitions with a reduction to a constant number of modes is depicted in Figure 3.15.

This yields the prior density for the IMM algorithm, since it holds that

$$p(\mathbf{x}_k | \mathcal{Z}^{k-1}) = \sum_{i_k=1}^{M} p(\mathbf{x}_k, i_k | \mathcal{Z}^{k-1}) \tag{3.324}$$

$$= \sum_{i_k=1}^{M} p_{k|k-1}^{i_k} \; \mathcal{N}(\mathbf{x}_k;\, \mathbf{x}_{k|k-1}^{i_k},\, \mathbf{P}_{k|k-1}^{i_k}). \tag{3.325}$$

The update step of the IMM is obtained via the Bayes theorem. It should be noted ahead that actually the measurement data provides the information on which model

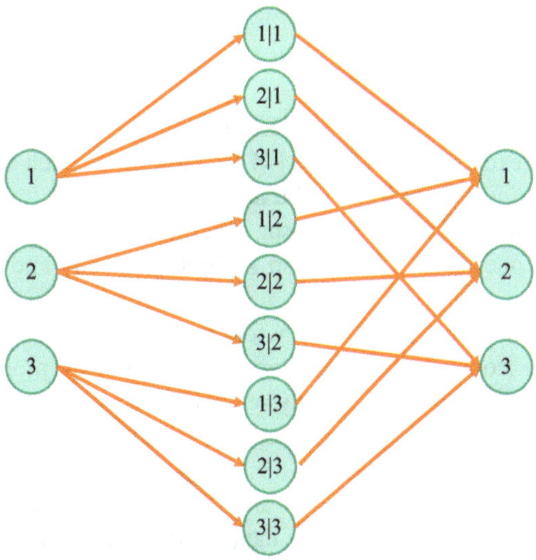

Figure 3.15 IMM prediction: all possible mode transitions are computed and reduced afterwards to a constant number of modes via moment matching

actually is in effect. Eventually, the density of the expected measurement updates the weights such that the best matching mode obtains the highest contribution. One has that

$$p(\mathbf{x}_k, i_k|\mathcal{Z}^k) = \frac{p_{k|k-1}^{i_k}\mathcal{N}(\mathbf{z}_k;\ \mathbf{H}_k\mathbf{x}_k,\ \mathbf{R}_k)\mathcal{N}(\mathbf{x}_k;\ \mathbf{x}_{k|k-1}^{i_k},\ \mathbf{P}_{k|k-1}^{i_k})}{\sum_{i_k}\int \mathrm{d}\mathbf{x}_k\, p_{k|k-1}^{i_k}\mathcal{N}(\mathbf{z}_k;\ \mathbf{H}_k\mathbf{x}_k,\ \mathbf{R}_k)\mathcal{N}(\mathbf{x}_k;\ \mathbf{x}_{k|k-1}^{i_k},\ \mathbf{P}_{k|k-1}^{i_k})} \tag{3.326}$$

$$= \frac{p_{k|k-1}^{i_k}\mathcal{N}(\mathbf{z}_k;\ \mathbf{H}_k\mathbf{x}_{k|k-1}^{i_k},\ \mathbf{S}_k^{i_k})\mathcal{N}(\mathbf{x}_k;\ \mathbf{x}_{k|k}^{i_k},\ \mathbf{P}_{k|k}^{i_k})}{\sum_{i_k}\int \mathrm{d}\mathbf{x}_k\, p_{k|k-1}^{i_k}\mathcal{N}(\mathbf{z}_k;\ \mathbf{H}_k\mathbf{x}_{k|k-1}^{i_k},\ \mathbf{S}_k^{i_k})\mathcal{N}(\mathbf{x}_k;\ \mathbf{x}_{k|k}^{i_k},\ \mathbf{P}_{k|k}^{i_k})} \tag{3.327}$$

$$= p_{k|k}^{i_k}\mathcal{N}(\mathbf{x}_k;\ \mathbf{x}_{k|k}^{i_k},\ \mathbf{P}_{k|k}^{i_k}), \tag{3.328}$$

where the filtering parameters $\mathbf{x}_{k|k}^{i_k}$ and $\mathbf{P}_{k|k}^{i_k}$ are given by the Kalman filter update formulae and the weights are updated according to the evaluation of the expected measurement density with

$$p_{k|k}^{i_k} = \frac{p_{k|k-1}^{i_k}\mathcal{N}(\mathbf{z}_k;\ \mathbf{H}_k\mathbf{x}_{k|k-1}^{i_k},\ \mathbf{S}_k^{i_k})}{\sum_{i_k} p_{k|k-1}^{i_k}\mathcal{N}(\mathbf{z}_k;\ \mathbf{H}_k\mathbf{x}_{k|k-1}^{i_k},\ \mathbf{S}_k^{i_k})} \tag{3.329}$$

$$\mathbf{S}_k^{i_k} = \mathbf{H}_k\mathbf{P}_{k|k-1}^{i_k}\mathbf{H}_k^\top + \mathbf{R}_k. \tag{3.330}$$

The parallel use of different evolution models in combination with an MHT scheme is a very effective combination for target tracking [32,42] due to the high degree of modeling in both, the sensor and the object behavior. In the literature, one can find sophisticated methods in order to derive the appropriate motion model for target tracking [43].

3.10 Sequential likelihood ratio test

The *Sequential Likelihood Ratio Test* (SLRT) is a stochastically optimal algorithm to decide between to exclusive hypotheses [44]. It has been proven an excellent means for *track confirmation*, where it has to be decided whether the track exists (that is it refers to a target) or not. Since new sensor data may be coming in, there is no final decision, instead, one updates a decision score sequentially and decides based on thresholds.

By tracking a tentative target even before its existence is confirmed, the SLRT allows an iterative computation of a score value $\mathrm{LR}(k)$ by processing new measurements. Since its derivation is closely related to the MHT (see Section 3.8), the combination of MHT and SLRT is particularly popular as many computations are identical and can be reused.

The problem of track confirmation and *track-before-detect* is to confirm the existence of a target based on ambiguous data. In particular when the detection probability is quite low, simple thresholds for the given number of associated observations within a certain time window would yield insufficient results. Based on more accurate stochastic models, it is possible to formulate a statistical test to decide between the two hypotheses

- h_1: There is a target.
- h_0: There is no target.

The decision is inferred from the information contained in the set of measurements $\mathcal{Z}^k$. The measurements at a given instant of time t_l is denoted by $Z_l = \{\mathbf{z}_l^1, \ldots, \mathbf{z}_l^{m_l}\}$, where m_l is the number of measurements. Among this data set, the target – given that it exists – may have produced (at most) one measurement with probability p_D. Note that in general the probability of detection p_D may also be state dependent, which would not change the general concept of the SLRT.

All other measurements are assumed to be caused by clutter or false detections. As in the above sections, the parameter ρ_F represents a fixed and constant clutter density and a spatial uniform distribution for the false alarms is assumed in the field of view. The number of false measurements m is modeled as a Poisson distribution as in (3.243) with mean value λ which is a fixed modeling parameter:

$$p_F(m) = \left(\frac{\lambda^m}{m!}\right) \mathrm{e}^{-\lambda}. \tag{3.331}$$

The decision whether a target is present in the data set $\mathcal{Z}^k$ can be solved optimally by SLRT in the sense that given a certain level of confidence the minimum number of update steps are required to compute a decision that matches this confidence.

For the SLRT, the likelihood ratio (LR) at the current time t_k which is given by the fraction

$$\mathrm{LR}(k) = \frac{p(h_1|\mathcal{Z}^k)}{p(h_0|\mathcal{Z}^k)} \tag{3.332}$$

is computed recursively. At each time step, one can check if the LR exceeds or falls below one of two thresholds A and B:

- $\mathrm{LR}(k) < A$: accept h_0, i.e. delete track
- $\mathrm{LR}(k) > B$: accept h_1, i.e. confirm track
- $A < \mathrm{LR}(k) < B$: continue processing.

The thresholds A and B depend on the level of confidence to confirm only true targets and reject noise only. A typical course of the LR score for a target is shown in Figure 3.16.

A recursive computation of the LR score is obtained by an application of the Bayes theorem on the numerator and on the denominator:

$$\mathrm{LR}(k) = \frac{p(\mathcal{Z}^k|h_1)}{p(\mathcal{Z}^k|h_0)} \cdot \frac{p(h_1)}{p(h_0)} \tag{3.333}$$

$$= \frac{p(\mathcal{Z}^k|h_1)}{p(\mathcal{Z}^k|h_0)} \tag{3.334}$$

where $\frac{p(h_1)}{p(h_0)} = 1$ since $p(h_1) = p(h_0)$ models equal chances for both hypotheses prior to the data processing.

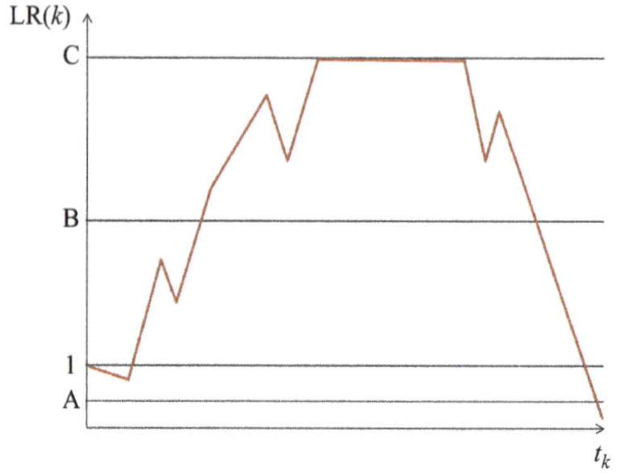

Figure 3.16 Example of a typical course of the LR score over time for a track, which is spawn at the beginning and lives for some time. The track is marked as tentative until the curve tops the threshold B. Since the score may increase exponentially, a cutoff value C is defined to keep it limited. As soon as the track vanishes, the LR score decreases and will fall below A at some time afterwards.

Now due to the definition of the conditional density we have that for both hypotheses $i = 0, 1$

$$p(\mathcal{Z}^k|h_i) = p(Z_k, \mathcal{Z}^{k-1}|h_i) \tag{3.335}$$

$$= p(Z_k|\mathcal{Z}^{k-1}, h_i)\, p(\mathcal{Z}^{k-1}|h_i). \tag{3.336}$$

Furthermore, we use marginalization of the target state $\mathbf{x}_k \in \mathbb{R}^n$ at time t_k for the numerator of $\mathrm{LR}(k)$ which yields

$$p(Z_k, \mathcal{Z}^{k-1}|h_1) = \int \mathrm{d}\mathbf{x}_k\, p(Z_k, \mathcal{Z}^{k-1}, \mathbf{x}_k|h_1) \tag{3.337}$$

$$= \int \mathrm{d}\mathbf{x}_k\, p(Z_k|\mathbf{x}_k, h_1)\, p(\mathbf{x}_k|\mathcal{Z}^{k-1}, h_1)\, p(\mathcal{Z}^{k-1}|h_1). \tag{3.338}$$

Therefore, one obtains

$$\mathrm{LR}(k) = \frac{\int \mathrm{d}\mathbf{x}_k\, p(Z_k|\mathbf{x}_k, h_1)\, p(\mathbf{x}_k|\mathcal{Z}^{k-1}, h_1)}{p(Z_k|h_0)} \cdot \mathrm{LR}(k-1) \tag{3.339}$$

$$=: \Lambda(k) \cdot \mathrm{LR}(k-1). \tag{3.340}$$

For the problem of track extraction, a decision based on the LR score has to be computed. For the recursive update, a multiplicative parameter $\Lambda(k)$ needs to be known:

$$\Lambda(k) = \frac{\int \mathrm{d}\mathbf{x}_k\, p(Z_k|\mathbf{x}_k, h_1)\, p(\mathbf{x}_k|\mathcal{Z}^{k-1}, h_1)}{p(Z_k|h_0)}. \tag{3.341}$$

In this section, a recursive computation of $\Lambda(k)$ will be derived based on a single target likelihood function for sensor data with non-perfect detection and false measurements.

Denominator

The denominator is the probability density of observing $Z_k = \mathbf{z}_k^1, \ldots, \mathbf{z}_k^{m_k}$ at time t_k if there is no target. The spatial distribution of a single false measurement can be modeled as

$$p(\mathbf{z}_k^i | h_0) = \frac{1}{|\text{FoV}|}, \tag{3.342}$$

where |FoV| is the spatial size of the field of view. This reflects the fact that no information is available on the spatial distribution of false alarms ("maximum entropy"). Now the event also consists of the fact that m_k measurements are observed. Together, we model the probability to observe m_k mutually independent false measurements. Therefore

$$p(Z_k | h_0) = |\text{FoV}|^{-m_k} p_F(m_k), \tag{3.343}$$

where $p_F(m_k)$ is the Poisson distribution with mean λ as introduced in (3.243).

Numerator

The likelihood in the numerator is the union of the events that the target was detected (D) and that it was not detected ($\neg D$):

$$\begin{aligned} p(Z_k | \mathbf{x}_k, h_1) &= p(Z_k | \mathbf{x}_k, h_1, D)\, p(D) \\ &\quad + p(Z_k | \mathbf{x}_k, h_1, \neg D) p(\neg D). \end{aligned} \tag{3.344}$$

In this simplified model, it is assumed that the target detection probability of the sensor is a constant p_D, that is it is independent of the state of the target. Thus, we also directly obtain

$$p(D) = p_D, \tag{3.345}$$

$$p(\neg D) = (1 - p_D). \tag{3.346}$$

Now, the probability of the sensor likelihood in the case of a detection is the union of the events that the jth measurement is from the target, where $j = 1, \ldots, m_k$. If $\mathbf{z}_k^j$ is from the target, then still all other measurements are false measurements. Therefore, one obtains:

$$p(Z_k | \mathbf{x}_k, h_1, D) = \sum_{j=1}^{m_k} p(Z_k | \mathbf{x}_k, h_1, D, j) p(j) \tag{3.347}$$

$$= \sum_{j=1}^{m_k} p(\{\mathbf{z}_k^l\}_{l \neq j} | \text{clutter})\, p(j)\, p(\mathbf{z}_k^j | \mathbf{x}_k, h_1, D, j) \tag{3.348}$$

$$= \sum_{j=1}^{m_k} |\text{FoV}|^{-(m_k - 1)} p_F(m_k - 1) \frac{1}{m_k} \mathcal{N}(\mathbf{z}_k^j;\ \mathbf{H}_k \mathbf{x}_k,\ \mathbf{R}_k), \tag{3.349}$$

where each hypothesis has the same probability $p(j) = \frac{1}{m_k}$. Now for the number of clutter measurements, the following equation holds:

$$p_F(m) = \frac{\lambda}{m} \cdot p_F(m-1). \tag{3.350}$$

Therefore, by means of the clutter density $\rho_F = \frac{\lambda}{|\mathrm{FoV}|}$, we have that

$$|\mathrm{FoV}|^{-(m_k-1)} p_F(m_k - 1)\frac{1}{m_k} = |\mathrm{FoV}|\frac{1}{\lambda}|\mathrm{FoV}|^{-m_k} p_F(m_k) \tag{3.351}$$

$$= \frac{1}{\rho_F}|\mathrm{FoV}|^{-m_k} p_F(m_k). \tag{3.352}$$

For the likelihood function, one obtains

$$p(Z_k|\mathbf{x}_k, h_1) = (|\mathrm{FoV}|^{-m_k} p_F(m_k))((1 - p_D) + \frac{p_D}{\rho_F}\sum_{j=1}^{m_k} \mathcal{N}(\mathbf{z}_k^j;\ \mathbf{H}_k\mathbf{x}_k,\ \mathbf{R}_k)) \tag{3.353}$$

Concluding the numerator and the denominator, one obtains

$$\Lambda(k) = \int \mathrm{d}\mathbf{x}_k \left((1 - p_D) + \frac{p_D}{\rho_F}\sum_{j=1}^{m_k} \mathcal{N}(\mathbf{z}_k^j;\ \mathbf{H}_k\mathbf{x}_k,\ \mathbf{R}_k)\right) p(\mathbf{x}_k|\mathcal{Z}^{k-1}). \tag{3.354}$$

Inserting the prior density given by the Gaussian $p(\mathbf{x}_k|\mathcal{Z}^{k-1}) = \mathcal{N}(\mathbf{x}_k;\ \mathbf{x}_{k|k-1},\ \mathbf{P}_{k|k-1})$ yields when applying the product formula (3.123)

$$\Lambda(k) = \sum_{j=0}^{m_k} \lambda_j \tag{3.355}$$

$$\lambda_j = \begin{cases} (1 - p_D), & \text{if } j = 0 \\ \frac{p_D}{\rho_F}\mathcal{N}(\mathbf{z}_k^j;\ \mathbf{H}_k\mathbf{x}_{k|k-1},\ \mathbf{S}_k), & \text{else} \end{cases} \tag{3.356}$$

where $\mathbf{S}_k = \mathbf{H}_k^\top \mathbf{P}_{k|k-1}\mathbf{H}_k^\top + \mathbf{R}_k$ is the current innovation covariance matrix [34]. It can be seen that those parameters are calculated within the MHT process and therefore do not require any additional complex computations.

Discussion

While this algorithm is optimal in order to decide *initially* on the track existence, it does not reflect the fact that it may stop to exist.¶¶ Actually, there is a probabilistic approach,*** which models the state transition between existing and non-existing as a

¶¶We are avoiding the term "birth" and "death" within this context.

***It is called *Joint Integrated Probabilistic Data Association* (JIPDA).

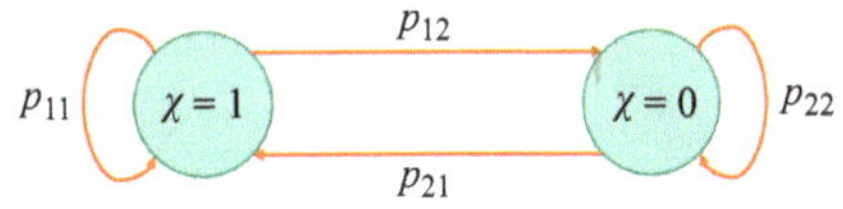

Figure 3.17 Markov chain state machine for the track existence ($\chi = 1$) or non-existence ($\chi = 0$) and the corresponding transition probabilities

Bernoulli distributed random variable [45]. The track existence variable χ^t for track t is then updated in the filtering step either by a probabilistic data association [46] or in a multi-scan MHT fashion [47]. The evolution of the state existence in the prediction step is modeled as a Markov process such that transition probabilities can be modeled:

$$P(\chi_k^t) = p_{11}P(\chi_{k-1}^t) + p_{21}(1 - P(\chi_{k-1}^t)). \tag{3.357}$$

Here, the Markov process and its transitions are shown in Figure 3.17.

This makes the Bernoulli approach beneficial when it comes to detect the end of track existence for long-lived tracks. However, the SLRT also can well be applied to detect the end of track existence by limiting the LR-score or its logarithm. The reason we focus on the SLRT is that it can be computed as a by-product of the MHT and that a distributed computation is well possible as will be presented in Chapter 8.

3.11 Conclusion

This chapter laid the foundation for state estimation and target tracking. Probabilities and probability density function provide wonderful means to represent knowledge on the state of a system, considered as a random variable. The simple but yet powerful tool of the Bayes theorem enables us to integrate the information of a time series of measurements into the posterior and therefore into the estimate and its error covariance matrix. The key for this process is the models, in particular for the sensor measurements and the object dynamics or state evolution. Here, the engineer has degrees of freedom to adapt the filter to a given scenario. We have seen that this becomes more complex as soon as ambiguities in the measurements arise. Hypothesising the data is a powerful approach in probability theory but will almost always lead to a growing memory disaster, which is to be handled based on approximations.

The upcoming chapter extends the described methods for state estimation in order to consider full trajectories of targets, where a trajectory is represented as a collection of states for discrete instants of time.

Chapter 4

Trajectory estimation

In the previous chapter, methods to compute the posterior density $p(\mathbf{x}_k|\mathcal{Z}^k)$ were presented, which contains the full information on the current state $\mathbf{x}_k$ at time t_k conditioned on the set of measurements $\mathcal{Z}^k$. If the full trajectory in the sense of a sequence of states $\mathbf{x}_1, \mathbf{x}_2, \ldots, \mathbf{x}_k$ is of interest, it is useful to update past states whenever new measurements have arrived. At a past time $t_l < t_k$ for instance, the posterior $p(\mathbf{x}_l|\mathcal{Z}^l)$ was computed. Under Kalman filter conditions, its parameters $\mathbf{x}_{l|l}$ and $\mathbf{P}_{l|l}$ might have been stored for further usage. In order to exploit the full information from the data, later measurements $\mathbf{z}_{l+1}, \ldots, \mathbf{z}_k$ have to be incorporated. Thus, we are interested in the density $p(\mathbf{x}_l|\mathcal{Z}^k)$ which is the goal of the first section in this chapter. Afterwards, joint densities of multiple states from the same object of the form

$$p(\mathbf{x}_k, \mathbf{x}_{k-1}, \ldots, \mathbf{x}_n|\mathcal{Z}^k) \tag{4.1}$$

are considered. It will be shown that these contain the full information on the individual states, and furthermore cross-covariance matrices may be inferred:

$$\begin{aligned}\operatorname{cov}\left[\mathbf{x}_l, \mathbf{x}_m|\mathcal{Z}^k\right] &= \int \mathrm{d}\mathbf{x}_k \int \mathrm{d}\mathbf{x}_{k-1} \cdots \int \mathrm{d}\mathbf{x}_n \\ &(\mathbf{x}_l - \mathrm{E}\left[\mathbf{x}_l|\mathcal{Z}^k\right])(\mathbf{x}_m - \mathrm{E}\left[\mathbf{x}_m|\mathcal{Z}^k\right])^\top p(\mathbf{x}_k, \mathbf{x}_{k-1}, \ldots, \mathbf{x}_n|\mathcal{Z}^k).\end{aligned} \tag{4.2}$$

These cross-covariances and the joint densities of full trajectories play a major role in some of the track-to-track fusion algorithms presented in Chapter 6.

4.1 Retrodiction

The term *"retrodiction"* is the counterpart to "prediction," since for the computation of updated past densities $p(\mathbf{x}_l|\mathcal{Z}^k)$ conditioned on current data ($t_l < t_k$), one is going backwards in time from the succeeding step:

$$p(\mathbf{x}_l|\mathcal{Z}^k) \quad \xleftarrow[\text{Retrodiction}]{} \quad p(\mathbf{x}_{l+1}|\mathcal{Z}^k). \tag{4.3}$$

This allows an iterative computation of all updated densities which refer to a past instant of time. By marginalization, the relation to $p(\mathbf{x}_{l+1}|\mathcal{Z}^k)$ can be explored:

$$p(\mathbf{x}_l|\mathcal{Z}^k) = \int \mathrm{d}\mathbf{x}_{l+1} \underbrace{p(\mathbf{x}_l|\mathbf{x}_{l+1}, \mathcal{Z}^k)}_{=p(\mathbf{x}_l|\mathbf{x}_{l+1}, \mathcal{Z}^l)} p(\mathbf{x}_{l+1}|\mathcal{Z}^k). \tag{4.4}$$

For $l+1=k$, this iterative computation can be initialized by means of the current posterior at time t_k. It should be noted that this process does not provide any additional information on the current state $\mathbf{x}_k$.

For the integration kernel in (4.4), the Bayes rule may be applied. This yields

$$p(\mathbf{x}_l|\mathbf{x}_{l+1}, \mathcal{Z}^l) = \frac{p(\mathbf{x}_{l+1}|\mathbf{x}_l)\, p(\mathbf{x}_l|\mathcal{Z}^l)}{\int \mathrm{d}\mathbf{x}_l\, p(\mathbf{x}_{l+1}|\mathbf{x}_l)\, p(\mathbf{x}_l|\mathcal{Z}^l)}. \tag{4.5}$$

By using the linear Gaussian model for the evolution model $p(\mathbf{x}_{l+1}|\mathbf{x}_l)$ and the Gaussian for the posterior $p(\mathbf{x}_l|\mathcal{Z}^l)$ at time t_l, one obtains

$$p(\mathbf{x}_l|\mathbf{x}_{l+1}, \mathcal{Z}^l) = \frac{\mathcal{N}(\mathbf{x}_{l+1};\, \mathbf{F}_{l+1|l}\mathbf{x}_l,\, \mathbf{Q}_{l+1|l})\, \mathcal{N}(\mathbf{x}_l;\, \mathbf{x}_{l|l},\, \mathbf{P}_{l|l})}{\int \mathrm{d}\mathbf{x}_l\, \mathcal{N}(\mathbf{x}_{l+1};\, \mathbf{F}_{l+1|l}\mathbf{x}_l,\, \mathbf{Q}_{l+1|l})\, \mathcal{N}(\mathbf{x}_l;\, \mathbf{x}_{l|l},\, \mathbf{P}_{l|l})} \tag{4.6}$$

$$= \frac{\mathcal{N}(\mathbf{x}_{l+1};\, \mathbf{x}_{l+1|l},\, \mathbf{P}_{l+1|l})}{\mathcal{N}(\mathbf{x}_{l+1};\, \mathbf{x}_{l+1|l},\, \mathbf{P}_{l+1|l})}$$

$$\cdot \mathcal{N}(\mathbf{x}_l;\, \mathbf{x}_{l|l} + \mathbf{W}_{l|l+1}(\mathbf{x}_{l+1} - \mathbf{x}_{l+1|l}),\, \mathbf{P}_{l|l} - \mathbf{W}_{l|l+1}\mathbf{P}_{l+1|l}\mathbf{W}_{l|l+1}^{\top}), \tag{4.7}$$

where the last equation is achieved by an application of the product formula (3.123) on the numerator and the denominator and by the fact that a single Gaussian integrates to one. Here the matrix $\mathbf{W}_{l|l+1}$ controls the gain with respect to the correction term and is therefore called *"retrodiction gain matrix."* According to the product formula, it is given by

$$\mathbf{W}_{l|l+1} = \mathbf{P}_{l|l}\mathbf{F}_{l+1|l}^{\top}\mathbf{P}_{l+1|l}^{-1}. \tag{4.8}$$

Let the previous step in the retrodiction iteration be given by the Gaussian density

$$p(\mathbf{x}_{l+1}|\mathcal{Z}^k) = \mathcal{N}(\mathbf{x}_{l+1};\, \mathbf{x}_{l+1|k},\, \mathbf{P}_{l+1|k}). \tag{4.9}$$

Filling it into (4.4) yields

$$p(\mathbf{x}_l|\mathcal{Z}^k) = \int \mathrm{d}\mathbf{x}_{l+1}\, \mathcal{N}(\mathbf{x}_l;\, \mathbf{x}_{l|l} + \mathbf{W}_{l|l+1}(\mathbf{x}_{l+1} - \mathbf{x}_{l+1|l}),\, \mathbf{P}_{l|l} - \mathbf{W}_{l|l+1}\mathbf{P}_{l+1|l}\mathbf{W}_{l|l+1}^{\top})$$

$$\mathcal{N}(\mathbf{x}_{l+1};\, \mathbf{x}_{l+1|k},\, \mathbf{P}_{l+1|k}) \tag{4.10}$$

$$= \mathcal{N}(\mathbf{x}_l;\, \mathbf{x}_{l|k},\, \mathbf{P}_{l|k}). \tag{4.11}$$

Here, again the Product Formula (3.123) was used in order to have the Gaussian in $\mathbf{x}_l$ independent of the integration variable. Obviously, the remaining term integrates to one due to the normalization property of densities.

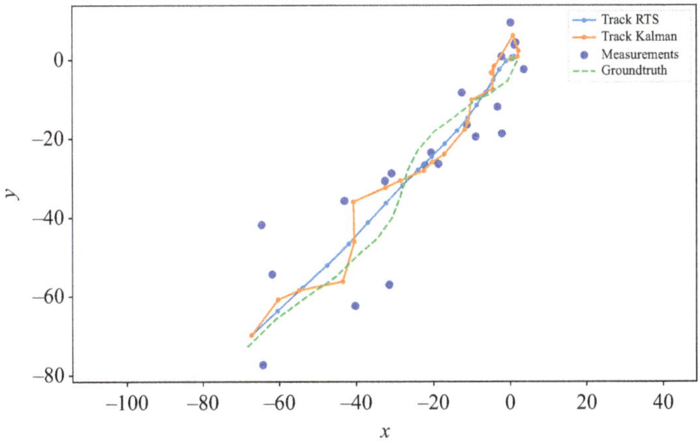

Figure 4.1 Simulation example of target measurements being processed with a Kalman filter ("Track Kalman") and with the RTS-retrodiction ("Track RTS")

The resulting parameters for the retrodicted density $p(\mathbf{x}_l|\mathcal{Z}^k) = \mathcal{N}(\mathbf{x}_l;\, \mathbf{x}_{l|k}, \mathbf{P}_{l|k})$ are known as the *Rauch–Tung–Striebel* (RTS) [48] equations:

$$\mathbf{x}_{l|k} = \mathbf{x}_{l|l} + \mathbf{W}_{l|l+1}(\mathbf{x}_{l+1|k} - \mathbf{x}_{l+1|l}) \tag{4.12}$$

$$\mathbf{P}_{l|k} = \mathbf{P}_{l|l} + \mathbf{W}_{l|l+1}\left(\mathbf{P}_{l+1|k} - \mathbf{P}_{l+1|l}\right)\mathbf{W}_{l|l+1}^{\top}, \tag{4.13}$$

where the *Retrodiction Gain Matrix* $\mathbf{W}_{l|l+1}$ is given by

$$\mathbf{W}_{l|l+1} = \mathbf{P}_{l|l}\mathbf{F}_{l+1|l}^{\top}\mathbf{P}_{l+1|l}^{-1}. \tag{4.14}$$

An example of a retrodicted trajectory is shown in Figure 4.1, where a target was tracked in a 2D numerical example using a simulated range-bearing sensor ("MSR"). The Kalman filter estimates ("Track Kalman") seen as a retrospective trajectory are a bit volatile, whereas the result of the Rauch–Tung–Striebel retrodiction ("Track RTS") is very smooth. This motivates the name *"smoothing,"* which is also a common name for retrodiction.

4.2 Notion of accumulated state densities

All information on a target trajectory within a time window $t_k, t_{k-1}, \ldots, t_n$ of length $k - n + 1$ with the *accumulated state*

$$\mathbf{x}_{k:n} := (\mathbf{x}_k^{\top}, \ldots, \mathbf{x}_n^{\top})^{\top} \tag{4.15}$$

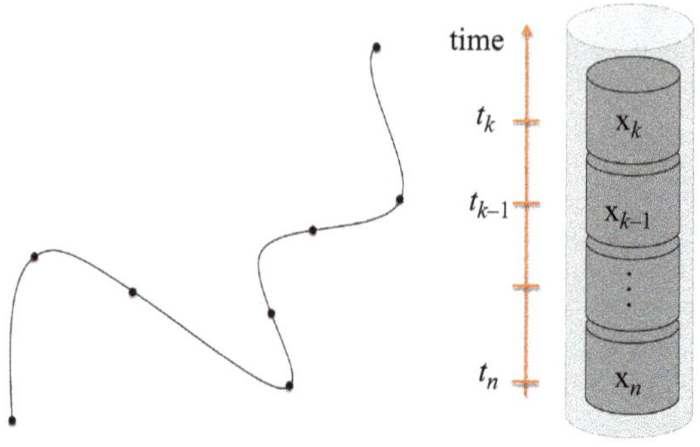

Figure 4.2 Left: A trajectory is represented in terms of its states at different instants of time. Right: Concept of the accumulated state.

that can be extracted from the time series of accumulated sensor data $\mathcal{Z}^k$ up to and including time t_k is contained in a joint density function $p(\mathbf{x}_{k:n}|\mathcal{Z}^k)$, which may be called *accumulated state density (ASD)*. Here t_k typically denotes the current time and $t_n \leq t_k$ is the time of initialization or the lower bound of a sliding time window. This scheme is visualized in Figure 4.2.

Via marginalizing over $\mathbf{x}_k, \ldots, \mathbf{x}_{l+1}, \mathbf{x}_{l-1}, \ldots, \mathbf{x}_n$,

$$p(\mathbf{x}_l|\mathcal{Z}^k) = \int \mathrm{d}\mathbf{x}_k \ldots \int \mathrm{d}\mathbf{x}_{l+1} \int \mathrm{d}\mathbf{x}_{l-1} \ldots \int \mathrm{d}\mathbf{x}_n \; p(\mathbf{x}_k, \ldots, \mathbf{x}_n|\mathcal{Z}^k), \tag{4.16}$$

the filtering density $p(\mathbf{x}_k|\mathcal{Z}^k)$ for $l = k$ and the retrodiction densities $p(\mathbf{x}_l|\mathcal{Z}^k)$ for $l < k$ result from the ASD. ASDs thus in a way unify the notions of filtering and retrodiction. In addition, ASDs also contain all mutual correlations between the individual object states at different instants of time. Bayes' theorem provides a recursion formula for updating ASDs:

$$p(\mathbf{x}_{k:n}|\mathcal{Z}^k) = \frac{p(\mathbf{z}_k|\mathbf{x}_k)\, p(\mathbf{x}_k|\mathbf{x}_{k-1})\, p(\mathbf{x}_{k-1:n}|\mathcal{Z}^{k-1})}{\int \mathrm{d}\mathbf{x}_{k:n}\, p(\mathbf{z}_k|\mathbf{x}_k)\, p(\mathbf{x}_k|\mathbf{x}_{k-1})\, p(\mathbf{x}_{k-1:n}|\mathcal{Z}^{k-1})}. \tag{4.17}$$

4.3 ASD posterior

Under conditions where Kalman filtering is applicable (see Section 3.2), a closed-form representation of $p(\mathbf{x}_{k:n}|\mathcal{Z}^k)$ can be derived. This section presents this calculation following [49]. By a repeated use of Bayes' theorem and the Markov property, we get from (4.17):

$$
\begin{aligned}
p(\mathbf{x}_{k:n}|\mathcal{Z}^k) &= \frac{p(\mathbf{z}_k|\mathbf{x}_k)\,p(\mathbf{x}_k|\mathbf{x}_{k-1})\,p(\mathbf{x}_{k-1:n}|\mathcal{Z}^{k-1})}{\int \mathrm{d}\mathbf{x}_{k:n}\,p(\mathbf{z}_k|\mathbf{x}_k)\,p(\mathbf{x}_k|\mathbf{x}_{k-1})\,p(\mathbf{x}_{k-1:n}|\mathcal{Z}^{k-1})} && (4.18)\\
&= \frac{\mathcal{N}(\mathbf{z}_k;\,\mathbf{H}_k\mathbf{x}_k,\,\mathbf{R}_k)\;\mathcal{N}(\mathbf{x}_k;\,\mathbf{F}_{k|k-1}\mathbf{x}_{k-1},\,\mathbf{Q}_{k|k-1})\,p(\mathbf{x}_{k-1:n}|\mathcal{Z}^{k-1})}{\int \mathrm{d}\mathbf{x}_{k:n}\;\mathcal{N}(\mathbf{z}_k;\,\mathbf{H}_k\mathbf{x}_k,\,\mathbf{R}_k)\;\mathcal{N}(\mathbf{x}_k;\,\mathbf{F}_{k|k-1}\mathbf{x}_{k-1},\,\mathbf{Q}_{k|k-1})\,p(\mathbf{x}_{k-1:n}|\mathcal{Z}^{k-1})} && (4.19)\\
&= \frac{\prod_{l=n+1}^{k}\mathcal{N}(\mathbf{z}_l;\,\mathbf{H}_l\mathbf{x}_l,\,\mathbf{R}_l)\;\mathcal{N}(\mathbf{x}_l;\,\mathbf{F}_{l|l-1}\mathbf{x}_{l-1},\,\mathbf{Q}_{l|l-1})\;\mathcal{N}(\mathbf{x}_n;\,\mathbf{x}_{n|n},\,\mathbf{P}_{n|n})}{\int \mathrm{d}\mathbf{x}_{k:n}\prod_{l=n+1}^{k}\mathcal{N}(\mathbf{z}_l;\,\mathbf{H}_l\mathbf{x}_l,\,\mathbf{R}_l)\;\mathcal{N}(\mathbf{x}_l;\,\mathbf{F}_{l|l-1}\mathbf{x}_{l-1},\,\mathbf{Q}_{l|l-1})\;\mathcal{N}(\mathbf{x}_n;\,\mathbf{x}_{n|n},\,\mathbf{P}_{n|n})}. && (4.20)
\end{aligned}
$$

A successive use of the product formula for Gaussian densities (3.123) now directly yields a factorized representation of the augmented density up to a factor of normalization, which can be omitted for the moment:

$$
p(\mathbf{x}_{k:n}|\mathcal{Z}^k) \propto \mathcal{N}(\mathbf{x}_k;\,\mathbf{x}_{k|k},\,\mathbf{P}_{k|k})\prod_{l=n}^{k-1}\mathcal{N}(\mathbf{x}_l;\,\mathbf{h}_{l|l+1}(\mathbf{x}_{l+1}),\,\mathbf{R}_{l|l+1}), \tag{4.21}
$$

where the auxiliary quantities $\mathbf{h}_{l|l+1}$, $\mathbf{R}_{l|l+1}$, $l \leq k$, are defined by:

$$
\begin{aligned}
\mathbf{h}_{l|l+1}(\mathbf{x}_{l+1}) &= \mathbf{x}_{l|l} + \mathbf{W}_{l|l+1}(\mathbf{x}_{l+1} - \mathbf{x}_{l+1|l}) && (4.22)\\
\mathbf{R}_{l|l+1} &= \mathbf{P}_{l|l} - \mathbf{W}_{l|l+1}\mathbf{P}_{l|l+1}\mathbf{W}_{l|l+1}^{\top} && (4.23)
\end{aligned}
$$

and $\mathbf{W}_{l|l+1}$ is the retrodiction gain matrix from (4.14).

Note that $\mathcal{N}(\mathbf{x}_l;\,\mathbf{h}_{l|l+1}(\mathbf{x}_{l+1}),\,\mathbf{R}_{l|l+1})$ can be interpreted by analogy to a Gaussian likelihood function with a linear measurement function $\mathbf{h}_{l|l+1}(\mathbf{x}_{l+1})$. $\mathbf{h}_{l|l+1}$, $\mathbf{R}_{l|l+1}$ are defined by the parameters of $p(\mathbf{x}_l|\mathcal{Z}^l) = \mathcal{N}(\mathbf{x}_l;\,\mathbf{x}_{l|l},\,\mathbf{P}_{l|l})$.

With $\mathbf{x}_{l|k}$, and $\mathbf{P}_{l|k}$ known from the RTS recursion (see Section 4.1), it is possible to rewrite the posterior as

$$
p(\mathbf{x}_{k:n}|\mathcal{Z}^k) = \mathcal{N}(\mathbf{x}_k;\,\mathbf{x}_{k|k},\,\mathbf{P}_{k|k}) \prod_{l=n}^{k-1}\mathcal{N}(\mathbf{x}_l - \mathbf{W}_{l|l+1}\mathbf{x}_{l+1};\,\mathbf{x}_{l|k} - \mathbf{W}_{l|l+1}\mathbf{x}_{l+1|k},\,\mathbf{U}_{l|k}), \tag{4.24}
$$

where the abbreviation for the auxiliary covariance matrices is used:

$$
\mathbf{U}_{l|k} = \mathbf{P}_{l|k} - \mathbf{W}_{l|l+1}\mathbf{P}_{l+1|k}\mathbf{W}_{l|l+1}^{\top}. \tag{4.25}
$$

Due to elementary matrix manipulations, one can show that this product can be represented by a single Gaussian,

$$
p(\mathbf{x}_{k:n}|\mathcal{Z}^k) = \mathcal{N}(\mathbf{x}_{k:n};\,\mathbf{x}_{k:n|k},\,\mathbf{P}_{k:n|k}), \tag{4.26}
$$

with a joint expectation vector $\mathbf{x}_{k:n|k}$ defined by:

$$
\mathbf{x}_{k:n|k} = (\mathbf{x}_{k|k}^{\top}, \mathbf{x}_{k-1|k}^{\top}, \ldots, \mathbf{x}_{n|k}^{\top})^{\top}, \tag{4.27}
$$

while the corresponding joint covariance matrix $\mathbf{P}_{k:n|k}$ can be written as an inverse of a tridiagonal block matrix:

$$\mathbf{P}_{k:n|k}^{-1} = \begin{pmatrix} \mathbf{T}_{k|k} & -\mathbf{W}_{k-1|k}^{\top}\mathbf{U}_{k-1|k}^{-1} & \mathbf{O} & \cdots & \mathbf{O} \\ -\mathbf{U}_{k-1|k}^{-1}\mathbf{W}_{k-1|k} & \mathbf{T}_{k-1|k} & -\mathbf{W}_{k-2|k}^{\top}\mathbf{U}_{k-2|k}^{-1} & \ddots & \vdots \\ \mathbf{O} & -\mathbf{U}_{k-2|k}^{-1}\mathbf{W}_{k-2|k} & \ddots & \ddots & \mathbf{O} \\ \vdots & \ddots & \ddots & \mathbf{T}_{n+1|k} & -\mathbf{W}_{n|k}^{\top}\mathbf{U}_{n|k} \\ \mathbf{O} & \cdots & \mathbf{O} & -\mathbf{U}_{n|k}\mathbf{W}_{n|k} & \mathbf{T}_{n|k} \end{pmatrix} \tag{4.28}$$

where the auxiliary quantities $\mathbf{T}_{l|k}$, $n \leq l \leq k$ are defined by:

$$\mathbf{T}_{l|k} = \begin{cases} \mathbf{P}_{k|k}^{-1} + \mathbf{W}_{l-1|l}^{\top}\mathbf{U}_{l-1|k}^{-1}\mathbf{W}_{l-1|l} & \text{for } l = k \\ \mathbf{U}_{l|k}^{-1} + \mathbf{W}_{l-1|l}^{\top}\mathbf{U}_{l-1|k}^{-1}\mathbf{W}_{l-1|l} & \text{for } n < l < k\,. \\ \mathbf{U}_{n|k}^{-1} & \text{for } l = n \end{cases} \tag{4.29}$$

This can be seen by considering projections $\mathbf{\Pi}_l$ defined by:

$$\mathbf{\Pi}_l \mathbf{x}_{k:n} = \begin{cases} (\mathbf{I}, \mathbf{O}, \ldots, \mathbf{O})\mathbf{x}_{k:n}, & l = k \\ (\mathbf{O}, \ldots, -\mathbf{W}_{l|l+1}, \mathbf{I}, \ldots, \mathbf{O})\,\mathbf{x}_{k:n}, & n \leq l < k \end{cases}$$

$$= \begin{cases} \mathbf{x}_k, l = k \\ \mathbf{x}_l - \mathbf{W}_{l|l+1}\mathbf{x}_{l+1}, \quad n \leq l < k. \end{cases} \tag{4.30}$$

Using $\mathbf{\Pi}_l$ and $\mathbf{U}_{l|k}$, $l = 1, \ldots, k$, the ASD can be rewritten:

$$p(\mathbf{x}_{k:n}|\mathcal{Z}^k) = \prod_{l=n}^{k} \mathcal{N}(\mathbf{\Pi}_l\mathbf{x}_{k:n};\ \mathbf{\Pi}_l\mathbf{x}_{k:n|k},\ \mathbf{U}_{l|k}) \tag{4.31}$$

$$\propto \prod_{l=n}^{k} \mathrm{e}^{-\frac{1}{2}\left(\mathbf{\Pi}_l\mathbf{x}_{k:n}-\mathbf{\Pi}_l\mathbf{x}_{k:n|k}\right)^{\top}\mathbf{U}_{l|k}^{-1}\left(\mathbf{\Pi}_l\mathbf{x}_{k:n}-\mathbf{\Pi}_l\mathbf{x}_{k:n|k}\right)} \tag{4.32}$$

$$= \mathrm{e}^{-\frac{1}{2}\left(\mathbf{x}_{k:n}-\mathbf{x}_{k:n|k}\right)^{\top}\left(\sum_{l=n}^{k}\mathbf{\Pi}_l^{\top}\mathbf{U}_{l|k}^{-1}\mathbf{\Pi}_l\right)\left(\mathbf{x}_{k:n}-\mathbf{x}_{k:n|k}\right)} \tag{4.33}$$

$$= \mathcal{N}(\mathbf{x}_{k:n};\ \mathbf{x}_{k:n|k},\ \mathbf{P}_{k:n|k}). \tag{4.34}$$

Thus, one can see that the resulting joint density is a Gaussian with mean $\mathbf{x}_{k:n|k}$ and covariance matrix $\mathbf{P}_{k:n|k}$, which is given by an the formula:

$$\mathbf{P}_{k:n|k} = \left(\sum_{l=n}^{k}\mathbf{\Pi}_l^{\top}\mathbf{U}_{l|k}^{-1}\mathbf{\Pi}_l\right)^{-1}. \tag{4.35}$$

The summation of the matrices $\mathbf{\Pi}_l^{\top}\mathbf{U}_{l|k}^{-1}\mathbf{\Pi}_l$ directly yields the inverse ASD covariance matrix as a tridiagonal block matrix as in (4.28). The tridiagonal structure is a consequence of the Markov property of the underlying evolution model. This representation of the inverse of $\mathbf{P}_{k:n|k}$ is useful in practical calculations.

By a repeated use of the matrix inversion lemma (Appendix, Section A.1) and an induction argument, the inverse of this tridiagonal block matrix can be calculated. The resulting block matrix is given by

$$\mathbf{P}_{k:n|k} = \begin{pmatrix} \mathbf{P}_{k|k} & \mathbf{P}_{k|k}\mathbf{W}_{k-1|k}^{\top} & \mathbf{P}_{k|k}\mathbf{W}_{k-2|k}^{\top} & \cdots & \mathbf{P}_{k|k}\mathbf{W}_{n|k}^{\top} \\ \mathbf{W}_{k-1|k}\mathbf{P}_{k|k} & \mathbf{P}_{k-1|k} & \mathbf{P}_{k-1|k}\mathbf{W}_{k-2|k-1}^{\top} & * & \mathbf{P}_{k-1|k}\mathbf{W}_{n|k-1}^{\top} \\ \mathbf{W}_{k-2|k}\mathbf{P}_{k|k} & \mathbf{W}_{k-2|k-1}\mathbf{P}_{k-1|k} & \mathbf{P}_{k-2|k} & * & \vdots \\ \vdots & * & * & * & \mathbf{P}_{n+1|k}\mathbf{W}_{n|n+1}^{\top} \\ \mathbf{W}_{n|k}\mathbf{P}_{k|k} & \mathbf{W}_{n|k-1}\mathbf{P}_{k-1|k} & \cdots & \mathbf{W}_{n|n+1}\mathbf{P}_{n+1|k} & \mathbf{P}_{n|k} \end{pmatrix} \tag{4.36}$$

where the following abbreviations were used:

$$\mathbf{W}_{l|k} = \prod_{\lambda=l}^{k-1} \mathbf{W}_{\lambda|\lambda+1} = \prod_{\lambda=l}^{k-1} \mathbf{P}_{\lambda|\lambda}\mathbf{F}_{\lambda+1|\lambda}^{\top}\mathbf{P}_{\lambda+1|\lambda}^{-1}. \tag{4.37}$$

It should be noted that the cross-covariance matrices $\mathrm{cov}\left[\mathbf{x}_l, \mathbf{x}_m | \mathcal{Z}^k\right]$ of different states in time are directly obtained as the corresponding off-diagonal elements.

The posterior ASD covariance matrix can be expressed recursively as

$$\mathbf{P}_{l:n|k} = \begin{pmatrix} \mathbf{P}_{l|k} & \mathbf{P}_{l|k}\mathbf{W}_{l-1:n}^{\top} \\ \mathbf{W}_{l-1:n}\mathbf{P}_{l|k} & \mathbf{P}_{l-1:n|k} \end{pmatrix}, \tag{4.38}$$

$n+1 \leq l \leq k$, with $\mathbf{P}_{n:n|k} = \mathbf{P}_{n|k}$ and $\mathbf{W}_{l:n}$ given by:

$$\mathbf{W}_{l:n} = \begin{pmatrix} \mathbf{W}_{l|l+1} \\ \mathbf{W}_{l-1:n}\mathbf{W}_{l|l+1} \end{pmatrix} \tag{4.39}$$

and $\mathbf{W}_{n:n} = \mathbf{W}_{n|n+1}$.

This statement directly follows from a straightforward induction argument, though the necessary calculations are perhaps somewhat tedious. The proposition holds for $k = n$. If it is assumed that it is true at time t_k, the ASD at time t_{k+1} can be represented by:

$$p(\mathbf{x}_{k+1:n}|\mathcal{Z}^{k+1}) = \frac{p(\mathbf{z}_{k+1}|\mathbf{x}_{k+1})\, p(\mathbf{x}_{k+1}|\mathbf{x}_k)\, p(\mathbf{x}_{k:n}|\mathcal{Z}^k)}{\int d\mathbf{x}_{k+1:n}\, p(\mathbf{z}_{k+1}|\mathbf{x}_{k+1})\, p(\mathbf{x}_{k+1}|\mathbf{x}_k)\, p(\mathbf{x}_{k:n}|\mathcal{Z}^k)}. \tag{4.40}$$

Using the projection matrices $\mathbf{\Pi}_k = (\mathbf{I}, \mathbf{O}, \ldots, \mathbf{O})$ defined by:

$$\mathbf{\Pi}_k \mathbf{x}_{k:n} = \mathbf{x}_k \tag{4.41}$$

and $\mathbf{\Pi}_{k:n} = (-\mathbf{W}_{k:n}, \mathbf{I})$ defined by:

$$\mathbf{\Pi}_{k:n}(\mathbf{x}_{k+1}^{\top}, \mathbf{x}_{k:n}^{\top})^{\top} = -\mathbf{W}_{k:n}\mathbf{x}_{k+1} + \mathbf{x}_{k:n}, \tag{4.42}$$

a repeated use of the product formula (3.123) yields:

$$\begin{aligned}
&p(\mathbf{z}_{k+1}|\mathbf{x}_{k+1})\, p(\mathbf{x}_{k+1}|\mathbf{x}_k)\, p(\mathbf{x}_{k:n}|\mathcal{Z}^k) = \\
&\mathcal{N}(\mathbf{z}_{k+1};\, \mathbf{H}_{k+1}\mathbf{x}_{k+1},\, \mathbf{R}_{k+1})\, \mathcal{N}(\mathbf{x}_{k+1};\, \mathbf{x}_{k+1|k},\, \mathbf{P}_{k+1|k}) \\
&\quad \cdot \mathcal{N}(\mathbf{x}_{k:n};\, \mathbf{x}_{k:n|k} + \mathbf{W}_{k:n}(\mathbf{x}_{k+1} - \mathbf{x}_{k+1|k}),\, \mathbf{R}_{k:n}) \\
&= \mathcal{N}(\mathbf{z}_{k+1};\, \mathbf{H}_{k+1}\mathbf{x}_{k+1|k},\, \mathbf{S}_{k+1|k})\, \mathcal{N}(\mathbf{x}_{k+1};\, \mathbf{x}_{k+1|k+1},\, \mathbf{P}_{k+1|k+1}) \\
&\quad \cdot \mathcal{N}(\mathbf{\Pi}_{k:n}\mathbf{x}_{k+1:n};\, \mathbf{\Pi}_{k:n}(\mathbf{x}_{k+1|k}^\top, \mathbf{x}_{k:n|k}^\top),\, \mathbf{R}_{k:n})
\end{aligned} \tag{4.43}$$

with $\mathbf{W}_{k:n}$ and $\mathbf{R}_{k:n}$ given by:

$$\mathbf{W}_{k:n} = \mathbf{P}_{k:n|k}\mathbf{\Pi}_k^\top \mathbf{F}_{k+1|k}^\top \mathbf{P}_{k+1|k}^{-1} \tag{4.44}$$

$$= \begin{pmatrix} \mathbf{W}_{k|k+1} \\ \mathbf{W}_{k-1:n}\mathbf{W}_{k|k+1} \end{pmatrix} \tag{4.45}$$

$$\mathbf{R}_{k:n} = \mathbf{P}_{k:n|k} - \mathbf{W}_{k:n}\mathbf{P}_{k+1|k}\mathbf{W}_{k:n}^\top. \tag{4.46}$$

In particular,

$$\mathbf{\Pi}_{k:n}^\top \mathbf{R}_{k:n}^{-1}\mathbf{\Pi}_{k:n} = \begin{pmatrix} \mathbf{W}_{k:n}^\top \mathbf{R}_{k:n}^{-1}\mathbf{W}_{k:n} & -\mathbf{W}_{k:n}^\top \mathbf{R}_{k:n}^{-1} \\ -\mathbf{R}_{k:n}^{-1}\mathbf{W}_{k:n} & \mathbf{R}_{k:n}^{-1} \end{pmatrix}. \tag{4.47}$$

By an another application of the product formula, the following term is obtained, which is proportional to a constant factor which is independent of the state vectors:

$$\begin{aligned}
p(\mathbf{z}_{k+1}|\mathbf{x}_{k+1}) \quad & p(\mathbf{x}_{k+1}|\mathbf{x}_k)\, p(\mathbf{x}_{k:n}|\mathcal{Z}^k) \\
& \propto \mathcal{N}(\mathbf{z}_{k+1};\, \mathbf{H}_{k+1}\mathbf{x}_{k+1|k},\, \mathbf{S}_{k+1|k})\, \mathcal{N}(\mathbf{x}_{k+1:n};\, \mathbf{x}_{k+1:n|k+1},\, \mathbf{P}_{k+1:n|k+1}),
\end{aligned} \tag{4.48}$$

where the covariance matrix $\mathbf{P}_{k+1:n|k+1}$ is given by:

$$\mathbf{P}_{k+1:n|k+1} = (\mathbf{\Pi}_{k+1}^\top \mathbf{P}_{k+1|k+1}^{-1}\mathbf{\Pi}_{k+1} + \mathbf{\Pi}_{k:n}^\top \mathbf{R}_{k:n}^{-1}\mathbf{\Pi}_{k:n})^{-1} \tag{4.49}$$

$$= \begin{pmatrix} \mathbf{P}_{k+1|k+1}^{-1} + \mathbf{W}_{k:n}^\top \mathbf{R}_{k:n}^{-1}\mathbf{W}_{k:n} & -\mathbf{W}_{k:n}^\top \mathbf{R}_{k:n}^{-1} \\ -\mathbf{R}_{k:n}^{-1}\mathbf{W}_{k:n} & \mathbf{R}_{k:n}^{-1} \end{pmatrix}^{-1}. \tag{4.50}$$

This block matrix can directly be inverted by using a matrix inversion formula (Appendix, Section A.1). The corresponding Schur-complement is given by:

$$\mathbf{T} = \mathbf{P}_{k+1|k+1}^{-1} + \mathbf{W}_{k:n}^\top \mathbf{R}_{k:n}^{-1}\mathbf{W}_{k:n} - \mathbf{W}_{k:n}^\top \mathbf{R}_{k:n}^{-1}\mathbf{R}_{k:n}\mathbf{R}_{k:n}^{-1}\mathbf{W}_{k:n} = \mathbf{P}_{k+1|k+1}^{-1}. \tag{4.51}$$

The covariance matrix of the ASD can therefore be expressed as:

$$\mathbf{P}_{k+1:n|k+1} = \begin{pmatrix} \mathbf{P}_{k+1|k+1} & \mathbf{P}_{k+1|k+1}\mathbf{W}_{k:n}^\top \\ \mathbf{W}_{k:n}\mathbf{P}_{k+1|k+1} & \mathbf{R}_{k:n} + \mathbf{W}_{k:n}\mathbf{P}_{k+1|k+1}\mathbf{W}_{k:n}^\top \end{pmatrix} \tag{4.52}$$

$$= \begin{pmatrix} \mathbf{P}_{k+1|k+1} & \mathbf{P}_{k+1|k+1}\mathbf{W}_{k:n}^\top \\ \mathbf{W}_{k:n}\mathbf{P}_{k+1|k+1} & \mathbf{P}_{k:n}^k + \mathbf{W}_{k:n}(\mathbf{P}_{k+1|k+1} - \mathbf{P}_{k+1|k})\mathbf{W}_{k:n}^\top \end{pmatrix}. \tag{4.53}$$

Using the identity

$$\mathbf{W}_{k|k+1}(\mathbf{P}_{k+1|k+1} - \mathbf{P}_{k+1|k})\mathbf{W}_{k|k+1}^\top = \mathbf{P}_{k|k+1} - \mathbf{P}_{k|k}, \tag{4.54}$$

which results from the RTS equations, one can see that

$$\begin{aligned}\mathbf{W}_{k:n} \quad & (\mathbf{P}_{k+1|k+1} - \mathbf{P}_{k+1|k})\mathbf{W}_{k:n}^{\top} \\ &= \begin{pmatrix} \mathbf{P}_{k|k+1} - \mathbf{P}_{k|k} & (\mathbf{P}_{k|k+1} - \mathbf{P}_{k|k})\mathbf{W}_{k-1:n}^{\top} \\ \mathbf{W}_{k-1:n}(\mathbf{P}_{k|k+1} - \mathbf{P}_{k|k}) & \mathbf{W}_{k-1:n}(\mathbf{P}_{k|k+1} - \mathbf{P}_{k|k})\mathbf{W}_{k-1:n}^{\top} \end{pmatrix}. \end{aligned} \tag{4.55}$$

With this result, the block matrix $\mathbf{P}_{k:n}^{k} + \mathbf{W}_{k:n}(\mathbf{P}_{k+1|k+1} - \mathbf{P}_{k+1|k})\mathbf{W}_{k:n}^{\top}$ on the right-lower corner on the right side of (4.53) is given by:

$$\begin{aligned}\mathbf{P}_{k:n|k} + \quad & \mathbf{W}_{k:n}(\mathbf{P}_{k+1|k+1} - \mathbf{P}_{k+1|k})\mathbf{W}_{k:n}^{\top} \\ &= \begin{pmatrix} \mathbf{P}_{k|k+1} & \mathbf{P}_{k|k+1}\mathbf{W}_{k-1:n}^{\top} \\ \mathbf{W}_{k-1:n}\mathbf{P}_{k|k+1} & \mathbf{P}_{k-1:n|k} + \mathbf{W}_{k-1:n}(\mathbf{P}_{k|k+1} - \mathbf{P}_{k|k})\mathbf{W}_{k-1:n}^{\top} \end{pmatrix}. \end{aligned} \tag{4.56}$$

An induction argument for the block matrix on the right-lower corner directly yields:

$$\mathbf{P}_{k:n|k} + \mathbf{W}_{k:n}(\mathbf{P}_{k+1|k+1} - \mathbf{P}_{k+1|k})\mathbf{W}_{k:n}^{\top} = \mathbf{P}_{k:n|k+1}. \tag{4.57}$$

According to the product formula (3.123), $\mathbf{x}_{k+1:n|k+1}$ is the sum of the following vectors:

$$\mathbf{P}_{k+1:n|k+1}\mathbf{\Pi}_{k:n}^{\top}\mathbf{R}_{k:n}^{-1}\mathbf{\Pi}_{k:n}(\mathbf{x}_{k+1|k}^{\top}, \mathbf{x}_{k:n|k}^{\top})^{\top} = \begin{pmatrix} \mathbf{O} \\ -\mathbf{W}_{k:n}\mathbf{x}_{k+1|k} + \mathbf{x}_{k:n|k} \end{pmatrix} \tag{4.58}$$

$$\mathbf{P}_{k+1:n|k+1}\mathbf{\Pi}_{k+1}^{\top}\mathbf{P}_{k+1|k+1}^{-1}\mathbf{\Pi}_{k+1}\mathbf{x}_{k+1|k+1} = \begin{pmatrix} \mathbf{x}_{k+1|k+1} \\ \mathbf{W}_{k:n}\mathbf{x}_{k+1|k+1} \end{pmatrix}. \tag{4.59}$$

By using an induction argument, we thus obtain:

$$\mathbf{x}_{k+1:n|k+1} = \begin{pmatrix} \mathbf{x}_{k+1|k+1} \\ \mathbf{x}_{k|k+1} \\ \mathbf{x}_{k-1:n|k} + \mathbf{W}_{k-1:n}(\mathbf{x}_{k|k+1} - \mathbf{x}_{k|k}) \end{pmatrix}. \tag{4.60}$$

An induction argument concludes the proof.

The densities $\{\mathcal{N}(\mathbf{x}_l;\, \mathbf{x}_{l|k},\, \mathbf{P}_{l|k})\}_{l=n,\ldots,k}$ are directly obtained via marginalizing, since the covariance matrices $\mathbf{P}_{l|k}$, $n \leq l \leq k$, appear on the diagonal of this block matrix. It should be mentioned that the ASD is completely defined by the results of prediction, filtering, and retrodiction obtained for the time window $t_k, \ldots, t_n$, that is, it is a by-product for Kalman filtering and RTS smoothing.

4.4 Recursive measurement fusion using ASDs

Using the ASD-posterior density given in the previous section, it is possible to derive an iterative filtering scheme using an ASD with a fixed window length, which is illustrated in Figure 4.3.

This section addresses the problem of how to process new measurements upon arrival. The proposed method is to apply standard Kalman filter equations to the accumulated state by using projections. Via linear projections, it is possible to directly access a state of any time instant within the window. This fact has a very important consequence:

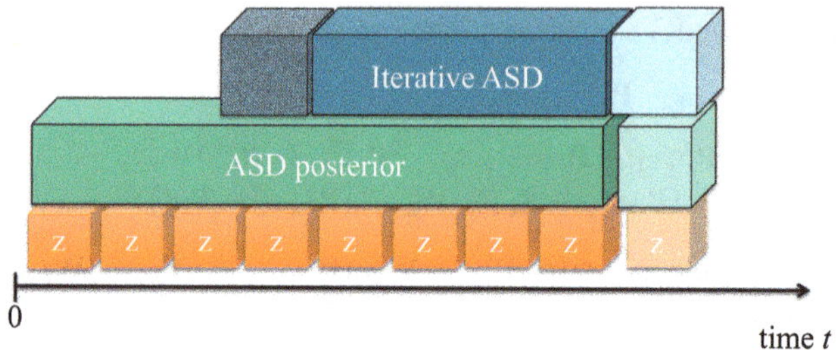

Figure 4.3 Schematic illustration of an iterative ASD filter and the ASD posterior. The iterative filtering scheme includes the calculation of a prior ASD density as well as a fixed window size. It is based on an ASD posterior, which yields the joint density conditioned on the entire set of measurements.

> For the iterative ASD filter, all measurements within the time window can be processed by means of a Kalman update together with projections. Therefore, this also holds for *Out-of-Sequence* (OoS) data, which refer to time instants before the current time.

To become more concrete, it is assumed that a measurement $\mathbf{z}_m$ is given which was produced at time t_m with $t_n \leq t_m \leq t_k$, that is possibly before the "present" time t_k, where the time series $\mathcal{Z}^k$ is available and has been processed. The goal is to calculate impact ("gain") of this new, but late sensor information on the accumulated state $\mathbf{x}_{k:n}$. Let $\mathbf{z}_m$ be a measurement of the target state $\mathbf{x}_m$ at time t_m characterized by a Gaussian likelihood function, which is defined by a measurement matrix $\mathbf{H}_m$ and a corresponding measurement error covariance matrix $\mathbf{R}_m$. At first, it is useful to renumber the target states $\mathbf{x}_k, \ldots, \mathbf{x}_n$ such that $\mathbf{x}_k, \ldots, \mathbf{x}_m, \ldots, \mathbf{x}_n =: \mathbf{x}_{k:m:n}$ are consistent with their time stamps $(t_l)_{l=k,\ldots,m,\ldots,n}$.

In order to process the measurement $\mathbf{z}_m$, it is necessary to calculate the prior density $p(\mathbf{x}_{k:m:n}|\mathcal{Z}^k)$ where $\mathbf{z}_m \notin \mathcal{Z}^k$. This new joint probability density function can be calculated by the same method used to derive the posterior in Section 4.3. In the first case, it is assumed that $m < k$. Then:

$$p(\mathbf{x}_{k:m:n}|\mathcal{Z}^k) \quad \propto p(\mathbf{z}_k|\mathbf{x}_k)\, p(\mathbf{x}_k|\mathbf{x}_{k-1}) \cdots$$

$$p(\mathbf{x}_{m+1}|\mathbf{x}_m)\, p(\mathbf{x}_m|\mathbf{x}_{m-1})\, p(\mathbf{z}_{m-1}|\mathbf{x}_{m-1}) \cdots p(\mathbf{x}_n|\mathcal{Z}^n) \tag{4.61}$$

$$= \prod_{l \neq m} \mathcal{N}(\mathbf{z}_l;\, \mathbf{H}_l\mathbf{x}_l,\, \mathbf{R}_l)\; \mathcal{N}(\mathbf{x}_l;\, \mathbf{F}_{l|l-1}\mathbf{x}_{l-1},\, \mathbf{Q}_{l|l-1})$$

$$\cdot \mathcal{N}(\mathbf{x}_m;\, \mathbf{F}_{m|m-1}\mathbf{x}_{m-1},\, \mathbf{Q}_{m|m-1})\; \mathcal{N}(\mathbf{x}_n;\, \mathbf{x}_{n|n},\, \mathbf{P}_{n|n}). \tag{4.62}$$

Applying the product formula to all terms from n up to $m-1$, it is possible to proceed as in (4.20). From there on, one should note the following fact:

$$
\begin{aligned}
&\mathcal{N}(\mathbf{x}_{m-1};\ \mathbf{x}_{m-1|m-1},\ \mathbf{P}_{m-1|m-1})\,\mathcal{N}(\mathbf{x}_m;\ \mathbf{F}_{m|m-1}\mathbf{x}_{m-1},\ \mathbf{Q}_{m|m-1})\\
&\cdot\mathcal{N}(\mathbf{x}_{m+1};\ \mathbf{F}_{m+1|m}\mathbf{x}_m,\ \mathbf{Q}_{m+1|m})\\
&= \mathcal{N}(\mathbf{x}_{m-1};\ \mathbf{h}_{m-1|m}(\mathbf{x}_m),\ \mathbf{R}_{m-1|m})\,\mathcal{N}(\mathbf{x}_m;\ \mathbf{x}_{m|m-1},\ \mathbf{P}_{m|m-1})\\
&\cdot\mathcal{N}(\mathbf{x}_{m+1};\ \mathbf{F}_{m+1|m}\mathbf{x}_m,\ \mathbf{Q}_{m+1|m})\\
&= \mathcal{N}(\mathbf{x}_{m-1};\ \mathbf{h}_{m-1|m}(\mathbf{x}_m),\ \mathbf{R}_{m-1|m})\,\mathcal{N}(\mathbf{x}_m;\ \mathbf{h}_{m|m+1}(\mathbf{x}_{m+1}),\ \mathbf{R}_{m|m+1})\\
&\cdot\mathcal{N}(\mathbf{x}_{m+1};\ \mathbf{x}_{m+1|m-1},\ \mathbf{P}_{m+1|m-1}).
\end{aligned}
\tag{4.63}
$$

This is due to the fact that $\mathbf{F}_{.}$ and $\mathbf{Q}_{.}$ describe a linear flow:

$$\mathbf{F}_{m+1|m-1} = \mathbf{F}_{m+1|m}\mathbf{F}_{m|m-1}, \tag{4.64}$$

$$\mathbf{Q}_{m+1|m-1} = \mathbf{F}_{m+1|m}\mathbf{Q}_{m|m-1}\mathbf{F}^{\top}_{m+1|m} + \mathbf{Q}_{m+1|m}. \tag{4.65}$$

Particularly, the function $\mathbf{h}_{m|m+1}$ and the covariance matrix $\mathbf{R}_{m|m+1}$ are given by

$$\mathbf{h}_{m|m+1}(\mathbf{x}_{m+1}) = \mathbf{x}_{m|m-1} + \mathbf{W}_{m|m+1}(\mathbf{x}_{m+1} - \mathbf{x}_{m+1|m-1}) \tag{4.66}$$

$$\mathbf{W}_{m|m+1} = \mathbf{P}_{m|m-1}\mathbf{F}^{\top}_{m+1|m}\mathbf{P}^{-1}_{m+1|m-1} \tag{4.67}$$

$$\mathbf{R}_{m|m+1} = \mathbf{P}_{m|m-1} - \mathbf{W}_{m|m+1}\mathbf{P}_{m+1|m-1}\mathbf{W}^{\top}_{m|m+1}. \tag{4.68}$$

This type of reasoning is also known as *continuous time retrodiction*, see [50] for a more detailed discussion. The resulting terms fit into the product representation and therefore a continued use of the product formula yields

$$p(\mathbf{x}_{k:m:n}|\mathcal{Z}^k) = \mathcal{N}(\mathbf{x}_k;\ \mathbf{x}_{k|k},\ \mathbf{P}_{k|k})\prod_{l=n}^{k-1}\mathcal{N}(\mathbf{x}_l;\ \mathbf{h}_{l|l+1}(\mathbf{x}_{l+1}),\ \mathbf{R}_{l|l+1}). \tag{4.69}$$

For the second case $m = k+1$, that is $t_m > t_l$ for all times t_l of measurement set $\mathcal{Z}^k$, a prediction factor as it is known from the standard Kalman equations is obtained. The predicted joint probability density function then is:

$$p(\mathbf{x}_{k:m:n}|\mathcal{Z}^k) = \mathcal{N}(\mathbf{x}_m;\ \mathbf{x}_{m|m-1},\ \mathbf{P}_{m|m-1})\prod_{l=n}^{k}\mathcal{N}(\mathbf{x}_l;\ \mathbf{h}_{l|l+1}(\mathbf{x}_{l+1}),\ \mathbf{R}_{l|l+1}). \tag{4.70}$$

In both cases, the same reasoning as above now leads to the extended ASD as a single Gaussian density function:

$$p(\mathbf{x}_{k:m:n}|\mathcal{Z}^k) = \mathcal{N}(\mathbf{x}_{k:m:n};\ \mathbf{x}_{k:m:n|k},\ \mathbf{P}_{k:m:n|k}). \tag{4.71}$$

For the filtering step of $\mathbf{z}_m$, a projection matrix $\mathbf{\Pi}_m$ is introduced, which is defined by

$$\mathbf{\Pi}_m\mathbf{x}_{k:m:n} = \mathbf{x}_m, \tag{4.72}$$

$$\mathbf{\Pi}_m = (\mathbf{O},\ldots,\mathbf{O},\mathbf{I},\mathbf{O},\ldots,\mathbf{O}) \tag{4.73}$$

such that it extracts the target state $\mathbf{x}_m$ from the accumulated state vector $\mathbf{x}_{k:m:n}$. The likelihood function of the OoS measurement with respect to the accumulated target state is thus given by:

$$p(\mathbf{z}_m|\mathbf{x}_{k:m:n}) = \mathcal{N}(\mathbf{z}_m;\ \mathbf{H}_m\mathbf{\Pi}_m\mathbf{x}_{k:m:n},\ \mathbf{R}_m). \tag{4.74}$$

Standard Bayesian reasoning and the product formula for Gaussians (3.123) directly yield for the accumulated state density:

$$p(\mathbf{x}_{k:m:n}|\mathbf{z}_m, \mathcal{Z}^k) = \frac{p(\mathbf{z}_m|\mathbf{x}_{k:m:n})\, p(\mathbf{x}_{k:m:n}|\mathcal{Z}^k)}{\int \mathrm{d}\mathbf{x}_{k:m:n}\, p(\mathbf{z}_m|\mathbf{x}_{k:m:n})\, p(\mathbf{x}_{k:m:n}|\mathcal{Z}^k)} \tag{4.75}$$

$$= \mathcal{N}(\mathbf{x}_{k:m:n};\ \mathbf{x}_{k:m:n|k,m},\ \mathbf{P}_{k:m:n|k,m}) \tag{4.76}$$

with parameters obtained by a version of the Kalman update equations:

$$\mathbf{x}_{k:m:n|k,m} = \mathbf{x}_{k:m:n|k} + \mathbf{W}_{k:m:n}(\mathbf{z}_m - \mathbf{H}_m\mathbf{\Pi}_m\mathbf{x}_{k:m:n|k}) \tag{4.77}$$

$$\mathbf{P}_{k:m:n|k,m} = \mathbf{P}_{k:m:n|k} - \mathbf{W}_{k:m:n}\mathbf{S}_{k:m:n}\mathbf{W}_{k:m:n}^\top, \tag{4.78}$$

where the corresponding Kalman gain and innovation matrices are given by:

$$\mathbf{S}_{k:m:n} = \mathbf{H}_m\mathbf{\Pi}_m\mathbf{P}_{k:m:n|k}\mathbf{\Pi}_m^\top\mathbf{H}_m^\top + \mathbf{R}_m \tag{4.79}$$

$$\mathbf{W}_{k:m:n} = \mathbf{P}_{k:m:n|k}\mathbf{\Pi}_m^\top\mathbf{H}_m^\top\mathbf{S}_{k:m:n}^{-1}. \tag{4.80}$$

Note that the innovation covariance matrix $\mathbf{S}_{k:m:n}$ to be inverted when calculating the Kalman gain matrix has the same dimension as the measurement vector $\mathbf{z}_m$, that is it is a low-dimensional matrix, just as in standard Kalman filtering. Moreover, it is easy to see that the innovation covariance of the ASD filter is equivalent to the standard innovation covariance of a Kalman filter due to

$$\mathbf{\Pi}_m\mathbf{P}_{k:m:n|k}\mathbf{\Pi}_m^\top = \mathbf{P}_{m|k}. \tag{4.81}$$

Nevertheless, the processing of an OoS measurement $\mathbf{z}_m$ has impact on all state estimates and the related estimate error covariance matrices in the considered time window. The strongest impact is observed for the estimate of the state at time t_m, which is the time instant the measurement was produced, while it declines the further one proceeds to the present time $t_k \geq t_l > t_m$ or to the past $t_m > t_l \geq t_n$. Accumulated state densities are therefore well suited to quantitatively discuss the question to what extent an OoS measurement is still useful or not, a phenomenon that is sometimes called "information aging." The posterior estimates of all states within the time window can be obtained by an application of the projections $\mathbf{\Pi}_l$, $n \leq l \leq k$:

$$p(\mathbf{x}_l|\mathbf{z}_m, \mathcal{Z}^k) = \mathcal{N}(\mathbf{x}_l;\ \mathbf{\Pi}_l\mathbf{x}_{k:m:n|k,m},\ \mathbf{\Pi}_l\mathbf{P}_{k:m:n|k,m}\mathbf{\Pi}_l^\top). \tag{4.82}$$

In a practical application, the focus is usually on the estimates of states and on their covariances. It is therefore sufficient to consider accumulated state densities $p(\mathbf{x}_{k:n}|\mathcal{Z}^k)$ characterized by lower dimensional parameters $\mathbf{x}_{l|k}$, $\mathbf{P}_{l|k}$, Section 4.5 will address the processing scheme, which is based on saving these parameters only. To determine the actual size of n to be taken into account is an important task in designing an ASD filter for sensor networks and depends on local storage and processing capacity as well as on link bandwidths.

Recursive ASD filter implementation

This paragraph shows how to calculate the parameters $\mathbf{x}_{k:n|k}$ and $\mathbf{P}_{k:n|k}$ *recursively*. This yields the most efficient way to implement an iterative ASD filter which processes a time series of measurements. For the prediction, it is assumed that the posterior ASD at time t_{k-1} is given in terms of $\mathbf{x}_{k-1:n|k-1}$ and $\mathbf{P}_{k-1:n|k-1}$. The prediction of the state is straight forward due to the Markov proposition:

$$\mathbf{x}_{k:n|k-1} = \left(\mathbf{x}_{k|k-1}^{\top} \ \mathbf{x}_{k-1|k-1}^{\top} \ \cdots \ \mathbf{x}_{n|k-1}^{\top}\right)^{\top}, \tag{4.83}$$

where $\mathbf{x}_{k|k-1} = \mathbf{F}_{k|k-1}\mathbf{x}_{k-1|k-1}$ is equivalent to a Kalman filter prediction. For the ASD covariance prediction, a recursive formulation of the ASD covariance in (4.36) is used:

$$\mathbf{P}_{k:n|k-1} = \begin{pmatrix} \mathbf{P}_{k|k-1} & \mathbf{P}_{k|k-1}\mathbf{W}_{k-1:n}^{\top} \\ \mathbf{W}_{k-1:n}\mathbf{P}_{k|k-1} & \mathbf{P}_{k-1:n|k-1} \end{pmatrix}, \tag{4.84}$$

where

$$\mathbf{P}_{k|k-1} = \mathbf{F}_{k|k-1}\mathbf{P}_{k-1|k-1}\mathbf{F}_{k|k-1}^{\top} + \mathbf{Q}_{k|k-1}, \tag{4.85}$$

$$\mathbf{W}_{k-1:n} = \begin{pmatrix} \mathbf{W}_{k-1|k} \\ \mathbf{W}_{k-2:n}\mathbf{W}_{k-1|k} \end{pmatrix}. \tag{4.86}$$

The expression in (4.84) can be simplified to

$$\mathbf{P}_{k:n|k-1} = \begin{pmatrix} \mathbf{P}_{k|k-1} & \mathbf{F}_{k|k-1}(\mathbf{P}_{k-1:n|k-1}^{(k-1)})^{\top} \\ \mathbf{P}_{k-1:n|k-1}^{(k-1)}\mathbf{F}_{k|k-1}^{\top} & \mathbf{P}_{k-1:n|k-1} \end{pmatrix}, \tag{4.87}$$

where $\mathbf{P}_{k-1:n|k-1}^{(k-1)}$ represents the $(k-1)$th block column for $n = 1$ which refers to time t_{k-1} of the previous posterior covariance matrix. This concludes the recursive ASD prediction.

For the filtering step, the likelihood for the current measurement $\mathbf{z}_k$ is obtained via the application of projections $\mathbf{\Pi}_k$ onto the current state:

$$p(\mathbf{z}_k|\mathbf{x}_k) = p(\mathbf{z}_k|\mathbf{\Pi}_k\mathbf{x}_{k:n}) \tag{4.88}$$

$$= \mathcal{N}(\mathbf{z}_k;\ \mathbf{H}_k\mathbf{\Pi}_k\mathbf{x}_{k:n},\ \mathbf{R}_k), \tag{4.89}$$

where

$$\mathbf{\Pi}_k = (\mathbf{I}, \mathbf{O}, \ldots, \mathbf{O}). \tag{4.90}$$

Then, the posterior parameters are obtained by the multiplication of the local prior density and the likelihood function. An application of the Kalman filter formulae yields

$$\mathbf{x}_{k:n|k} = \mathbf{x}_{k:n|k-1} + \mathbf{W}_{k:n|k-1}(\mathbf{z}_k - \mathbf{H}_k\mathbf{\Pi}_k\mathbf{x}_{k:n|k-1}), \tag{4.91}$$

$$\mathbf{P}_{k:n|k} = \mathbf{P}_{k:n|k-1} - \mathbf{W}_{k:n|k-1}\mathbf{S}_k\mathbf{W}_{k:n|k-1}^{\top} \tag{4.92}$$

$$\mathbf{W}_{k:n|k-1} = \mathbf{P}_{k:n|k-1}\mathbf{\Pi}_k^{\top}\mathbf{H}_k^{\top}\mathbf{S}_k^{-1}, \tag{4.93}$$

$$\mathbf{S}_k = \mathbf{H}_k\mathbf{\Pi}_k\mathbf{P}_{k:n|k-1}\mathbf{\Pi}_k^{\top}\mathbf{H}_k^{\top} + \mathbf{R}_k. \tag{4.94}$$

Table 4.1 Recursive ASD algorithm

Init.	set $\mathbf{x}_{0\|0}$, $\mathbf{P}_{0\|0}$.
Prediction $t_0 \to t_1$	$\mathbf{x}_{1\|0} = \mathbf{F}_{1\|0}\mathbf{x}_{0\|0}$ $\mathbf{P}_{1\|0} = \mathbf{F}_{1\|0}\mathbf{P}_{0\|0}\mathbf{F}_{1\|0}^\top + \mathbf{Q}_{1\|0}$ $\mathbf{x}_{1:0\|0} = (\mathbf{x}_{1\|0}^\top \ \mathbf{x}_{0\|0}^\top)^\top$ $\mathbf{P}_{1:0\|0} = \begin{pmatrix} \mathbf{P}_{1\|0} & \mathbf{F}_{1\|0}\mathbf{P}_{0\|0} \\ \mathbf{P}_{0\|0}\mathbf{F}_{1\|0}^\top & \mathbf{P}_{0\|0} \end{pmatrix}$
Filtering $\mathbf{z}_1$ at t_1	$\Pi_1 = (\mathbf{I}\ 0)$ $\mathbf{S}_1 = \mathbf{H}_1\mathbf{\Pi}_1\mathbf{P}_{1:0\|0}\mathbf{\Pi}_1^\top\mathbf{H}_1^\top + \mathbf{R}_1$ $\mathbf{W}_{1:0\|0} = \mathbf{P}_{1:0\|0}\mathbf{\Pi}_1^\top\mathbf{H}_1^\top\mathbf{S}_1^{-1}$ $\mathbf{x}_{1:0\|1} = \mathbf{x}_{1:0\|0} + \mathbf{W}_{1:0\|0}(\mathbf{z}_1 - \mathbf{H}_1\mathbf{\Pi}_1\mathbf{x}_{1:0\|0})$ $\mathbf{P}_{1:0\|1} = \mathbf{P}_{1:0\|0} - \mathbf{W}_{1:0\|0}\mathbf{S}_1\mathbf{W}_{1:0\|0}^\top$
Prediction $t_{k-1} \to t_k$	$\mathbf{x}_{k\|k-1} = \mathbf{F}_{k\|k-1}\mathbf{x}_{k-1\|k-1}$ $\mathbf{P}_{k\|k-1} = \mathbf{F}_{k\|k-1}\mathbf{P}_{k-1\|k-1}\mathbf{F}_{k\|k-1}^\top + \mathbf{Q}_{k\|k-1}$ $\mathbf{x}_{k:0\|k-1} = (\mathbf{x}_{k\|k-1}^\top \ \mathbf{x}_{k-1:0\|k-1}^\top)^\top$ $\mathbf{P}_{k-1:n\|k-1}^{(k-1)} = \begin{pmatrix} \mathbf{P}_{k-1\|k-1} \\ \mathbf{W}_{k-2\|k-1}\mathbf{P}_{k-1\|k-1} \\ \vdots \\ \mathbf{W}_{0\|k-1}\mathbf{P}_{k-1\|k-1} \end{pmatrix}$ $\mathbf{P}_{k:n\|k-1} = \begin{pmatrix} \mathbf{P}_{k\|k-1} & \mathbf{F}_{k\|k-1}(\mathbf{P}_{k-1:n\|k-1}^{(k-1)})^\top \\ \mathbf{P}_{k-1:n\|k-1}^{(k-1)}\mathbf{F}_{k\|k-1}^\top & \mathbf{P}_{k-1:n\|k-1} \end{pmatrix}$
Filtering $\mathbf{z}_k$ at t_k	$\Pi_k = (\mathbf{I}\ \mathbf{O}\ \dots\ \mathbf{O})$ $\mathbf{S}_k = \mathbf{H}_k\mathbf{\Pi}_k\mathbf{P}_{k:0\|k-1}\mathbf{\Pi}_k^\top\mathbf{H}_k^\top + \mathbf{R}_k$ $\mathbf{W}_{k:0\|k-1} = \mathbf{P}_{k:0\|k-1}\mathbf{\Pi}_k^\top\mathbf{H}_k^\top\mathbf{S}_k^{-1}$ $\mathbf{x}_{k:0\|k} = \mathbf{x}_{k:0\|k-1} + \mathbf{W}_{k:0\|k-1}(\mathbf{z}_k - \mathbf{H}_k\mathbf{\Pi}_k\mathbf{x}_{k:0\|k-1})$ $\mathbf{P}_{k:0\|k} = \mathbf{P}_{k:0\|k-1} - \mathbf{W}_{k:0\|k-1}\mathbf{S}_k\mathbf{W}_{k:0\|k-1}^\top$
Sliding window	Prune estimate for t_n from $\mathbf{x}_{k:n\|n}$ Prune column and row for t_n from $\mathbf{P}_{k:n\|n}$

As in the OoS case above, the dimension of $\mathbf{S}_k$ is in the dimension of $\mathbf{z}_k$, which is small dimensional in most applications. Moreover, as stated above, the smoothed states and covariances, respectively, are obtained by a single update step. All implementation details are outlined in Table 4.1.

4.5 Generalized smoothing and out-of-sequence processing

This section presents preliminary steps for generalizing the OoS processing to a single state filter, which does neither compute a full ASD density nor apply retrodiction necessarily. The crucial step hereby is a generalized smoother, which is based on the RTS equations.

As derived in (4.13) and (4.13), the RTS smoother is given by

$$\mathbf{x}_{l|l+1} = \mathbf{x}_{l|l} + \mathbf{W}_{l|l+1}(\mathbf{x}_{l+1|l+1} - \mathbf{x}_{l+1|l}) \quad (4.95)$$

$$\mathbf{P}_{l|l+1} = \mathbf{P}_{l|l} + \mathbf{W}_{l|l+1}(\mathbf{P}_{l+1|l+1} - \mathbf{P}_{l+1|l})\mathbf{W}_{l|l+1}^{\top}, \quad (4.96)$$

where $\mathbf{W}_{l|l+1}$ is the retrodiction gain matrix known from (4.14)

$$\mathbf{W}_{l|l+1} = \mathbf{P}_{l|l}\mathbf{F}_{l+1|l}^{\top}\mathbf{P}_{l+1|l}^{-1}. \quad (4.97)$$

A successive application of the RTS equations yields

$$\mathbf{x}_{k-1|k} = \mathbf{x}_{k-1|k-1} + \mathbf{W}_{k-1|k}(\mathbf{x}_{k|k} - \mathbf{x}_{k|k-1}) \quad (4.98)$$

$$\mathbf{x}_{k-2|k-1} = \mathbf{x}_{k-2|k-2} + \mathbf{W}_{k-2|k-1}(\mathbf{x}_{k-1|k-1} - \mathbf{x}_{k-1|k-2}) \quad (4.99)$$

$$\mathbf{x}_{k-2|k} = \mathbf{x}_{k-2|k-2} + \mathbf{W}_{k-2|k-1}(\mathbf{x}_{k-1|k} - \mathbf{x}_{k-1|k-2}) \quad (4.100)$$

Combining these equations and an induction argument directly yields the retrodicted state $\mathbf{x}_{l|k}$ for $n \leq l \leq k-1$:

$$\mathbf{x}_{l|k} = \mathbf{x}_{l|k-1} + \mathbf{W}_{l|k}(\mathbf{x}_{k|k} - \mathbf{x}_{k|k-1}), \quad (4.101)$$

where

$$\mathbf{W}_{r|s} = \begin{cases} \mathbf{I}, & \text{if } r \geq s, \\ \prod_{i=r}^{s-1} \mathbf{W}_{i|i+1} & \text{else.} \end{cases} \quad (4.102)$$

The same reasoning derives the retrodicted covariance matrix $\mathbf{P}_{l|k}$

$$\mathbf{P}_{l|k} = \mathbf{P}_{l|k-1} + \mathbf{W}_{l|k}(\mathbf{P}_{k|k} - \mathbf{P}_{k|k-1})\mathbf{W}_{l|k}^{\top}. \quad (4.103)$$

If the Kalman filter equations

$$\mathbf{x}_{k|k} = \mathbf{x}_{k|k-1} + \mathbf{W}_{k|k-1}(\mathbf{z}_k - \mathbf{H}_k\mathbf{x}_{k|k-1}), \quad (4.104)$$

$$\mathbf{P}_{k|k} = \mathbf{P}_{k|k-1} - \mathbf{W}_{k|k-1}\mathbf{S}_k\mathbf{W}_{k|k-1}^{\top} \quad (4.105)$$

are applied to (4.101) and (4.103), the following update rule for past states based on the current measurement $\mathbf{z}_k$ is obtained:

$$\mathbf{x}_{l|k} = \mathbf{x}_{l|k-1} + \mathbf{W}_{l|k}\mathbf{W}_{k|k-1}(\mathbf{z}_k - \mathbf{H}_k\mathbf{x}_{k|k-1}), \quad (4.106)$$

$$\mathbf{P}_{l|k} = \mathbf{P}_{l|k-1} - \mathbf{W}_{l|k}\mathbf{W}_{k|k-1}\mathbf{S}_k\mathbf{W}_{k|k-1}^{\top}\mathbf{W}_{l|k}^{\top}. \quad (4.107)$$

This result will be called *Generalized-RTS* in the following paragraphs. A comparison of both smoothing schemes is illustrated in Figure 4.4.

Generalization of OoS processing

Using the generalized RTS equations from above, it is possible to derive the cross-covariances of states in the past. The cross-covariance is needed to derive an optimal update of a current state $\mathbf{x}_k$ when a delayed measurement $\mathbf{z}_m$ is processed where $t_m < t_k$.

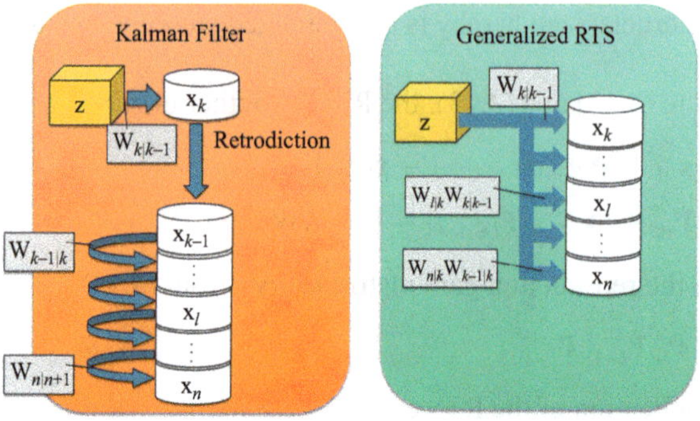

Figure 4.4 A comparison of the standard Kalman filter scheme including the Rauch-Tung-Striebel (RTS) recursion to the generalized RTS. The grey legend boxes indicate the individual gain matrices used for each step. As one can see, the generalized scheme uses combined gain matrices which saves numerical resources and computation time.

An application of the fundamental equations of linear estimation on the generalized retrodiction (4.106) and (4.107) yields:

$$\mathbf{W}_{l|k}\mathbf{P}_{k|k-1}\mathbf{H}_k^\top = \mathrm{cov}\left[\mathbf{x}_l, \mathbf{z}_k | \mathcal{Z}^{k-1}\right] \tag{4.108}$$

$$= \mathrm{cov}\left[\mathbf{x}_l, \mathbf{H}_k\mathbf{x}_k + v_k | \mathcal{Z}^{k-1}\right] \tag{4.109}$$

$$= \mathrm{cov}\left[\mathbf{x}_l, \mathbf{x}_k | \mathcal{Z}^{k-1}\right]\mathbf{H}_k^\top \tag{4.110}$$

As this is valid for an arbitrary matrix $\mathbf{H}_k$, it holds

$$\mathrm{cov}\left[\mathbf{x}_l, \mathbf{x}_k | \mathcal{Z}^{k-1}\right] = \mathbf{W}_{l|k}\mathbf{P}_{k|k-1}. \tag{4.111}$$

As a consequence of (4.111), it can be seen that

$$\mathrm{cov}\left[\mathbf{x}_k, \mathbf{z}_m | \mathcal{Z}^{k-1}\right] = \left(\mathrm{cov}\left[\mathbf{H}_m\mathbf{x}_m + v_m, \mathbf{x}_k | \mathcal{Z}^{k-1}\right]\right)^\top \tag{4.112}$$

$$= \mathbf{P}_{k|k-1}\mathbf{W}_{m|k}^\top\mathbf{H}_m^\top. \tag{4.113}$$

However, this term is conditioned on the sensor data up to time t_{k-1}, therefore, it is necessary to prove that this can be applied to the cross-covariance conditioned on $\mathcal{Z}^k$. Let the prior cross-covariance of a state $\mathbf{x}_k$ and a measurement $\mathbf{z}_m$ be given by

$$\mathrm{cov}\left[\mathbf{x}_k, \mathbf{z}_m | \mathcal{Z}^{k-1}\right] = \mathbf{P}_{k|k-1}\mathbf{W}_{m|k}^\top\mathbf{H}_m^\top. \tag{4.114}$$

Then, the posterior cross-covariance is given by

$$\mathrm{cov}\left[\mathbf{x}_k, \mathbf{z}_m | \mathcal{Z}^k\right] = \mathbf{P}_{k|k}\mathbf{W}_{m|k}^\top\mathbf{H}_m^\top. \tag{4.115}$$

This can be seen as follows. Due to Bayes' theorem, it holds

$$p(\mathbf{x}_k, \mathbf{z}_m | \mathcal{Z}^k) = \frac{p(\mathbf{z}_k | \mathbf{x}_k, \mathbf{z}_m)\, p(\mathbf{x}_k, \mathbf{z}_m | \mathcal{Z}^{k-1})}{\int\int \mathrm{d}\mathbf{x}_k\, \mathrm{d}\mathbf{z}_m\, p(\mathbf{z}_k | \mathbf{x}_k)\, p(\mathbf{x}_k, \mathbf{z}_m | \mathcal{Z}^{k-1})}. \tag{4.116}$$

If projections $\tilde{\mathbf{H}}_k$ are introduced, which are defined in the following line

$$\tilde{\mathbf{H}}_k = (\mathbf{H}_k, \mathbf{O}) \tag{4.117}$$

such that

$$\tilde{\mathbf{H}}_k(\mathbf{x}_k^\top, \mathbf{z}_m^\top)^\top = \mathbf{H}_k \mathbf{x}_k. \tag{4.118}$$

the likelihood function is

$$p(\mathbf{z}_k | \mathbf{x}_k, \mathbf{z}_m) = \mathcal{N}((\mathbf{x}_k^\top, \mathbf{z}_m^\top)^\top; \tilde{\mathbf{H}}_k(\mathbf{x}_k^\top, \mathbf{z}_m^\top)^\top, \mathbf{R}_k). \tag{4.119}$$

Due to the assumptions above, the prior joint pdf in (4.116) is given by

$$p(\mathbf{x}_k, \mathbf{z}_m | \mathcal{Z}^{k-1}) = \mathcal{N}((\mathbf{x}_k^\top, \mathbf{z}_m^\top)^\top; (\mathbf{x}_{k|k-1}^\top, \mathbf{z}_{m|k-1}^\top)^\top, \mathbf{C}_{xz|k-1}), \tag{4.120}$$

where

$$\mathbf{C}_{xz|k-1} = \begin{pmatrix} \mathbf{P}_{k|k-1} & \mathbf{P}_{k|k-1}\mathbf{W}_{m|k}^\top \mathbf{H}_m^\top \\ \mathbf{H}_m \mathbf{W}_{m|k} \mathbf{P}_{k|k-1} & \mathbf{S}_{m|k-1} \end{pmatrix}. \tag{4.121}$$

Applying the product formula for Gaussians on these densities yields

$$p(\mathbf{x}_k, \mathbf{z}_m | \mathcal{Z}^{k}) = \mathcal{N}((\mathbf{x}_k^\top, \mathbf{z}_m^\top)^\top; (\mathbf{x}_{k|k}^\top, \mathbf{z}_{m|k}^\top)^\top, \mathbf{C}_{xz|k}), \tag{4.122}$$

where the following abbreviations are used:

$$\tilde{\mathbf{W}}_k = \begin{pmatrix} \mathbf{W}_{k|k-1} \\ \mathbf{H}_m \mathbf{W}_{m|k} \mathbf{W}_{k|k-1} \end{pmatrix} \tag{4.123}$$

$$\mathbf{C}_{xz|k} = \mathbf{C}_{xz|k-1} - \tilde{\mathbf{W}}_k \mathbf{S}_{k|k-1} \tilde{\mathbf{W}}_k^\top. \tag{4.124}$$

By calculating the entries of $\mathbf{C}_{xz|k}$ the posterior* cross-covariance is directly obtained:

$$\operatorname{cov}\left[\mathbf{x}_k, \mathbf{z}_m | \mathcal{Z}^k\right] = \left(\mathbf{P}_{k|k-1} - \mathbf{W}_{k|k-1}\mathbf{S}_{k|k-1}\mathbf{W}_{k|k-1}^\top\right) \mathbf{W}_{m|k}^\top \mathbf{H}_m^\top \tag{4.125}$$

$$= \mathbf{P}_{k|k} \mathbf{W}_{m|k}^\top \mathbf{H}_m^\top. \tag{4.126}$$

This leads us to the following result. If a time-delayed measurement $\mathbf{z}_m$ and its prior expectation vector $\mathbf{z}_{m|k}$ and covariance $\mathbf{S}_{m|k}$ are given, the parameters of an estimation density for $\mathbf{x}_k$ given the data set $\mathcal{Z}^k$ where $\mathbf{z}_m$ is not contained in $\mathcal{Z}^k$ and $t_m < t_k$ are optimally updated as follows.

$$p(\mathbf{x}_k | \mathcal{Z}^{k,m}) = \mathcal{N}(\mathbf{x}_k;\, \mathbf{x}_{k|k,m}, \mathbf{P}_{k,m|k}), \tag{4.127}$$

$$\mathbf{W}_{k,m|k} = \mathbf{P}_{k|k} \mathbf{W}_{m|k}^\top \mathbf{H}_m^\top \mathbf{S}_{m|k}^{-1}. \tag{4.128}$$

$$\mathbf{x}_{k|k,m} = \mathbf{x}_{k|k} + \mathbf{W}_{k,m|k}\left(\mathbf{z}_m - \mathbf{z}_{m|k}\right), \tag{4.129}$$

$$\mathbf{P}_{k,m|k} = \mathbf{P}_{k|k} - \mathbf{W}_{k,m|k} \mathbf{S}_{m|k} \mathbf{W}_{k,m|k}^\top. \tag{4.130}$$

*For the proof, the term "posterior" refers to the measurement $\mathbf{z}_k$, not to the OoS measurement $\mathbf{z}_m$.

This can be seen by an application of the fundamental equations (3.45) and (3.46) by using the derived cross-covariance of (4.126).

4.5.1 Batch processing

If initial conditions are given by the density function

$$p(\mathbf{x}_n) = \mathcal{N}(\mathbf{x}_n;\ \mathbf{x}_{n|n},\ \mathbf{P}_{n|n}) \tag{4.131}$$

and a set of measurements

$$\mathbf{z}_{k:n} = \{\mathbf{z}_k, \mathbf{z}_{k-1}, \ldots, \mathbf{z}_n\} \tag{4.132}$$

are observations of the corresponding process, the ASD formulae directly provide the means to estimate the trajectory $\mathbf{x}_{k:n}$ based on the sensor data as shown in Figure 4.5.

This can easily be seen by an application of the Bayes' theorem:

$$p(\mathbf{x}_{k:n}|\mathcal{Z}^k) = \frac{p(\mathbf{z}_{k:n}|\mathbf{x}_{k:n})\, p(\mathbf{x}_{k:n})}{\int \mathrm{d}\mathbf{x}_{k:n}\, p(\mathbf{z}_{k:n}|\mathbf{x}_{k:n})\, p(\mathbf{x}_{k:n})}, \tag{4.133}$$

where $p(\mathbf{x}_{k:n})$ is conditioned on the initial knowledge on $\mathbf{x}_n$. Therefore, it can be obtained by a successive application of the ASD prediction (4.83) and (4.87) on the initial pdf. This yields a Gaussian density

$$p(\mathbf{x}_{k:n}) = \mathcal{N}(\mathbf{x}_{k:n};\ \mathbf{x}_{k:n|n},\ \mathbf{P}_{k:n|n}) \tag{4.134}$$

where the mean is given by

$$\mathbf{x}_{k:n|n} = (\mathbf{x}_{k|n}^\top\ \mathbf{x}_{k-1|n}^\top\ \cdots\ \mathbf{x}_{n|n}^\top)^\top \tag{4.135}$$

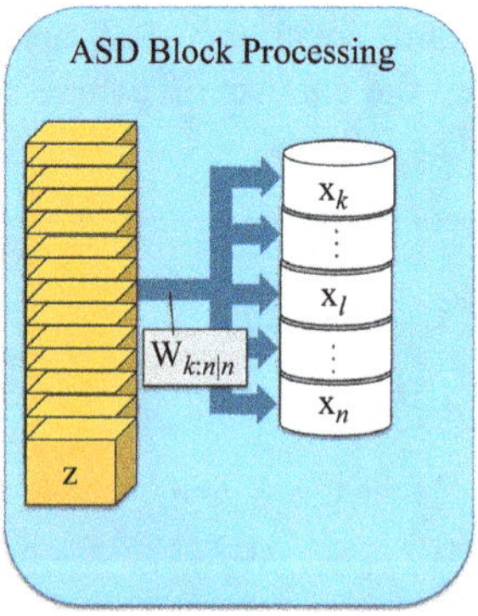

Figure 4.5 Scheme of the ASD block processing. A prior for the trajectory is computed, then it is updated in a single step using the full batch of measurements within a time series.

and

$$\mathbf{P}_{k:n|n} = \begin{pmatrix} \mathbf{P}_{k|n} & \mathbf{F}_{k|k-1}\mathbf{P}_{k-1|n} & \cdots & & \mathbf{F}_{k|k-1}\cdots\mathbf{F}_{n+1|n}\mathbf{P}_{n|n} \\ \mathbf{P}_{k-1|n}\mathbf{F}_{k|k-1}^{\top} & \ddots & & & \\ \vdots & & \mathbf{P}_{n+2|n} & \mathbf{F}_{n+2|n+1}\mathbf{P}_{n+1|n} & \mathbf{F}_{n+2|n+1}\mathbf{F}_{n+1|n}\mathbf{P}_{n|n} \\ \vdots & & \mathbf{P}_{n+1|n}\mathbf{F}_{n+2|n+1}^{\top} & \mathbf{P}_{n+1|n} & \mathbf{F}_{n+1|n}\mathbf{P}_{n|n} \\ \mathbf{P}_{n|n}\mathbf{F}_{n+1|n}^{\top}\cdots\mathbf{F}_{k|k-1}^{\top} & \cdots & \mathbf{P}_{n|n}\mathbf{F}_{n+1|n}^{\top}\mathbf{F}_{n+2|n+1}^{\top} & \mathbf{P}_{n|n}\mathbf{F}_{n+1|n}^{\top} & \mathbf{P}_{n|n} \end{pmatrix} \tag{4.136}$$

and the following abbreviations

$$\mathbf{x}_{l|n} = \mathbf{F}_{l|n}\mathbf{x}_{n|n} = \mathbf{F}_{l|l-1}\mathbf{x}_{l-1|n} \tag{4.137}$$

$$\mathbf{P}_{l|n} = \mathbf{F}_{l|n}\mathbf{P}_{n|n}\mathbf{F}_{l|n}^{\top} + \mathbf{Q}_{l|n} \tag{4.138}$$

$$= \mathbf{F}_{l|l-1}\mathbf{P}_{l-1|n}\mathbf{F}_{l|l-1}^{\top} + \mathbf{Q}_{l|l-1} \tag{4.139}$$

were used.

Since the measurements are mutually conditionally independent, the likelihood of the accumulated measurement set $\mathbf{z}_{k:n}$ is given by

$$p(\mathbf{z}_{k:n}|\mathbf{x}_{k:n}) = \mathcal{N}(\mathbf{z}_{k:n};\, \mathbf{H}_{k:n}\mathbf{x}_{k:n},\, \mathbf{R}_{k:n}) \tag{4.140}$$

where

$$\mathbf{z}_{k:n} = \begin{pmatrix} \mathbf{z}_k \\ \vdots \\ \mathbf{z}_n \end{pmatrix}, \tag{4.141}$$

$$\mathbf{R}_{k:n} = \begin{pmatrix} \mathbf{R}_k & & \\ & \ddots & \\ & & \mathbf{R}_n \end{pmatrix}, \tag{4.142}$$

$$\mathbf{H}_{k:n} = \begin{pmatrix} \mathbf{H}_k & & \\ & \ddots & \\ & & \mathbf{H}_n \end{pmatrix}. \tag{4.143}$$

Table 4.2 Batch processing algorithm

Init.	set $\mathbf{x}_{n\|n}$, $\mathbf{P}_{n\|n}$, gather $\{\mathbf{z}_l\}_{l=n}^{k}$
Prior up to time t_k	For $l = n+1, \ldots, k$ compute $\mathbf{x}_{l\|n} = \mathbf{F}_{l\|l-1}\mathbf{x}_{l-1\|n}$ $\mathbf{P}_{l\|n} = \mathbf{F}_{l\|l-1}\mathbf{P}_{l-1\|n}\mathbf{F}_{l\|l-1}^{\top} + \mathbf{Q}_{l\|l-1}$ End For $\mathbf{x}_{k:n\|n} = (\mathbf{x}_{k\|n}^{\top}\ \mathbf{x}_{k-1\|n}^{\top}\ \cdots\ \mathbf{x}_{n\|n}^{\top})^{\top}$ $\mathbf{P}_{k:n\|n}$ as in (4.136).
Batch likelihood	$\mathbf{z}_{k:n} = \left(\mathbf{z}_k^{\top}\ \ldots\ \mathbf{z}_n^{\top}\right)^{\top}$, $\mathbf{R}_{k:n} = \text{blkdiag}\left(\mathbf{R}_k\ \ldots\ \mathbf{R}_n\right)$, $\mathbf{H}_{k:n} = \text{blkdiag}\left(\mathbf{H}_k\ \ldots\ \mathbf{H}_n\right)$.
Processing	$\mathbf{x}_{k:n\|k} = \mathbf{x}_{k:n\|n} + \mathbf{W}_{k:n\|n}(\mathbf{z}_{k:n} - \mathbf{H}_{k:n}\mathbf{x}_{k:n\|n})$, $\mathbf{P}_{k:n\|k} = \mathbf{P}_{k:n\|n} - \mathbf{W}_{k:n\|n}\mathbf{S}_{k:n}\mathbf{W}_{k:n\|n}^{\top}$, $\mathbf{W}_{k:n\|n} = \mathbf{P}_{k:n\|n}\mathbf{H}_{k:n}^{\top}\mathbf{S}_{k:n}^{-1}$, $\mathbf{S}_{k:n} = \mathbf{H}_{k:n}\mathbf{P}_{k:n\|n}\mathbf{H}_{k:n}^{\top} + \mathbf{R}_{k:n}$.

According to Bayes' theorem, an application of the product formula directly yields the fully *filtered and smoothed* trajectory is given by the posterior ASD

$$p(\mathbf{x}_{k:n}|Z^k) = \mathcal{N}(\mathbf{x}_{k:n};\ \mathbf{x}_{k:n|k}, \mathbf{P}_{k:n|k}), \tag{4.144}$$

where

$$\mathbf{x}_{k:n|k} = \mathbf{x}_{k:n|n} + \mathbf{W}_{k:n|n}(\mathbf{z}_{k:n} - \mathbf{H}_{k:n}\mathbf{x}_{k:n|n}), \tag{4.145}$$

$$\mathbf{P}_{k:n|k} = \mathbf{P}_{k:n|n} - \mathbf{W}_{k:n|n}\mathbf{S}_{k:n}\mathbf{W}_{k:n|n}^{\top}, \tag{4.146}$$

$$\mathbf{W}_{k:n|n} = \mathbf{P}_{k:n|n}\mathbf{H}_{k:n}^{\top}\mathbf{S}_{k:n}^{-1}, \tag{4.147}$$

$$\mathbf{S}_{k:n} = \mathbf{H}_{k:n}\mathbf{P}_{k:n|n}\mathbf{H}_{k:n}^{\top} + \mathbf{R}_{k:n}. \tag{4.148}$$

The full algorithm on batch processing using ASDs is summarized in Table 4.2.

4.5.2 Block-line combined estimator and smoother

We have seen in Chapter 3 that the Kalman filter update equations are equivalent to the Information filter. Actually, the fact that the batch update can be written as the result of the Bayes' theorem with a prior and a single likelihood function, yields the most general form of the updated and retrodicted state estimates possible

In the following paragraph, it is shown that for a given prior estimate $\mathbf{x}_{n|n}$ for time t_n and for each instant of time t_l there exists a weighting matrix $\mathbf{K}_{l\leftarrow n}$ and a set of matrices $\{\mathbf{L}_{l\leftarrow j}\}_{j=n}^{k}$ such that the filtered and smoothed estimate for time t_l is given by

$$\mathbf{x}_{l|k} = \mathbf{K}_{l\leftarrow n}\mathbf{x}_{n|n} + \sum_{j=n}^{k} \mathbf{L}_{l\leftarrow j}\mathbf{z}_j. \tag{4.149}$$

The resulting estimate $\mathbf{x}_{l|k}$ is equivalent to a standard prediction-filtering recursion up to time t_k with a succeeding backwards iteration of the RTS smoother. This can be considered the block line version of the ASD update in (4.145) as schematically depicted in Figure 4.6. It can be seen by using the Information filter form with an inverted prior ASD covariance matrix, which has to be computed at first.

According to the Markov assumption and the linear-Gaussian models, it holds that

$$p(\mathbf{x}_{k:n}) = \prod_{l=n+1}^{k} p(\mathbf{x}_l|\mathbf{x}_{l-1})\,p(\mathbf{x}_n) \tag{4.150}$$

$$= \prod_{l=n+1}^{k} \mathcal{N}(\mathbf{x}_l;\, \mathbf{F}_{l|l-1}\mathbf{x}_{l-1},\, \mathbf{Q}_{l|l-1}) \cdot \mathcal{N}(\mathbf{x}_n;\, \mathbf{x}_{n|n},\, \mathbf{P}_{n|n}) \tag{4.151}$$

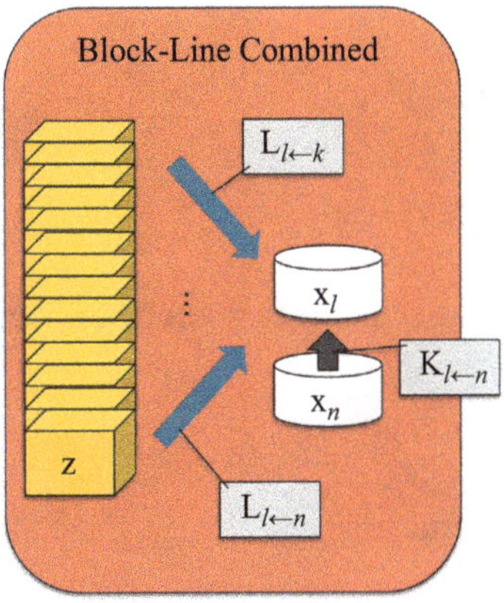

Figure 4.6 Block-line solution of the combined filtering and smoothing. The estimate for an arbitrary instant of time may be computed as a linear combination of a prior in the past and all measurements.

A successive application of the product formula (3.123) yields that the prior ASD density can be written as

$$p(\mathbf{x}_{k:n}) = \mathcal{N}(\mathbf{x}_k;\, \mathbf{x}_{k|n},\, \mathbf{P}_{k|n}) \cdot \prod_{l=n}^{k-1} \mathcal{N}(\mathbf{x}_l;\, \mathbf{x}_{l|n} + \mathbf{P}_{l|n}\mathbf{F}_{l+1|l}^{\top}\mathbf{P}_{l+1|n}^{-1}(\mathbf{x}_{l+1} - \mathbf{x}_{l+1|n}),\, \tilde{\mathbf{S}}_l), \tag{4.152}$$

where

$$\tilde{\mathbf{S}}_l = \begin{cases} \mathbf{P}_{l|n} - \mathbf{P}_{l|n}\mathbf{F}_{l+1|l}^{\top}\mathbf{P}_{l+1|n}^{-1}\mathbf{F}_{l+1|l}\mathbf{P}_{l|n}, & \text{if } l < k \\ \mathbf{P}_{k|n}, & \text{if } l = k, \end{cases} \tag{4.153}$$

This product can be rewritten to

$$p(\mathbf{x}_{k:n}) = \prod_{l=n}^{k} \mathcal{N}(\tilde{\Pi}_l \mathbf{x}_{k:n};\, \tilde{\Pi}_l \mathbf{x}_{k:n|n},\, \tilde{\mathbf{S}}_l), \tag{4.154}$$

by introducing the following affine projections:

$$\tilde{\Pi}_l = \begin{cases} \Pi_l - \mathbf{P}_{l|n}\mathbf{F}_{l+1|l}^{\top}\mathbf{P}_{l+1|n}^{-1}\Pi_{l+1}, & \text{if } l < k \\ \Pi_k, & \text{if } l = k \end{cases} \tag{4.155}$$

where

$$\Pi_l \mathbf{x}_{k:n} = \mathbf{x}_l \tag{4.156}$$

as above. The product in (4.154) yields a single Gaussian:

$$p(\mathbf{x}_{k:n}) = \mathcal{N}(\mathbf{x}_{k:n};\, \mathbf{x}_{k:n|n},\, \mathbf{P}_{k:n|n}) \tag{4.157}$$

with

$$\mathbf{x}_{k:n|n} = \begin{pmatrix} \mathbf{F}_{k|n}\mathbf{x}_{n|n} \\ \mathbf{F}_{k-1|n}\mathbf{x}_{n|n} \\ \vdots \\ \mathbf{F}_{n+1|n}\mathbf{x}_{n|n} \\ \mathbf{x}_{n|n} \end{pmatrix} \tag{4.158}$$

and the inverse of the prior ASD matrix is given by the sum

$$\mathbf{P}_{k:n|n}^{-1} = \sum_{l=n}^{k} \tilde{\Pi}_l^{\top}\tilde{\mathbf{S}}_l^{-1}\tilde{\Pi}_l, \tag{4.159}$$

which yields the following structure:

$$\mathbf{P}_{k:n|n}^{-1} = \begin{pmatrix} \mathbf{D}_k & \mathbf{M}_k & & & \\ \mathbf{M}_k^{\top} & \mathbf{D}_{k-1} & \mathbf{M}_{k-1} & & \\ & \ddots & \ddots & \ddots & \\ & & \mathbf{M}_{n+2}^{\top} & \mathbf{D}_{n+1} & \mathbf{M}_{n+1} \\ & & & \mathbf{M}_{n+1}^{\top} & \mathbf{D}_n \end{pmatrix}, \tag{4.160}$$

where

$$\mathbf{D}_l = \begin{cases} \mathbf{S}_l^{-1} - \mathbf{P}_{l|n}^{-1}\mathbf{F}_{l|l-1}\mathbf{P}_{l-1|n}\mathbf{S}_{l-1}^{-1}\mathbf{P}_{l-1|n}\mathbf{F}_{l|l-1}^\top\mathbf{P}_{l|n}^{-1}, & \text{if } l > n \\ \mathbf{S}_n^{-1}, & \text{if } l = n, \end{cases} \tag{4.161}$$

$$\mathbf{M}_l = -\mathbf{S}_{l-1}^{-1}\mathbf{P}_{l-1|n}\mathbf{F}_{l|l-1}^\top\mathbf{P}_{l|n}^{-1}. \tag{4.162}$$

The inverted prior in (4.160) can now be used for the Information filter form of the ASD block update:

$$\mathbf{x}_{k:n|k} = \mathbf{P}_{k:n|k}(\mathbf{P}_{k:n|n}^{-1}\mathbf{x}_{k:n|n} + \mathbf{H}_{k:n}^\top\mathbf{R}_{k:n}^{-1}\mathbf{z}_{k:n}). \tag{4.163}$$

The information contribution from the measurements is easy to derive, since it holds that

$$\mathbf{R}_{k:n}^{-1} = \begin{pmatrix} \mathbf{R}_k^{-1} & & \\ & \ddots & \\ & & \mathbf{R}_n^{-1} \end{pmatrix}. \tag{4.164}$$

Therefore,

$$\mathbf{H}_{k:n}^\top\mathbf{R}_{k:n}^{-1}\mathbf{z}_{k:n} = \begin{pmatrix} \mathbf{H}_k^\top\mathbf{R}_k^{-1}\mathbf{z}_k \\ \vdots \\ \mathbf{H}_n^\top\mathbf{R}_n^{-1}\mathbf{z}_n \end{pmatrix}. \tag{4.165}$$

As a consequence, the equation from above can be rewritten as follows:

$$\mathbf{x}_{k:n|k} = \mathbf{P}_{k:n|k}\mathbf{d}_{k:n}, \tag{4.166}$$

with $\mathbf{d}_{k:n}$ given by

$$\mathbf{d}_{k:n} = \begin{pmatrix} \mathbf{D}_k\mathbf{x}_{k|n} + \mathbf{M}_k\mathbf{x}_{k-1|n} + \mathbf{H}_k^\top\mathbf{R}_k^{-1}\mathbf{z}_k \\ \mathbf{M}_k^\top\mathbf{x}_{k|n} + \mathbf{D}_{k-1}\mathbf{x}_{k-1|n} + \mathbf{M}_{k-1}\mathbf{x}_{k-2|n} + \mathbf{H}_{k-1}^\top\mathbf{R}_{k-1}^{-1}\mathbf{z}_{k-1} \\ \vdots \\ \mathbf{M}_{n+2}^\top\mathbf{x}_{n+2|n} + \mathbf{D}_{n+1}\mathbf{x}_{n+1|n} + \mathbf{M}_{n+1}\mathbf{x}_{n|n} + \mathbf{H}_{n+1}^\top\mathbf{R}_{n+1}^{-1}\mathbf{z}_{n+1} \\ \mathbf{M}_{n+1}^\top\mathbf{x}_{n+1|n} + \mathbf{D}_n\mathbf{x}_{n|n} + \mathbf{H}_n^\top\mathbf{R}_n^{-1}\mathbf{z}_n \end{pmatrix}. \tag{4.167}$$

By exploiting the structure of the posterior ASD covariance matrix $\mathbf{P}_{k:n|k}$ in (4.28), it follows that the lth block entry of $\mathbf{x}_{k:n|k}$ is given by

$$\begin{aligned}
\mathbf{x}_{l|k} &= \Pi_l \mathbf{P}_{k:n|k} \mathbf{d}_{k:n} \qquad (4.168)\\
&= \mathbf{W}_{l|k}\mathbf{P}_{k|k}(\mathbf{D}_k\mathbf{x}_{k|n} + \mathbf{M}_k\mathbf{x}_{k-1|n} + \mathbf{H}_k^\top\mathbf{R}_k^{-1}\mathbf{z}_k)\\
&\quad + \mathbf{W}_{l|k-1}\mathbf{P}_{k-1|k}(\mathbf{M}_k^\top\mathbf{x}_{k|n} + \mathbf{D}_{k-1}\mathbf{x}_{k-1|n}\\
&\quad + \mathbf{M}_{k-1}\mathbf{x}_{k-2|n} + \mathbf{H}_{k-1}^\top\mathbf{R}_{k-1}^{-1}\mathbf{z}_{k-1})\\
&\quad + \ldots\\
&\quad + \mathbf{P}_{l|k}(\mathbf{M}_{l+1}^\top\mathbf{x}_{l+1|n} + \mathbf{D}_l\mathbf{x}_{l|n} + \mathbf{M}_l\mathbf{x}_{l-1|n}\\
&\quad + \mathbf{H}_l^\top\mathbf{R}_l^{-1}\mathbf{z}_l)\\
&\quad + \mathbf{P}_{l|k}\mathbf{W}_{l-1|l}^\top(\mathbf{M}_l^\top\mathbf{x}_{l|n} + \mathbf{D}_{l-1}\mathbf{x}_{l-1|n}\\
&\quad + \mathbf{M}_{l-1}\mathbf{x}_{l-2|n} + \mathbf{H}_{l-1}^\top\mathbf{R}_{l-1}^{-1}\mathbf{z}_{l-1})\\
&\quad + \cdots\\
&\quad + \mathbf{P}_{l|k}\mathbf{W}_{n|l}^\top(\mathbf{M}_{n+1}^\top\mathbf{x}_{n+1|n} + \mathbf{D}_n\mathbf{x}_{n|n}\\
&\quad + \mathbf{H}_n^\top\mathbf{R}_n^{-1}\mathbf{z}_n). \qquad (4.169)
\end{aligned}$$

Therefore, the super position equation is given by

$$\mathbf{x}_{l|k} = \mathbf{K}_{l\leftarrow n}\mathbf{x}_{n|n} + \sum_{j=n}^{k} \mathbf{L}_{l\leftarrow j}\mathbf{z}_j, \qquad (4.170)$$

where

$$\begin{aligned}
\mathbf{K}_{l\leftarrow n} = &\ (\mathbf{W}_{l|k}\mathbf{P}_{k|k}\mathbf{D}_k + \mathbf{W}_{l|k-1}\mathbf{P}_{k-1|k}\mathbf{M}_k^\top)\mathbf{F}_{k|n}\\
&+(\mathbf{W}_{l|k}\mathbf{P}_{k|k}\mathbf{M}_k + \mathbf{W}_{l|k-1}\mathbf{P}_{k-1|k}\mathbf{D}_{k-1}\\
&+\mathbf{W}_{l|k-2}\mathbf{P}_{k-2|k}\mathbf{M}_{k-1}^\top)\mathbf{F}_{k-1|n}\\
&+\cdots\\
&+(\mathbf{W}_{l|l+1}\mathbf{P}_{l+1|k}\mathbf{M}_{l+1} + \mathbf{P}_{l|k}\mathbf{D}_l\\
&+\mathbf{P}_{l|k}\mathbf{W}_{l-1|l}^\top\mathbf{M}_l^\top)\mathbf{F}_{l|n}\\
&+\cdots\\
&+(\mathbf{P}_{l|k}\mathbf{W}_{n+2|l}^\top\mathbf{M}_{n+2} + \mathbf{P}_{l|k}\mathbf{W}_{n+1|l}^\top\mathbf{D}_{n+1}\\
&+\mathbf{P}_{l|k}\mathbf{W}_{n|l}^\top\mathbf{M}_{n+1}^\top)\mathbf{F}_{n+1|n}\\
&+\mathbf{P}_{l|k}\mathbf{W}_{n+1|l}^\top\mathbf{M}_{n+1} + \mathbf{P}_{l|k}\mathbf{W}_{n|l}^\top\mathbf{D}_n \qquad (4.171)
\end{aligned}$$

and

$$\mathbf{L}_{l\leftarrow j} = \begin{cases} \mathbf{W}_{l|j}\mathbf{P}_{j|k}\mathbf{H}_j^\top\mathbf{R}_j^{-1}, & \text{if } j > l \\ \mathbf{P}_{l|k}\mathbf{H}_l^\top\mathbf{R}_l^{-1}, & \text{if } j = l \\ \mathbf{P}_{l|k}\mathbf{W}_{j|l}^\top\mathbf{H}_j^\top\mathbf{R}_j^{-1}, & \text{if } j < l \end{cases} \tag{4.172}$$

This concludes the proof.

4.6 Quantum physical interpretation

Nowadays, it is now well known that the tracking problem can be formulated by means of a path integral (see [51] for instance) where the observations induce potential functions which act as an attracting force field. However, a closed form solution inherently requires the notion of the ASDs, since they provide exact formulae for the probability measures of paths, at least in for the discrete case. As a result, one obtains a Feynman kernel which depends on starting and end points together with the set of measurements. Once applied to a prior density for the starting point, the Chapman–Kolmogorov equation can be used to obtain the posterior density of the end point where the Feynman kernel acts as a Markov kernel. This section shows the connection to the target tracking problem based on the considerations from Section 2.4 in Chapter 2. As a result, the closed form solution for the case of linear-Gaussian measurement models is derived.

4.7 The free quantum particle

The free particle moves in space–time with constant velocity $\dot{\mathbf{x}}$ where no potential field acts on the particle, that is $\mathbf{V} = 0$. Here, the particle is of a quantum nature, therefore the velocity is unknown for the transition kernel $\mathbf{K}_0(\mathbf{x}_k, \mathbf{x}_n; (t_k - t_n))$. This is due to the fact that the initial position $\mathbf{x}_n$ is fixed and Heisenberg's uncertainty principle [52] holds:

$$\sigma_x\, \sigma_p \geq \frac{\hbar}{2}. \tag{4.173}$$

Here σ_x and σ_p are the standard deviation for the estimation errors for the position and the momentum $p = m \cdot \dot{\mathbf{x}}$. For a fixed position $\mathbf{x}_n$ one has that $\sigma_x = 0$ and $\sigma_p = \infty$.

In this basic setup, the Hamiltonian is given by

$$\underline{\mathbf{H}}_0 = -\frac{\hbar^2}{2m}\Delta, \tag{4.174}$$

and the corresponding Lagrangian function is the constant energy of the particle, which is given by

$$\mathcal{L}(\mathbf{x}, \dot{\mathbf{x}}) = \frac{m}{2}\dot{\mathbf{x}}^\top\dot{\mathbf{x}} \quad \forall \mathbf{x} \tag{4.175}$$

According to the definition of the Feynman kernel above, we have

$$\mathbf{K}_0(\mathbf{x}_k, \mathbf{x}_n; (t_k - t_n)) \propto \lim_{\epsilon \to 0} \int \mathrm{d}\mathbf{x}_{k-1:n+1}\, \mathrm{e}^{\frac{i}{\hbar} \sum_{l=n}^{k-1} \epsilon \mathcal{L}(\mathbf{x}_{l+1}, \frac{\mathbf{x}_{l+1} - \mathbf{x}_l}{\epsilon})} \tag{4.176}$$

$$= \lim_{\epsilon \to 0} \int \mathrm{d}\mathbf{x}_{k-1:n+1}\, \mathrm{e}^{\frac{im}{2\hbar} \sum_{l=n}^{k-1} \epsilon (\frac{\mathbf{x}_{l+1} - \mathbf{x}_l}{\epsilon})^2} \tag{4.177}$$

$$= \lim_{\epsilon \to 0} \int \mathrm{d}\mathbf{x}_{k-1:n+1}\, \mathrm{e}^{\frac{im}{2\hbar\epsilon} \sum_{l=n}^{k-1} (\mathbf{x}_{l+1} - \mathbf{x}_l)^\top (\mathbf{x}_{l+1} - \mathbf{x}_l)} \tag{4.178}$$

$$= \lim_{\epsilon \to 0} \int \mathrm{d}\mathbf{x}_{k-1:n+1} \prod_{l=n+1}^{k} \mathcal{N}(\mathbf{x}_l;\, \mathbf{x}_{l-1},\, \mathbf{D}_{l|l-1}), \tag{4.179}$$

where

$$\mathbf{D}_{l|l-1} = \frac{i\epsilon\hbar}{m}\mathbf{I}. \tag{4.180}$$

According to the product formula (3.123), the product of Gaussians can be rewritten as

$$\prod_{l=n+1}^{k} \mathcal{N}(\mathbf{x}_l;\, \mathbf{x}_{l-1},\, \mathbf{D}_{l|l-1}) = \mathcal{N}(\mathbf{x}_{k:n+1};\, \mathbf{x}_n^{(k)},\, \mathbf{D}_{k:n+1|n}) \tag{4.181}$$

where

$$\mathbf{D}_{k:n+1|n} = \begin{pmatrix} \mathbf{D}_{k|n} & \mathbf{D}_{k-1|n} & \dots & \mathbf{D}_{n+1|n} \\ \mathbf{D}_{k-1|n} & \mathbf{D}_{k-1|n} & \dots & \mathbf{D}_{n+1|n} \\ \vdots & & \ddots & \vdots \\ \mathbf{D}_{n+1|n} & \mathbf{D}_{n+1|n} & \dots & \mathbf{D}_{n+1|n} \end{pmatrix}, \tag{4.182}$$

$$\mathbf{x}_n^{(k)} = \begin{pmatrix} \mathbf{x}_n \\ \mathbf{x}_n \\ \vdots \\ \mathbf{x}_n \end{pmatrix}, \tag{4.183}$$

$$\mathbf{D}_{l|n} = \sum_{j=n+1}^{l} \mathbf{D}_{j|j-1} = \frac{i(t_l - t_n)\hbar}{m}\mathbf{I}. \tag{4.184}$$

Now it is crucial to note that the integration in some variables of a multi-variate density yields a marginalization of exactly those parameters. Therefore, we have that

$$\int \mathrm{d}\mathbf{x}_{k-1:n+1}\, \mathcal{N}(\mathbf{x}_{k:n+1};\, \mathbf{x}_n^{(k)},\, \mathbf{D}_{k:n+1|n}) = \mathcal{N}(\mathbf{x}_k;\, \mathbf{x}_n,\, \mathbf{D}_{k|n}), \tag{4.185}$$

and since this equation is independent of the choice of ϵ the Feynman kernel is given by the Gaussian

$$\mathbf{K}_0(\mathbf{x}_k, \mathbf{x}_n; (t_k - t_n)) = \left(\frac{m}{2\pi\hbar i(t_k - t_n)}\right)^{\frac{n}{2}} \mathrm{e}^{-\frac{1}{2}\frac{m}{\hbar i(t_k - t_n)}(\mathbf{x}_k - \mathbf{x}_n)^\top(\mathbf{x}_k - \mathbf{x}_n)} \tag{4.186}$$

$$= \mathcal{N}(\mathbf{x}_k;\ \mathbf{x}_n,\ \mathbf{D}_{k|n}). \tag{4.187}$$

Therefore, one can see that the Feynman kernel for the free particle can be interpreted as a diffusion process in *complex time* iT. Moreover, this implicates that the analytic continuation $K(\mathbf{x}_k, \mathbf{x}_n; -i(t_k - t_n))$ becomes the fundamental solution of the diffusion equation with diffusion constant $\frac{m}{i\hbar}$.

4.7.1 State estimation for macroscopic objects with path integrals

In this section, it is assumed that a sensor produces noisy observations of the state at discrete instants of time. The measurement model $p(\mathbf{z}_k|\mathbf{x}_k)$ therefore is the linear Gaussian likelihood function as above such that one has for time t_k that

$$p(\mathbf{z}_k|\mathbf{x}_k) = \mathcal{N}(\mathbf{z}_k;\ \mathbf{H}_k\mathbf{x}_k,\ \mathbf{R}_k). \tag{4.188}$$

By means of the conditional probability, the posterior at time t_k can be expressed by an integration kernel and the posterior at time t_{k-1}:

$$p(\mathbf{x}_k|\mathcal{Z}^k) \propto \int \mathrm{d}\mathbf{x}_{k-1}\, p(\mathbf{z}_k|\mathbf{x}_k)\, p(\mathbf{x}_k|\mathbf{x}_{k-1}) \cdot p(\mathbf{x}_{k-1}|\mathcal{Z}^{k-1}). \tag{4.189}$$

A successive use of the Markov property and the Bayes theorem shows that the Kernel can be extended to multiple time steps:

$$p(\mathbf{x}_k|\mathcal{Z}^k) \propto \int \mathrm{d}\mathbf{x}_n\, \mathrm{d}\mathbf{x}_{n+1}\, \ldots\, \mathrm{d}\mathbf{x}_{k-1}\, p(\mathbf{x}_k, \ldots, \mathbf{x}_n|\mathcal{Z}^k) \tag{4.190}$$

$$= \int \mathrm{d}\mathbf{x}_{k-1:n}\, p(\mathbf{x}_k|\mathbf{x}_{k-1})\, p(\mathbf{z}_k|\mathbf{x}_k)\, p(\mathbf{x}_{k-1:n}|\mathcal{Z}^{k-1}) \tag{4.191}$$

$$= \int \mathrm{d}\mathbf{x}_{k-1:n} \left\{\prod_{l=n+1}^{k} p(\mathbf{x}_l|\mathbf{x}_{l-1})\, p(\mathbf{z}_l|\mathbf{x}_l)\right\} p(\mathbf{x}_n|\mathcal{Z}^n) \tag{4.192}$$

It now becomes obvious that the posterior is obtained by means of a Feynman kernel:

$$p(\mathbf{x}_k|\mathcal{Z}^k) \propto \int \mathrm{d}\mathbf{x}_n\, \mathbf{K}(\mathbf{x}_k, \mathbf{x}_n; \mathbf{z}_{k:n+1})\, p(\mathbf{x}_n|\mathcal{Z}^n), \tag{4.193}$$

$$\mathbf{K}(\mathbf{x}_k, \mathbf{x}_n; \mathbf{z}_{k:n+1}) = \int \mathrm{d}\mathbf{x}_{k-1:n+1} \left\{\prod_{l=n+1}^{k} p(\mathbf{x}_l|\mathbf{x}_{l-1})\, p(\mathbf{z}_l|\mathbf{x}_l)\right\} \tag{4.194}$$

In case that linear-Gaussian models as described above are assumed, the Feynman kernel becomes

$$\mathbf{K}(\mathbf{x}_k, \mathbf{x}_n; \mathbf{z}_{k:n+1}) = \int \mathrm{d}\mathbf{x}_{k-1:n+1}$$

$$\left\{ \prod_{l=n+1}^{k} \mathcal{N}(\mathbf{x}_l;\ \mathbf{F}_{l|l-1}\mathbf{x}_{l-1},\ \mathbf{Q}_{l|l-1})\ \mathcal{N}(\mathbf{z}_l;\ \mathbf{H}_l\mathbf{x}_l,\ \mathbf{R}_l) \right\} \tag{4.195}$$

$$\propto \int \mathrm{d}\mathbf{x}_{k-1:n+1}\ \mathrm{e}^{-\frac{1}{2}\mathcal{S}[\mathbf{x}_{k:n}]} \tag{4.196}$$

where the "action" $\mathcal{S}[\mathbf{x}_{k:n}]$ of the path $\mathbf{x}_n, \ldots, \mathbf{x}_k$ is given by

$$\mathcal{S}[\mathbf{x}_{k:n}] = \int_{t_n}^{t_k} \mathrm{d}t\ \mathcal{L}(\mathbf{x}_l) \tag{4.197}$$

$$\mathcal{L}(\mathbf{x}_l) = \mathbf{T}(\mathbf{x}_l, \mathbf{x}_{l-1}) - \mathbf{V}(\mathbf{x}_l) \tag{4.198}$$

$$\int_{t_n}^{t_k} \mathrm{d}t\ \mathcal{L}(\mathbf{x}_l) = \sum_{l=n+1}^{k} (\mathbf{x}_l - \mathbf{F}_{l|l-1}\mathbf{x}_{l-1})^\top \mathbf{Q}_{l|l-1}^{-1}(\mathbf{x}_l - \mathbf{F}_{l|l-1}\mathbf{x}_{l-1})$$

$$+(\mathbf{z}_l - \mathbf{H}_l\mathbf{x}_l)^\top \mathbf{R}_l^{-1}(\mathbf{z}_l - \mathbf{H}_{l|l-1}\mathbf{x}_l) \tag{4.199}$$

$$\mathbf{T}(\mathbf{x}_l, \mathbf{x}_{l-1}) = (\mathbf{x}_l - \mathbf{F}_{l|l-1}\mathbf{x}_{l-1})^\top \mathbf{Q}_{l|l-1}^{-1}(\mathbf{x}_l - \mathbf{F}_{l|l-1}\mathbf{x}_{l-1})\frac{\mathrm{d}}{\mathrm{d}t} \tag{4.200}$$

$$\mathbf{V}(\mathbf{x}_l) = -(\mathbf{z}_l - \mathbf{H}_l\mathbf{x}_l)^\top \mathbf{R}_l^{-1}(\mathbf{z}_l - \mathbf{H}_{l|l-1}\mathbf{x}_l)\frac{\mathrm{d}}{\mathrm{d}t}. \tag{4.201}$$

As one can see, (4.196)–(4.201) show that inferred knowledge on the stochastic motion of an object which is observed by a noisy sensor obeys the laws of a Wick-rotated Schrödiger equation. In this "tracking" setup, the log-likelihood of the measurements becomes the potential function which are in force only at a single instant of time, but nonetheless they act on the complete trajectory. Since the measurement potential $\mathbf{V}$ is negative, one can see its attracting nature, that is, the trajectory is attracted towards the measurements. This is visualized in Figure 4.7.

4.7.2 The free dynamic particle

Here, a macroscopic object is considered. One has that Heisenberg's uncertainty principle does not hold, that is, the state transition is modeled by a stochastic process with given drift despite the fact that the initial state is fixed in the kernel $\mathbf{K}_0(\mathbf{x}_k, \mathbf{x}_n)$. We have seen above in Section 4.7 that the Feynman kernel for quantum particles corresponds to a diffusion process in the complex time. This is a different due to the fact that the drift (velocity) components may be known for classical objects with the state variables $\{\mathbf{x}_l\}_{l=n,\ldots,k}$. These two cases are shown schematically in Figure 4.8.

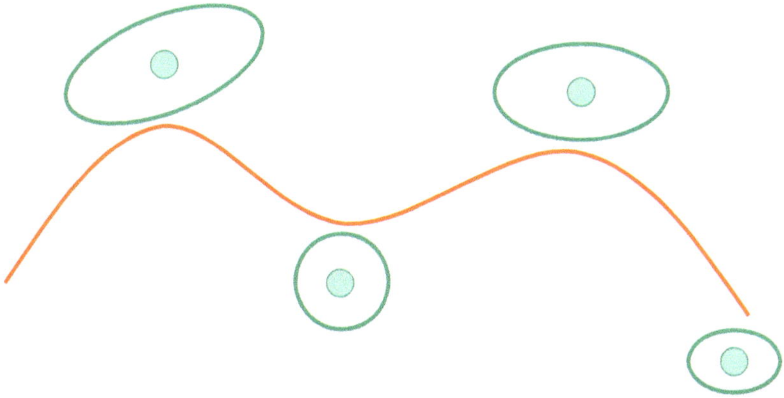

Figure 4.7 *A smoothed trajectory (orange) estimated based on measurements (green). The trajectory is attracted by the potential generated by the measurements.*

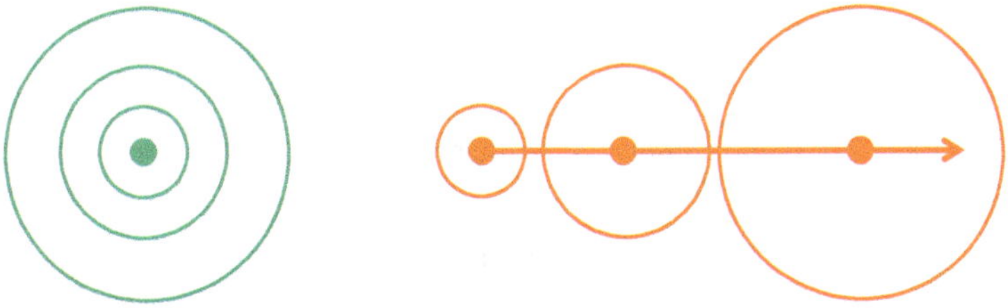

Figure 4.8 *The transition kernel for a free particle is a diffusion process due to the stochastic nature. In the quantum case (left), it is stationary due to the fixed initial position whereas in the classical case, the state may include a known initial velocity (right).*

To begin with, it is assumed that $\mathbf{V}_l(\mathbf{x}_l) = 0$ at all instants of time t_l which is equivalent to a stochastic diffusion without observations. As a first step, it is important to notice that

$$\mathbf{K}_0(\mathbf{x}_k, \mathbf{x}_n) = \int d\mathbf{x}_{k-1:n+1} \left\{ \prod_{l=n+1}^{k} p(\mathbf{x}_l|\mathbf{x}_{l-1}) \right\} \tag{4.202}$$

$$= \int d\mathbf{x}_{k-1:n+1}\, p(\mathbf{x}_{k:n+1}|\mathbf{x}_n) \tag{4.203}$$

The multi-step propagation kernel $p(\mathbf{x}_{k:n+1}|\mathbf{x}_n)$ is a Gaussian and therefore fully defined by its mean and covariance. The mean is given by taking the expectation which yields

$$\mathrm{E}\left[\mathbf{x}_{k:n+1}|\mathbf{x}_n\right] = (\mathrm{E}\left[\mathbf{x}_k|\mathbf{x}_n\right]^\top \dots \mathrm{E}\left[\mathbf{x}_{n+1}|\mathbf{x}_n\right]^\top)^\top \tag{4.204}$$

$$= \begin{pmatrix} \mathbf{F}_{k|n} \\ \vdots \\ \mathbf{F}_{n+2|n} \\ \mathbf{F}_{n+1|n} \end{pmatrix} \mathbf{x}_n \tag{4.205}$$

$$=: \mathbf{F}_{k:n+1|n}\mathbf{x}_n. \tag{4.206}$$

The covariance matrix $\mathbf{Q}_{k:n+1|n}$ can be computed by means of block entries

$$\mathbf{Q}_{k:n+1|n} = \left(\mathrm{cov}\left[\mathbf{x}_i, \mathbf{x}_j|\mathbf{x}_n\right]\right)_{i,j}. \tag{4.207}$$

The block diagonal entries $\mathrm{cov}\left[\mathbf{x}_i, \mathbf{x}_i|\mathbf{x}_n\right]$ are given by

$$\mathrm{cov}\left[\mathbf{x}_i, \mathbf{x}_i|\mathbf{x}_n\right] = \mathbf{Q}_{i|n}, \tag{4.208}$$

whereas for the off-diagonal entries $\mathrm{cov}\left[\mathbf{x}_i, \mathbf{x}_j|\mathbf{x}_n\right]$, it can be assumed without loss of generality that $i > j$. Then

$$\mathrm{cov}\left[\mathbf{x}_i, \mathbf{x}_j|\mathbf{x}_n\right] = \mathrm{cov}\left[\mathbf{F}_{i|j}\mathbf{x}_j + \mathbf{w}_{i|j}, \mathbf{x}_j|\mathbf{x}_n\right] \tag{4.209}$$

$$= \mathbf{F}_{i|j}\mathbf{Q}_{j|n}. \tag{4.210}$$

since $\mathbf{w}_{i|j}$ is zero-mean and orthogonal to the process noise up to time t_j. Now putting it all together yields

$$p(\mathbf{x}_{k:n+1}|\mathbf{x}_n) = \mathcal{N}(\mathbf{x}_{k:n+1}; \mathbf{F}_{k:n+1|n}\mathbf{x}_n, \mathbf{Q}_{k:n+1|n}) \tag{4.211}$$

and the structure of the accumulated process noise covariance $\mathbf{Q}_{k:n+1|n}$ is given by

$$\mathbf{Q}_{k:n+1|n} = \begin{pmatrix} \mathbf{Q}_{k|n} & \mathbf{F}_{k|k-1}\mathbf{Q}_{k-1|n} & \cdots & \mathbf{F}_{k|n+2}\mathbf{Q}_{n+2|n} & \mathbf{F}_{k|n+1}\mathbf{Q}_{n+1|n} \\ \mathbf{Q}_{k-1|n}\mathbf{F}_{k|k-1}^\top & \mathbf{Q}_{k-1|n} & \cdots & \mathbf{F}_{k-1|n+2}\mathbf{Q}_{n+2|n} & \mathbf{F}_{k-1|n+1}\mathbf{Q}_{n+1|n} \\ \vdots & \ddots & & & \vdots \\ \mathbf{Q}_{n+1|n}\mathbf{F}_{k|n+1}^\top & \mathbf{Q}_{n+1|n}\mathbf{F}_{k-1|n+2}^\top & \cdots & \mathbf{Q}_{n+1|n}\mathbf{F}_{n+2|n+1} & \mathbf{Q}_{n+1|n} \end{pmatrix} \tag{4.212}$$

which also can be expressed recursively:

$$\mathbf{Q}_{l:n+1|n} = \begin{pmatrix} \mathbf{Q}_{l|n} & \mathbf{F}_{l|l-1}\Pi_{l-1}\mathbf{Q}_{l-1:n+1|n} \\ \mathbf{Q}_{l-1:n+1|n}\Pi_{l-1}^\top\mathbf{F}_{l|l-1}^\top & \mathbf{Q}_{l-1:n+1|n} \end{pmatrix} \tag{4.213}$$

where Π_{l-1} is the projection $(\mathbf{I}, \mathbf{O}, \dots, \mathbf{O})$ such that $\Pi_{l-1}\mathbf{Q}_{l-1:n+1|n}$ is the first block row of $\mathbf{Q}_{l-1:n+1|n}$.

Since

$$\mathbf{K}_0(\mathbf{x}_k, \mathbf{x}_n) = \int \mathrm{d}\mathbf{x}_{k-1:n+1}\ \mathcal{N}(\mathbf{x}_{k:n+1}; \mathbf{F}_{k:n+1|n}\mathbf{x}_n, \mathbf{Q}_{k:n+1|n}) \tag{4.214}$$

the Gaussian density is marginalized up to the instance referring to time t_k. Therefore, one obtains

$$\mathbf{K}_0(\mathbf{x}_k, \mathbf{x}_n) = \mathcal{N}(\mathbf{x}_k; \mathbf{F}_{k|n}\mathbf{x}_n, \mathbf{Q}_{k|n}). \tag{4.215}$$

4.7.3 Linear-Gaussian measurements

By means of the results of the previous section, it becomes obvious that the Feynman kernel for linear Gaussian measurements is given by

$$\mathbf{K}(\mathbf{x}_k, \mathbf{x}_n; \mathbf{z}_{k:n+1}) = \int d\mathbf{x}_{k-1:n+1}\ \mathcal{N}(\mathbf{x}_{k:n+1};\ \mathbf{F}_{k:n+1|n}\mathbf{x}_n,\ \mathbf{Q}_{k:n+1|n}) \prod_{l=n+1}^{k} \mathcal{N}(\mathbf{z}_l;\ \mathbf{H}_l\mathbf{x}_l,\ \mathbf{R}_l) \tag{4.216}$$

Now since the measurements are mutually independent it holds that

$$\prod_{l=n+1}^{k} p(\mathbf{z}_l|\mathbf{x}_l) = p(\mathbf{z}_{k:n+1}|\mathbf{x}_{k:n+1}) \tag{4.217}$$

where $\mathbf{z}_{k:n+1}$ is the accumulated vector of the set of measurements $\{\mathbf{z}_l\}_{l=n+1}^{k}$. In the linear Gaussian case, the right side of (4.217) equals the Gaussian

$$p(\mathbf{z}_{k:n+1}|\mathbf{x}_{k:n+1}) = \mathcal{N}(\mathbf{z}_{k:n+1};\ \mathbf{H}_{k:n+1}\mathbf{x}_{k:n+1},\ \mathbf{R}_{k:n+1}) \tag{4.218}$$

in which the following abbreviations were used:

$$\mathbf{z}_{k:n+1} = \left(\mathbf{z}_k^\top\ \mathbf{z}_{k-1}^\top\ \ldots\ \mathbf{z}_{n+1}^\top\right)^\top, \tag{4.219}$$

$$\mathbf{R}_{k:n+1} = \operatorname{diag}\left(\mathbf{R}_k\ \ldots\ \mathbf{R}_{n+1}\right), \tag{4.220}$$

$$\mathbf{H}_{k:n+1} = \operatorname{diag}\left(\mathbf{H}_k\ \ldots\ \mathbf{H}_{n+1}\right). \tag{4.221}$$

Inserting into (4.216) yields

$$\mathbf{K}(\mathbf{x}_k, \mathbf{x}_n; \mathbf{z}_{k:n+1}) = \int d\mathbf{x}_{k-1:n+1}\ \mathcal{N}(\mathbf{x}_{k:n+1};\ \mathbf{F}_{k:n+1|n}\mathbf{x}_n,\ \mathbf{Q}_{k:n+1|n})\,\mathcal{N}(\mathbf{z}_{k:n+1};\ \mathbf{H}_{k:n+1}\mathbf{x}_{k:n+1},\ \mathbf{R}_{k:n+1}) \tag{4.222}$$

on which the product formula (3.123) can be applied. One obtains

$$\mathbf{K}(\mathbf{x}_k, \mathbf{x}_n; \mathbf{z}_{k:n+1}) = \mathcal{N}(\mathbf{z}_{k:n+1};\ \mathbf{H}_{k:n+1}\mathbf{F}_{k:n+1|n}\mathbf{x}_n,\ \mathbf{U}_{k:n+1|n}) \int d\mathbf{x}_{k-1:n+1}\ \mathcal{N}(\mathbf{x}_{k:n+1};\ \mathbf{F}_{k:n+1|n}\mathbf{x}_n + \mathbf{J}_{k:n+1|n}\boldsymbol{\nu}_{k:n+1}(\mathbf{x}_n),\ \mathbf{G}_{k:n+1|n}) \tag{4.223}$$

where

$$\mathbf{U}_{k:n+1|n} = \mathbf{H}_{k:n+1|n}\mathbf{Q}_{k:n+1|n}\mathbf{H}_{k:n+1|n}^\top + \mathbf{R}_{k:n+1}, \tag{4.224}$$

$$\mathbf{J}_{k:n+1|n} = \mathbf{Q}_{k:n+1|n}\mathbf{H}_{k:n+1|n}^\top\mathbf{U}_{k:n+1|n}^{-1}, \tag{4.225}$$

$$\boldsymbol{\nu}_{k:n+1}(\mathbf{x}_n) = \mathbf{z}_{k:n+1} - \mathbf{H}_{k:n+1|n}\mathbf{F}_{k:n+1|n}\mathbf{x}_n, \tag{4.226}$$

$$\mathbf{G}_{k:n+1|n} = \mathbf{Q}_{k:n+1|n} - \mathbf{J}_{k:n+1|n}\mathbf{U}_{k:n+1|n}\mathbf{J}_{k:n+1|n}^\top. \tag{4.227}$$

As the integration in (4.223) is over a single Gaussian, all variables except for $\mathbf{x}_k$ are marginalized. What remains is a single Gaussian in $\mathbf{x}_k$ where the mean and the covariance are given by means of a projection Π_k such that $\Pi_k \mathbf{x}_{k:n+1} = \mathbf{x}_k$:

$$\int \mathrm{d}\mathbf{x}_{k-1:n+1}\ \mathcal{N}(\mathbf{x}_{k:n+1};\ \mathbf{F}_{k:n+1|n}\mathbf{x}_n + \mathbf{J}_{k:n+1|n}\boldsymbol{\nu}_{k:n+1}(\mathbf{x}_n),\ \mathbf{G}_{k:n+1|n})$$
$$= \mathcal{N}(\mathbf{x}_k;\ \Pi_k(\mathbf{F}_{k:n+1|n}\mathbf{x}_n + \mathbf{J}_{k:n+1|n}\boldsymbol{\nu}_{k:n+1}(\mathbf{x}_n)),\ \Pi_k \mathbf{G}_{k:n+1|n}\Pi_k^\top) \tag{4.228}$$

Putting it all together, the Feynman kernel for linear-Gaussian measurements is given by

$$\mathbf{K}(\mathbf{x}_k, \mathbf{x}_n; \mathbf{z}_{k:n+1}) = \mathcal{N}(\mathbf{z}_{k:n+1};\ \mathbf{H}_{k:n+1}\mathbf{F}_{k:n+1|n}\mathbf{x}_n,\ \mathbf{U}_{k:n+1|n})$$
$$\cdot \mathcal{N}(\mathbf{x}_k;\ \Pi_k(\mathbf{F}_{k:n+1|n}\mathbf{x}_n + \mathbf{J}_{k:n+1|n}\boldsymbol{\nu}_{k:n+1}(\mathbf{x}_n)),\ \Pi_k \mathbf{G}_{k:n+1|n}\Pi_k^\top). \tag{4.229}$$

4.8 Application on a Gaussian density

In case of a Gaussian boundary density

$$p(\mathbf{x}_n|\mathcal{Z}^n) = \mathcal{N}(\mathbf{x}_n;\ \mathbf{x}_{n|n},\ \mathbf{P}_{n|n}) \tag{4.230}$$

the posterior at time t_k can be obtained by two applications of the product formula since

$$p(\mathbf{x}_k|\mathcal{Z}^k) = \int \mathrm{d}\mathbf{x}_n\ \mathbf{K}(\mathbf{x}_k, \mathbf{x}_n; \mathbf{z}_{k:n+1})\, p(\mathbf{x}_n|\mathcal{Z}^n), \tag{4.231}$$

where the kernel in (4.229) consists of a product of two Gaussian densities. The computation of a posterior density based on a set of measurements and a prior in the past using the path integral formulation is visualized schematically in Figure 4.9.

A first usage yields

$$\mathcal{N}(\mathbf{z}_{k:n+1};\ \mathbf{H}_{k:n+1}\mathbf{F}_{k:n+1|n}\mathbf{x}_n,\ \mathbf{U}_{k:n+1|n})\ \mathcal{N}(\mathbf{x}_n;\ \mathbf{x}_{n|n},\ \mathbf{P}_{n|n})$$
$$= \mathcal{N}(\mathbf{z}_{k:n+1};\ \mathbf{H}_{k:n+1}\mathbf{x}_{k:n+1|n},\ \mathbf{S}_{k:n+1})\ \mathcal{N}(\mathbf{x}_n;\ \mathbf{m}_{n|k},\ \mathbf{M}_{n|k}) \tag{4.232}$$

where

$$\mathbf{x}_{k:n+1|n} = \left(\mathbf{x}_{k|n}^\top\ \mathbf{x}_{k-1|n}^\top\ \cdots\ \mathbf{x}_{n+1|n}^\top\right)^\top, \tag{4.233}$$
$$\mathbf{S}_{k:n+1} = \mathbf{H}_{k:n+1}\mathbf{F}_{k:n+1|n}\mathbf{P}_{n|n}\mathbf{F}_{k:n+1|n}^\top\mathbf{H}_{k:n+1}^\top + \mathbf{U}_{k:n+1|n}, \tag{4.234}$$
$$= \mathbf{H}_{k:n+1}\mathbf{P}_{k:n+1|n}\mathbf{H}_{k:n+1}^\top + \mathbf{R}_{k:n+1}, \tag{4.235}$$
$$\mathbf{P}_{k:n+1|n} = \mathbf{F}_{k:n+1|n}\mathbf{P}_{n|n}\mathbf{F}_{k:n+1|n}^\top + \mathbf{Q}_{k:n+1|n}, \tag{4.236}$$
$$\tilde{\mathbf{W}}_{k:n+1} = \mathbf{P}_{n|n}\mathbf{F}_{k:n+1|n}^\top\mathbf{H}_{k:n+1}^\top\mathbf{S}_{k:n+1}^{-1}, \tag{4.237}$$
$$\mathbf{m}_{n|k} = \mathbf{x}_{n|n} + \tilde{\mathbf{W}}_{k:n+1}(\mathbf{z}_{k:n+1} - \mathbf{H}_{k:n+1}\mathbf{x}_{k:n+1|n}), \tag{4.238}$$
$$\mathbf{M}_{n|k} = \mathbf{P}_{n|n} - \tilde{\mathbf{W}}_{k:n+1}\mathbf{S}_{k:n+1}\tilde{\mathbf{W}}_{k:n+1}^\top. \tag{4.239}$$

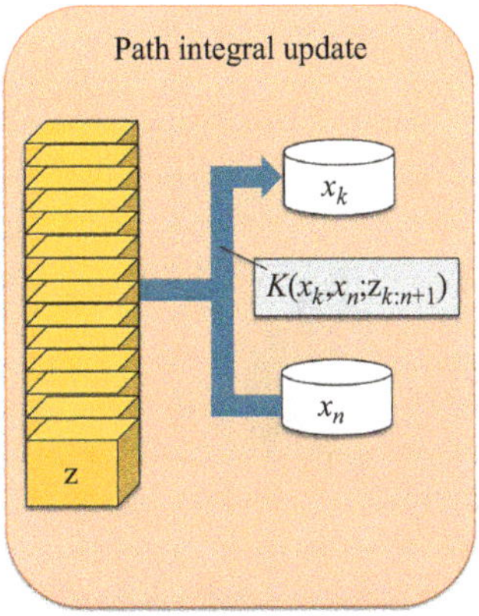

Figure 4.9 Schematic visualization of the path integral update, where a posterior is computed by means of the transition kernel $\mathbf{K}(\mathbf{x}_k, \mathbf{x}_n; \mathbf{z}_{k:n+1})$.

The second usage then is applied on

$$\mathcal{N}(\mathbf{x}_n;\ \mathbf{m}_{n|k},\ \mathbf{M}_{n|k}) \\ \cdot \mathcal{N}(\mathbf{x}_k;\ \Pi_k(\mathbf{F}_{k:n+1|n}\mathbf{x}_n + \mathbf{J}_{k:n+1|n}\boldsymbol{\nu}_{k:n+1}(\mathbf{x}_n)),\ \Pi_k\mathbf{G}_{k:n+1|n}\Pi_k^\top). \tag{4.240}$$

Since the Gaussian in $\mathbf{x}_n$ is integrated, only the density in $\mathbf{x}_k$ remains which then is the resulting posterior:

$$p(\mathbf{x}_k|\mathcal{Z}^k) = \mathcal{N}(\mathbf{x}_k;\ \mathbf{x}_{k|k},\ \mathbf{P}_{k|k}) \tag{4.241}$$

where the parameters are given by

$$\mathbf{x}_{k|k} = \Pi_k(\mathbf{F}_{k:n+1|n}\mathbf{m}_{n|k} + \mathbf{J}_{k:n+1|n}\boldsymbol{\nu}_{k:n+1}(\mathbf{m}_{n|k})) \tag{4.242}$$

and

$$\mathbf{P}_{k|k} = \Pi_k(\mathbf{T}_{k:n+1}\mathbf{M}_{n|k}\mathbf{T}_{k:n+1}^\top + \mathbf{G}_{k:n+1|n})\Pi_k^\top, \tag{4.243}$$

$$\mathbf{T}_{k:n+1} = \mathbf{F}_{k:n+1|n} - \mathbf{J}_{k:n+1|n}\mathbf{H}_{k:n+1|n}\mathbf{F}_{k:n+1|n}. \tag{4.244}$$

Of course, in the linear Gaussian case, these parameters are equal to a sequential processing of a Kalman filter. This is easily verified for instance by using the MATLAB® code provided in the Appendix Section B.

Chapter 5
Track-to-track fusion with known correlations

The theory of track-to-track fusion (T2TF) in target tracking and state estimation addresses the challenge of fusing a set of given tracks on a commonly observed target or system in terms of their means and covariance matrices such that a joint estimate is computed. In the optimal case, the result is equal to processing the measurements from all involved sensors themselves. Such scenarios occur for instance, if a distinguished fusion center receives the result of processed, that is fused sensor data, or if a platform in a mesh network of spatially distributed nodes is processing state estimates from its connected neighbors. In general, it is obvious that multiple sensors can provide more complementary information and better coverage either in geographical terms or in other parametrical spaces. In tracking applications, it is of particular interest to increase the detection probability and reduce false alarms such that the situational awareness is based on stable tracks with a high continuity. On the other hand, automotive driving and advanced driver assistance systems heavily rely on a robust recognition of the surrounding traffic, which in many cases is observed by multiple radar, lidar, and camera sensors mounted on the cars. The produced amount of data makes it infeasible to process in a centralized manner for the available computation hardware. As a consequence, one can find hierarchical architectures, where a close-sensor signal processing and filtering is followed by a fusion of sensors within a local group and finally the global picture is assembled by a third fusion layer by means of T2TF algorithms.

In general, one distinguishes between the following architectures:

- *Centralized fusion:* In the centralized fusion architecture, all sensors send their measurements at all instants of time to a distinguished *fusion center* (FC) node, which updates its tracks using a Bayesian prediction-filtering recursion. The data sets from the sensors may be processed iteratively, that is the posterior of processing the measurements from sensor s is used as the prior for sensor $s + 1$. Since the full set of measurements potentially includes a large amount of false alarms, this scheme is hardly feasible in practical applications. As indicated above, either the link bandwidth is too narrow to transfer the raw data at each instant of time or the processing capacity of the fusion center is insufficient to produce results in realtime. However, since the methods for single sensor processing may well be applied to measurements from multiple sensors to estimate a jointly observed target state, it often serves as a *golden standard* to compare T2TF algorithms against. This is due to the fact that often approximations of the "optimal" solution are applied in T2TF in order to deal with arising correlations of track estimates.

- *Decentralized fusion:* Wireless ad-hoc networks with agile nodes often have a dynamic topology, that is, the connectivity between platforms changes over time. Each instant typically has a processing unit in order to fuse the state estimates from its neighbors with its own. As a consequence, *"data incest"* may appear. Consider for instance that node A sends its estimate to node B, after fusing the result is transmitted to node C, who in turn is connected with A. If A fuses the received data with its estimate, part of the information is double accounted for. In this case, only approximate solutions exist [53], which increase the resulting covariance matrix artificially in order to reflect the reduced trust.
- *Distributed fusion:* As indicated in Figure 5.1(c), distributed fusion considers a similar architecture as in the centralized manner, but in this case the local processing nodes at the sensor sites fuse their data and produce an estimate for the jointly observed state. As a consequence, only the mean and its covariance matrix is transmitted to the FC. For the computation of the global estimate conditioned on the full information from all sensors, the FC applies T2TF algorithms. Since the local tracks are correlated, standard Kalman-based methods would not provide the optimal result. This is due to the fact that the Kalman filter assumes stochastic independence between the measurement and the prior in the update step.

It turns out that the distributed fusion scheme may well be applied in various architecture, including mesh-type sensor networks with limited variation in the topology and hierarchical multi-layer data fusion setups [11]. The focus for T2TF methods in this book therefore is on the distributed scheme.

As important as the estimate itself is a corresponding joint error covariance matrix, which reflects the estimation error in a consistent way. This implies that all

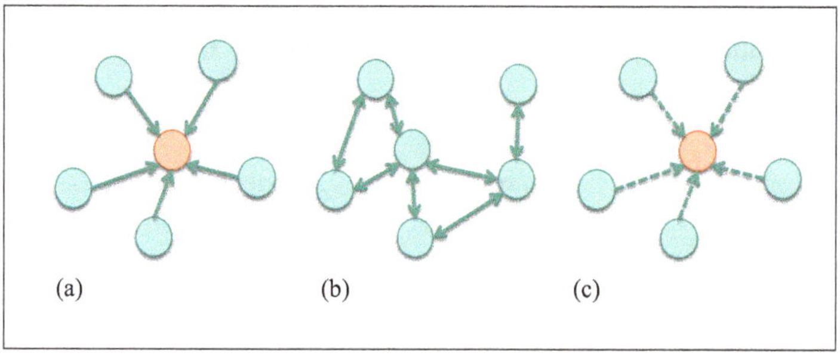

Figure 5.1 (a) Centralized Fusion: all sensors send the raw measurements to a distinguished fusion center. (b) Decentralized Fusion: dynamic platforms which perform sensing and track fusion incorporate the information form their connected neighbors. The topology may vary over time. (c) Distributed Fusion: there is a distinguished fusion center which receives the tracks from connected sensors. The topology is star-like.

instances in the set refer to the same target, that is, it is assumed that the association is solved already. In contrast to measurements, tracks may in general not be considered conditionally independent, which is due to the joint process noise of the observed target. This becomes evident, if a target with very high process noise is considered. Obviously, this noise will affect the estimation error of each track significantly. As a consequence, those errors are not mutually independent anymore.

This lack of mutual independence between two tracks $(\mathbf{x}^i_{k|k}, \mathbf{P}^i_{k|k})$ and $(\mathbf{x}^j_{k|k}, \mathbf{P}^j_{k|k})$, respectively, on a target with state $\mathbf{x}_k$ at time instant t_k is measured in terms of the cross-covariance matrix

$$\mathbf{P}^{i,j}_{k|k} = \mathrm{E}\left[(\mathbf{x}_k - \mathbf{x}^i_{k|k})(\mathbf{x}_k - \mathbf{x}^j_{k|k})\right] \tag{5.1}$$

$$= \mathbf{P}^{j,i\,\top}_{k|k}. \tag{5.2}$$

Here, i and j denote sensor indices from the set $\{1, \ldots, S\}$, where S is the number of processing platforms which have a track on the target at time t_k.

As we will see in Section 5.4 of this chapter, the cross-covariance matrices can be calculated iteratively when Kalman filter assumptions hold. This, however, often is infeasible in practical applications due to the large amount of required parameters. Approximative algorithms which avoid the explicit calculation of cross-covariances are the focus of the next chapter. For the theoretical foundation of how to fuse tracks optimally, algorithms with known cross-correlations are required. This chapter presents methods for T2TF, where it is assumed that the cross-covariance matrices $\mathbf{P}^{i,j}_{k|k}$ for all pairs $i \neq j$ are known.

5.1 Bar-Shalom–Campo formula

Let us initially assume that a T2TF scenario is given as it is schematically indicated in Figure 5.2 where two estimates of the state $\mathbf{x}_k$ for a common observed target at time t_k are given in terms of their mean and covariance $\{(\mathbf{x}^1_{k|k}, \mathbf{P}^1_{k|k}), (\mathbf{x}^2_{k|k}, \mathbf{P}^2_{k|k}\}$. The objective is then to compute a combined mean $\mathbf{x}_{k|k}$ together with a consistent estimation error covariance matrix $\mathbf{P}_{k|k}$. One may regard the "local" tracks as sensor information with

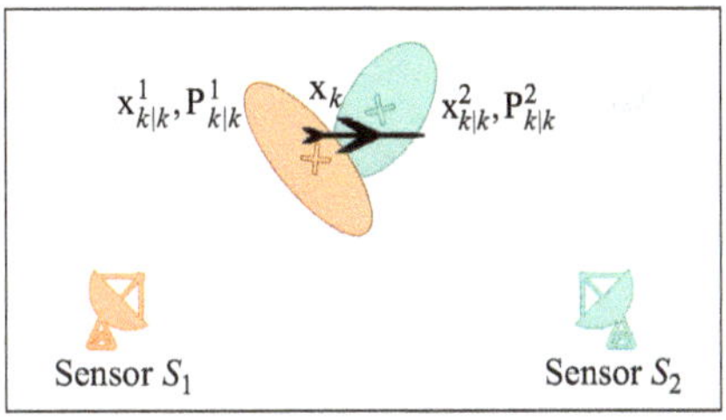

Figure 5.2 Scheme of the T2TF problem for two sensors, S_1 and S_2, which have produced individual tracks on a joint object with state $\mathbf{x}_k$

an obviously linear observation matrix (being the identity) and additive zero-mean Gaussian noise. As a consequence, the fundamental equations of linear estimation (3.45) and (3.46), respectively, can be applied, if the cross-covariance matrix is known. To this end, let us denote the respective estimation errors by

$$\tilde{\mathbf{x}}^1_{k|k} := \mathbf{x}_k - \mathbf{x}^1_{k|k} \quad \text{and} \quad \tilde{\mathbf{x}}^2_{k|k} := \mathbf{x}_k - \mathbf{x}^2_{k|k}. \tag{5.3}$$

Since the local estimators are assumed to be unbiased, their individual estimation errors are zero mean, that is $\mathrm{E}\left[\tilde{\mathbf{x}}^i_{k|k}\right] = \mathbf{O}$ for $i = 1, 2$. Let the cross-covariance of $\tilde{\mathbf{x}}^1_{k|k}$ and $\tilde{\mathbf{x}}^2_{k|k}$ be denoted by

$$\mathbf{P}^{1,2}_{k|k} := \operatorname{cov}\left[\tilde{\mathbf{x}}^1_{k|k}, \tilde{\mathbf{x}}^2_{k|k}\right] \tag{5.4}$$

$$= \mathrm{E}\left[\tilde{\mathbf{x}}^1_{k|k}\tilde{\mathbf{x}}^{2\top}_{k|k}\right] \tag{5.5}$$

and analogously

$$\mathbf{P}^{2,1}_{k|k} := \operatorname{cov}\left[\tilde{\mathbf{x}}^2_{k|k}, \tilde{\mathbf{x}}^1_{k|k}\right] \tag{5.6}$$

$$= \mathbf{P}^{1,2\top}_{k|k} \tag{5.7}$$

Consider the joint estimation error given by the block-vector:

$$\tilde{\mathbf{x}}^\star_{k|k} := \begin{pmatrix} \tilde{\mathbf{x}}^1_{k|k} \\ \tilde{\mathbf{x}}^2_{k|k} \end{pmatrix} \tag{5.8}$$

with error covariance

$$\operatorname{cov}\left[\tilde{\mathbf{x}}^\star_{k|k}\right] = \begin{pmatrix} \mathbf{P}^1_{k|k} & \mathbf{P}^{1,2}_{k|k} \\ \mathbf{P}^{2,1}_{k|k} & \mathbf{P}^2_{k|k} \end{pmatrix}. \tag{5.9}$$

The two local estimates $\mathbf{x}^1_{k|k}$ and $\mathbf{x}^2_{k|k}$ can be fused by means of the fundamental equation of linear estimation. To this end, $\mathbf{x}^1_{k|k}$ is used as the prior information and $\mathbf{x}^2_{k|k}$ is used as the "measurement." Regarding the measurement fusion notation, one would have that

$$\mathrm{E}\left[\mathbf{x}_k|\mathbf{x}^1_{k|k}, \mathbf{x}^2_{k|k}\right] =: \mathbf{x}_{k|k} \tag{5.10}$$

$$\mathbf{x}_{k|k-1} = \mathbf{x}^1_{k|k} \tag{5.11}$$

$$\mathbf{z}_k = \mathbf{x}^2_{k|k} \tag{5.12}$$

$$\mathbf{H}_k = \mathbf{I} \tag{5.13}$$

$$\mathbf{z}_{k|k-1} = \mathbf{x}^1_{k|k}. \tag{5.14}$$

Therefore, an application of the fundamental equation (3.45) yields

$$\mathbf{x}_{k|k} = \mathbf{x}^1_{k|k} + \mathbf{P}_{xz}\mathbf{P}_{zz}(\mathbf{x}^2_{k|k} - \mathbf{x}^1_{k|k}) \tag{5.15}$$

$$\mathbf{P}_{k|k} = \mathbf{P}^1_{k|k-1} - \mathbf{P}_{xz}\mathbf{P}^{-1}_{zz}\mathbf{P}_{zx}, \tag{5.16}$$

where $\mathbf{P}_{xz}$ is the *prior* cross-covariance matrix $\operatorname{cov}\left[\mathbf{x}_k, \mathbf{x}^2_{k|k}|\mathbf{x}^1_{k|k}\right]$ conditioned on the observed value of track 1. This, however, can be computed in closed form:

$$\begin{aligned}
\mathbf{P}_{xz} &= \operatorname{cov}\left[\mathbf{x}_k, \mathbf{x}^2_{k|k}|\mathbf{x}^1_{k|k}\right] && (5.17)\\
&= \mathrm{E}\left[(\mathbf{x}_k - \mathrm{E}\left[\mathbf{x}_k|\mathbf{x}^1_{k|k}\right])(\mathbf{x}^2_{k|k} - \mathrm{E}\left[\mathbf{x}^2_{k|k}|\mathbf{x}^1_{k|k}\right])^\top\right] && (5.18)\\
&= \mathrm{E}\left[(\mathbf{x}_k - \mathbf{x}^1_{k|k})(\mathbf{x}^2_{k|k} - \mathbf{x}^1_{k|k})^\top\right] && (5.19)\\
&= \mathrm{E}\left[(\mathbf{x}_k - \mathbf{x}^1_{k|k})((\mathbf{x}_k - \mathbf{x}^1_{k|k}) - (\mathbf{x}_k - \mathbf{x}^2_{k|k}))^\top\right] && (5.20)\\
&= \mathbf{P}^1_{k|k} - \mathbf{P}^{1,2}_{k|k} && (5.21)
\end{aligned}$$

Analogously, $\mathbf{P}_{zz}$ is the covariance of the *expected measurement* density function $p(\mathbf{x}^2_{k|k}|\mathbf{x}^1_{k|k})$. One has that

$$\begin{aligned}
\mathbf{P}_{zz} &= \operatorname{cov}\left[\mathbf{x}^2_{k|k}|\mathbf{x}^1_{k|k}\right] && (5.22)\\
&= \mathrm{E}\left[(\mathbf{x}^2_{k|k} - \mathrm{E}\left[\mathbf{x}^2_{k|k}|\mathbf{x}^1_{k|k}\right])(\mathbf{x}^2_{k|k} - \mathrm{E}\left[\mathbf{x}^2_{k|k}|\mathbf{x}^1_{k|k}\right])^\top\right] && (5.23)\\
&= \mathrm{E}\left[(\mathbf{x}^2_{k|k} - \mathbf{x}^1_{k|k})(\mathbf{x}^2_{k|k} - \mathbf{x}^1_{k|k})^\top\right] && (5.24)\\
&= \mathrm{E}\left[(\mathbf{x}_k - \mathbf{x}^1_{k|k}) - (\mathbf{x}_k - \mathbf{x}^2_{k|k}))((\mathbf{x}_k - \mathbf{x}^1_{k|k}) - (\mathbf{x}_k - \mathbf{x}^2_{k|k}))^\top\right] && (5.25)\\
&= \mathbf{P}^1_{k|k} - \mathbf{P}^{1,2}_{k|k} - \mathbf{P}^{2,1}_{k|k} + \mathbf{P}^2_{k|k} && (5.26)
\end{aligned}$$

Using these results in (5.15) and (5.16) yields

$$\begin{aligned}
\mathbf{x}_{k|k} &= \mathbf{x}^1_{k|k} + (\mathbf{P}^1_{k|k} - \mathbf{P}^{1,2}_{k|k})(\mathbf{P}^1_{k|k} - \mathbf{P}^{1,2}_{k|k} - \mathbf{P}^{2,1}_{k|k} + \mathbf{P}^2_{k|k})^{-1}(\mathbf{x}^2_{k|k} - \mathbf{x}^1_{k|k}) && (5.27)\\
\mathbf{P}_{k|k} &= \mathbf{P}^1_{k|k} - (\mathbf{P}^1_{k|k} - \mathbf{P}^{1,2}_{k|k})(\mathbf{P}^1_{k|k} - \mathbf{P}^{1,2}_{k|k} - \mathbf{P}^{2,1}_{k|k} + \mathbf{P}^2_{k|k})^{-1}(\mathbf{P}^1_{k|k} - \mathbf{P}^{1,2}_{k|k})^\top, && (5.28)
\end{aligned}$$

which are known as the *Bar-Shalom–Campo formulae*. Those can be applied for T2TF of two tracks with given cross-covariances.

Discussion

The Bar-Shalom–Campo formula is exact in the sense that no approximation is used to calculate the Bayesian posterior of the fused track. Therefore, it is useful for the purpose of comparison against other approaches in theoretical considerations. For practical scenarios, its application often is hindered by the fact that the cross-covariance $\mathbf{P}^{1,2}_{k|k} = \mathbf{P}^{2,1\,\top}_{k|k}$ is unknown. We will see below that there exists a recursive formula to compute it under Kalman filter conditions, this, however, hardly is easing the problem since a lot of communication resources would be required in order to transmit the necessary parameters.

5.2 T2TF for an arbitrary number of sensors with known cross-covariances

The approach of the previous section can well be generalized to an arbitrary number of sensors S. This is achieved by means of the Least Squares (LS) estimate, where the local tracks are seen as correlated observations, which all refer to the same instant of time. The local tracking results of S sensors are given in terms of

$$\left\{(\mathbf{x}^1_{k|k}, \mathbf{P}^1_{k|k}), \ldots, (\mathbf{x}^S_{k|k}, \mathbf{P}^S_{k|k})\right\}. \tag{5.29}$$

With respect to the fused estimate $\mathbf{x}_{k|k}$ and its error covariance matrix $\mathbf{P}_{k|k}$ the local tracks $\mathbf{x}^1_{k|k}, \ldots, \mathbf{x}^S_{k|k}$ are considered as correlated observations which are to be fused. Therefore, a pseudo measurement $\mathbf{z}_k$ is constructed by a vertical concatenation given by

$$\mathbf{z}_k = \begin{pmatrix} \mathbf{x}^1_{k|k} \\ \vdots \\ \mathbf{x}^S_{k|k} \end{pmatrix} \tag{5.30}$$

Analogously, the corresponding measurement model is to be formulated, which is rather simple since it is assumed that the local estimates $\{\mathbf{x}^s_{k|k}\}_s$ are unbiased and refer to the same state space. One obtains

$$\mathbf{H}_k = (\mathbf{I}, \ldots, \mathbf{I})^\top. \tag{5.31}$$

The measurement error covariance matrix $\mathbf{R}_k$ contains the individual track covariances as the diagonal entries and the cross-covariance matrices $\{\mathbf{P}^{(i,j)}_{k|k}\}_{i \neq j}$ as off-diagonals:

$$\mathbf{R}_k = \begin{pmatrix} \mathbf{P}^1_{k|k} & \cdots & \mathbf{P}^{1,S}_{k|k} \\ \vdots & \ddots & \vdots \\ \mathbf{P}^{S,1}_{k|k} & \cdots & \mathbf{P}^S_{k|k} \end{pmatrix} \tag{5.32}$$

An application of the LS estimate (3.82) and (3.83) yields

$$\mathbf{P}_{k|k} = (\mathbf{H}_k^\top \mathbf{R}_k^{-1} \mathbf{H}_k)^{-1} \tag{5.33}$$

$$\mathbf{x}_{k|k} = \mathbf{P}_{k|k} \mathbf{H}_k^\top \mathbf{R}_k^{-1} \mathbf{z}_k. \tag{5.34}$$

This provides the fused estimate and its error covariance for an arbitrary number of tracks with known cross-covariances.

It should be noted that due to the inversion lemma (see Section A in the Appendix), an alternative formula for the fusion of two tracks with given cross-covariance matrix $\mathbf{P}^{1,2}_{k|k}$ can be found. One has that

$$\mathbf{P}^{-1}_{k|k} = (\mathbf{I}, \mathbf{I}) \begin{pmatrix} \mathbf{P}^{1}_{k|k} & \mathbf{P}^{1,2}_{k|k} \\ (\mathbf{P}^{1,2}_{k|k})^{\top} & \mathbf{P}^{2}_{k|k} \end{pmatrix}^{-1} \begin{pmatrix} \mathbf{I} \\ \mathbf{I} \end{pmatrix} \tag{5.35}$$

$$= (\mathbf{I}, \mathbf{I}) \begin{pmatrix} \mathbf{S}^{-1} & -\mathbf{S}^{-1}\mathbf{P}^{1,2}_{k|k}(\mathbf{P}^{2}_{k|k})^{-1} \\ -(\mathbf{P}^{2}_{k|k})^{-1}(\mathbf{P}^{1,2}_{k|k})^{\top}\mathbf{S} & (\mathbf{P}^{2}_{k|k})^{-1}(\mathbf{P}^{1,2}_{k|k})^{\top}\mathbf{S}\mathbf{P}^{1,2}_{k|k}(\mathbf{P}^{2}_{k|k})^{-1} \end{pmatrix} \begin{pmatrix} \mathbf{I} \\ \mathbf{I} \end{pmatrix} \tag{5.36}$$

$$= \mathbf{S}^{-1} - \mathbf{S}^{-1}\mathbf{P}^{1,2}_{k|k}(\mathbf{P}^{2}_{k|k})^{-1} - (\mathbf{P}^{2}_{k|k})^{-1}(\mathbf{P}^{1,2}_{k|k})^{\top}\mathbf{S} + (\mathbf{P}^{2}_{k|k})^{-1}(\mathbf{P}^{1,2}_{k|k})^{\top}\mathbf{S}\mathbf{P}^{1,2}_{k|k}(\mathbf{P}^{2}_{k|k})^{-1}, \tag{5.37}$$

where the abbreviation $\mathbf{S} = \mathbf{P}^{1}_{k|k} - \mathbf{P}^{1,2}_{k|k}(\mathbf{P}^{2}_{k|k})^{-1}\mathbf{P}^{2\top}_{k|k}$ was used for the Schur-complement of the joint error covariance matrix $\mathbf{R}_k$.

5.3 Fusion with arbitrary gains

The T2TF rule with known cross-covariances from (5.34) corresponds to a linear combination of the track states $\{\mathbf{x}^{s}_{k|k}\}_s$ with a gain matrix $\mathbf{W}_{\mathrm{LS}}$, which is optimal in the least squares sense and is given by

$$\mathbf{x}_{k|k} = \mathbf{W}_{\mathrm{LS}} \begin{pmatrix} \mathbf{x}^{1}_{k|k} \\ \vdots \\ \mathbf{x}^{S}_{k|k} \end{pmatrix} \tag{5.38}$$

$$\mathbf{W}_{\mathrm{LS}} := \mathbf{P}_{k|k}\mathbf{H}^{\top}_{k}\mathbf{R}^{-1}_{k}, \tag{5.39}$$

where the notation for the pseudo observation model from the previous section was used. The matrix $\mathbf{W}_{\mathrm{LS}}$ is of block-dimension $1 \times S$ and can therefore be expressed in terms of block-components $\{\mathbf{K}^{s}\}_s = 1, \ldots, S$ such that $\mathbf{W}_{\mathrm{LS}} = (\mathbf{K}^{1}, \ldots, \mathbf{K}^{S})$. Due to the construction of $\mathbf{H}$ in (5.31), one has that

$$\sum_{s=1}^{S} \mathbf{K}^{s} = \mathbf{W}_{\mathrm{LS}}\mathbf{H} \tag{5.40}$$

$$= \mathbf{P}_{k|k}\mathbf{H}^{\top}_{k}(\mathbf{R}_{k})^{-1}\mathbf{H} \tag{5.41}$$

$$= \mathbf{P}_{k|k}\mathbf{P}^{-1}_{k|k} = \mathbf{I}. \tag{5.42}$$

Thus, the exact fusion rule is a matrix-valued convex combination where the estimate can be written as

$$\mathbf{x}_{k|k} = \sum_{s=1}^{S} \mathbf{K}^s \mathbf{x}_{k|k}^s. \tag{5.43}$$

One can see that the block-entries $\mathbf{K}^s$ act as individual *gain matrices* for the local estimates. Actually, it is well possible to compute the exact formula for the fused error covariance matrix for *arbitrary* gain matrices $\tilde{\mathbf{K}}^s$ which obey the normalization that $\sum \mathbf{K}^s = \mathbf{I}$, which is also called the *condition for unbiasedness*, since any deviation from it would lead to a bias in the fused estimate, conditioned that all local tracks are unbiased themselves. This is of particular interest, if the gain matrices are derived according to some specific optimization criterion or are even modeled based on heuristic arguments.

One has that

$$\mathbf{P}_{k|k} = \mathrm{E}\left[(\mathbf{x}_{k|k} - \mathbf{x}_k)(\mathbf{x}_{k|k} - \mathbf{x}_k)^{\top}\right] \tag{5.44}$$

$$= \mathrm{E}\left[\left(\sum_{s=1}^{S} \tilde{\mathbf{K}}^s \mathbf{x}_{k|k}^s - \mathbf{x}_k\right)\left(\sum_{s=1}^{S} \tilde{\mathbf{K}}^s \mathbf{x}_{k|k}^s - \mathbf{x}_k\right)^{\top}\right] \tag{5.45}$$

$$= \mathrm{E}\left[\left(\sum_{s=1}^{S} \tilde{\mathbf{K}}^s (\mathbf{x}_{k|k}^s - \mathbf{x}_k)\right)\left(\sum_{s=1}^{S} \tilde{\mathbf{K}}^s (\mathbf{x}_{k|k}^s - \mathbf{x}_k)\right)^{\top}\right] \tag{5.46}$$

$$= \sum_{i,j=1}^{S} \tilde{\mathbf{K}}^i \mathrm{E}\left[(\mathbf{x}_{k|k}^i - \mathbf{x}_k)(\mathbf{x}_{k|k}^j - \mathbf{x}_k)^{\top}\right] \tilde{\mathbf{K}}^{j\,\top} \tag{5.47}$$

$$= \sum_{i,j=1}^{S} \tilde{\mathbf{K}}^i \mathbf{P}_{k|k}^{(i,j)} \tilde{\mathbf{K}}^{j\,\top} \tag{5.48}$$

$$= \tilde{\mathbf{W}} \mathbf{R} \tilde{\mathbf{W}}^{\top}, \tag{5.49}$$

where $\tilde{\mathbf{W}} := (\tilde{\mathbf{K}}^1 \dots \tilde{\mathbf{K}}^S)$ is the generalized composed gain matrix and $\mathbf{R}$ is the joint error covariance matrix defined in (5.32).

Discussion

The computation of a consistent error covariance matrix for given gain matrices is of high importance. This is due to the fact that it suffices to optimize the latter in order to obtain a complete T2TF algorithm. Moreover, this also opens the door towards heuristically engineered gains, which reflect context knowledge such as the operating environment conditions or poor information on the tracks, that is for instance, if only "track quality" measures are given in terms of real numbers instead of consistent local error covariances.

5.4 Iterative calculation of cross-covariances

In the methods presented up to here, it was assumed that the tracks are correlated, that is, the cross-covariances are non-zero and that the joint covariance matrix is known. This motivates the question, if and under which circumstances they actually are correlated. A derivation of the cross-covariance matrices under Kalman filter conditions yields a better understanding of the conditions under which tracks become correlated and why that is the case. To this end, a recursive computation scheme is formulated as a result of the prediction-filtering cycle.

Initialization

We initialize the cross-covariance of the estimation error of sensor i and j at the initialization time t_0 with a zero matrix: $\mathbf{P}^{i,j}_{0|0} = \mathbf{O}$. This matches common conditions in many practical applications, since the local tracks might be initiated based on singular measurements, which are mutually independent. As a consequence, the initial tracks are mutually independent, too.

Prediction

The prediction considers the evolution of the cross-covariance from time t_{k-1} to t_k. It is assumed that the result after the previous filtering step is given in terms of $\mathbf{P}^{i,j}_{k-1|k-1} = \operatorname{cov}\left[\mathbf{x}^i_{k-1|k-1}, \mathbf{x}^j_{k-1|k-1}\right]$. After the prediction step of sensor i and sensor j to time t_k, the cross-covariance becomes

$$\begin{aligned}
\mathbf{P}^{i,j}_{k|k-1} &= \mathrm{E}\left[(\mathbf{x}_k - \mathbf{x}^i_{k|k-1})(\mathbf{x}_k - \mathbf{x}^j_{k|k-1})^\top\right] && (5.50)\\
&= \mathrm{E}\left[(\mathbf{F}_{k|k-1}\mathbf{x}_{k-1} + \mathbf{w}_{k|k-1} - \mathbf{F}_{k|k-1}\mathbf{x}^i_{k-1|k-1})(\mathbf{F}_{k|k-1}\mathbf{x}_{k-1} + \mathbf{w}_{k|k-1} - \mathbf{F}_{k|k-1}\mathbf{x}^j_{k-1|k-1})^\top\right]\\
&= \mathrm{E}\left[(\mathbf{F}_{k|k-1}\tilde{\mathbf{x}}^i_{k-1|k-1} + \mathbf{w}_{k|k-1})(\mathbf{F}_{k|k-1}\tilde{\mathbf{x}}^j_{k-1|k-1} + \mathbf{w}_{k|k-1})^\top\right]\\
&= \mathbf{F}_{k|k-1}\mathbf{P}^{i,j}_{k-1|k-1}\mathbf{F}^\top_{k|k-1} + \mathbf{Q}_{k|k-1}. && (5.51)
\end{aligned}$$

Here, the tilde-parameters $\tilde{\mathbf{x}}^i_{k-1|k-1} = \mathbf{x}_k - \mathbf{x}^i_{k-1|k-1}$ denote the corresponding estimation errors.

Filtering

In the filtering step of sensor i and j the respective measurements $\mathbf{z}_k^i$ and $\mathbf{z}_k^j$ are processed locally by means of the Kalman filter update equation. This yields for the posterior cross-covariance that

$$\mathbf{P}_{k|k}^{i,j} = \mathrm{E}\left[(\mathbf{x}_k - \mathbf{x}_{k|k}^i)(\mathbf{x}_k - \mathbf{x}_{k|k}^j)^\top\right] \tag{5.52}$$

$$= \mathrm{E}\left[(\mathbf{x}_k - \mathbf{x}_{k|k-1}^i - \mathbf{W}_{k|k-1}^i(\mathbf{z}_k^i - \mathbf{H}_k\mathbf{x}_{k|k-1}^i))(\mathbf{x}_k - \mathbf{x}_{k|k-1}^j - \mathbf{W}_{k|k-1}^j(\mathbf{z}_k^j - \mathbf{H}_k\mathbf{x}_{k|k-1}^j))^\top\right]$$

$$= \mathrm{E}[(\mathbf{x}_k - \mathbf{x}_{k|k-1}^i - \mathbf{W}_{k|k-1}^i(\mathbf{H}_k\mathbf{x}_k + \mathbf{v}_k^i - \mathbf{H}_k\mathbf{x}_{k|k-1}^i))$$

$$(\mathbf{x}_k - \mathbf{x}_{k|k-1}^j - \mathbf{W}_{k|k-1}^j(\mathbf{H}_k\mathbf{x}_k + \mathbf{v}_k^j - \mathbf{H}_k\mathbf{x}_{k|k-1}^j))^\top] \tag{5.53}$$

$$= \mathrm{E}\left[(\tilde{\mathbf{x}}_{k|k-1}^i - \mathbf{W}_{k|k-1}^i(\mathbf{H}_k\tilde{\mathbf{x}}_{k|k-1}^i + \mathbf{v}_k^i))(\tilde{\mathbf{x}}_{k|k-1}^j - \mathbf{W}_{k|k-1}^j(\mathbf{H}_k\tilde{\mathbf{x}}_{k|k-1}^j + \mathbf{v}_k^j))^\top\right] \tag{5.54}$$

$$= \mathrm{E}\left[((\mathbf{I}_\mathbf{x} - \mathbf{W}_{k|k-1}^i\mathbf{H}_k)\tilde{\mathbf{x}}_{k|k-1}^i - \mathbf{W}_{k|k-1}^i\mathbf{v}_k^i)((\mathbf{I}_\mathbf{x} - \mathbf{W}_{k|k-1}^j\mathbf{H}_k)\tilde{\mathbf{x}}_{k|k-1}^j - \mathbf{W}_{k|k-1}^j\mathbf{v}_k^j)^\top\right] \tag{5.55}$$

$$= (\mathbf{I}_\mathbf{x} - \mathbf{W}_{k|k-1}^i\mathbf{H}_k)\mathbf{P}_{k|k-1}^{i,j}(\mathbf{I}_\mathbf{x} - \mathbf{W}_{k|k-1}^j\mathbf{H}_k)^\top + \mathbf{W}_{k|k-1}^i \underbrace{\mathrm{E}\left[\mathbf{v}_k^i(\mathbf{v}_k^j)^\top\right]}_{=\mathbf{O}} \mathbf{W}_{k|k-1}^{j\,\top}$$

$$= (\mathbf{I}_\mathbf{x} - \mathbf{W}_{k|k-1}^i\mathbf{H}_k)\mathbf{P}_{k|k-1}^{i,j}(\mathbf{I}_\mathbf{x} - \mathbf{W}_{k|k-1}^j\mathbf{H}_k)^\top. \tag{5.56}$$

The cross-covariance $\mathrm{E}\left[\mathbf{v}_k^i(\mathbf{v}_k^j)^\top\right]$ of the zero-mean measurement noise is again zero since they are statistically independent.

Concluding the derivations from above, we have that the iterative computation of the cross-covariance $\mathbf{P}_{k|k}^{i,j}$ is given by

$$\mathbf{P}_{0|0}^{i,j} = \mathbf{O} \tag{5.57}$$

$$\mathbf{P}_{k|k-1}^{i,j} = \mathbf{F}_{k|k-1}\mathbf{P}_{k-1|k-1}^{i,j}\mathbf{F}_{k|k-1}^\top + \mathbf{Q}_{k|k-1} \tag{5.58}$$

$$\mathbf{P}_{k|k}^{i,j} = (\mathbf{I}_\mathbf{x} - \mathbf{W}_{k|k-1}^i\mathbf{H}_k)\mathbf{P}_{k|k-1}^{i,j}(\mathbf{I}_\mathbf{x} - \mathbf{W}_{k|k-1}^j\mathbf{H}_k)^\top. \tag{5.59}$$

Discussion

It is now interesting to see that cross-covariance matrix $\mathbf{P}_{k|k}^{i,j}$ remains zero throughout the process, if (and only if) the process noise covariance $\mathbf{Q}_{k|k-1}$ is zero for all time steps k. In this case, one may speak of a "deterministic target," since its dynamics are fully predictable by means of the transition matrices $\mathbf{F}_{k|k-1}$. In all other cases, it is obvious that local tracks quickly become correlated and that the correlation depends on the ratio between process noise and measurement noise. In particular, one can see that a full computation of all cross-covariances is infeasible in practical applications since it depends on the local gain matrices $\mathbf{W}_{k|k-1}^s$ and the situation becomes even more complex in case of algorithms which mitigate non-detections and false alarms such as *Probabilistic Data Association* and *Multi Hypotheses Tracker*.

5.5 Distributed calculation of the cross-covariances using the square root decomposition

We have seen above that T2TF can be solved, if the full joint covariance matrix or equivalently all the cross-covariances $\mathbf{P}^{i,j}_{k|k}$ are available. An application of the LS equations (5.33) and (5.34) yields the desired posterior parameters. However, in the previous section, it became obvious that the centralized computation of the cross-covariances is infeasible in most practical applications. But the structure of the recursive computation for $\mathbf{P}^{i,j}_{k|k}$ allows a distributed computation [54]. That is, each sensor i computes an auxiliary matrix $\mathbf{L}^i_k$ which encodes all relevant information for the fusion center to reconstruct a cross-covariance $\mathbf{P}^{i,j}_{k|k}$ by combining $\mathbf{L}^i_k$ with $\mathbf{L}^j_k$ from sensor j. This distributed calculation scheme is visualized in Figure 5.3.

In order to derive the computation of the local contributions $\mathbf{L}^i_k$, we note that the recursive computation of the cross-covariance in (5.58) and (5.59) can be combined:

$$\mathbf{P}^{i,j}_{k|k} = \mathbf{M}^i_k(\mathbf{F}_{k|k-1}\mathbf{P}^{i,j}_{k-1|k-1}\mathbf{F}^\top_{k|k-1} + \mathbf{Q}_{k|k-1})\mathbf{M}^{j\top}_k \tag{5.60}$$

$$= \mathbf{T}^i_{k|k-1}\mathbf{P}^{i,j}_{k-1|k-1}\mathbf{T}^{j\top}_{k|k-1} + \mathbf{T}^i_{k|k}\mathbf{Q}_{k|k-1}\mathbf{T}^{j\top}_{k|k} \tag{5.61}$$

$$\mathbf{M}^i_k = \mathbf{I}_\mathbf{x} - \mathbf{W}^i_{k|k-1}\mathbf{H}^i_k \tag{5.62}$$

$$\mathbf{T}^i_{k|k-1} = \mathbf{M}^i_k\mathbf{F}_{k|k-1} \tag{5.63}$$

$$\mathbf{T}^i_{k|k} = \mathbf{M}^i_k \tag{5.64}$$

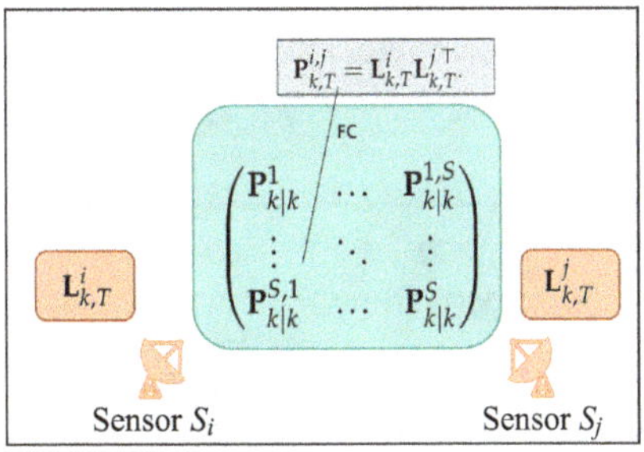

Figure 5.3 By means of a decomposition of the recursive computation of the cross-covariance matrices, it is possible to define local parameters for each sensor, which can be combined in the fusion center (FC) in order to reconstruct the approximative components of the full joint error covariance matrix

Another application of the same formulae reveals a recursive structure:

$$\mathbf{P}^{i,j}_{k|k} = \mathbf{T}^{i}_{k|k-2}\mathbf{P}^{i,j}_{k-2|k-2}\mathbf{T}^{j\top}_{k|k-1} + \sum_{l=k-1}^{k} \mathbf{T}^{i}_{k|l}\mathbf{Q}_{l|l-1}\mathbf{T}^{j\top}_{k|l} \tag{5.65}$$

$$\mathbf{T}^{i}_{k|l} = \mathbf{T}^{i}_{k|k}\mathbf{T}^{i}_{k|k-1}\mathbf{T}^{i}_{k-1|k-2}\cdots\mathbf{T}^{i}_{l+1|l}. \tag{5.66}$$

An iterative application therefore yields

$$\mathbf{P}^{i,j}_{k|k} = \mathbf{T}^{i}_{k|0}\mathbf{P}^{i,j}_{0|0}\mathbf{T}^{j\top}_{k|0} + \sum_{l=1}^{k} \mathbf{T}^{i}_{k|l}\mathbf{Q}_{l|l-1}\mathbf{T}^{j\top}_{k|l}. \tag{5.67}$$

In the next step, one is using a square root decomposition of a symmetric positive definite matrix $\mathbf{A}$ such that

$$\sqrt{\mathbf{A}}\sqrt{\mathbf{A}}^{\top} = \mathbf{A}. \tag{5.68}$$

The Cholesky decomposition is an example which may be used for convenience. One may now define auxiliary variables

$$\mathbf{A}^{i}_{l} = \begin{cases} \mathbf{T}^{i}_{k|l}\sqrt{\mathbf{P}^{i,j}_{0|0}}, & \text{if } l = 0 \\ \mathbf{T}^{i}_{k|l}\sqrt{\mathbf{Q}_{l|l-1}}, & \text{if } l \geq 1. \end{cases} \tag{5.69}$$

As a consequence, it is possible to rewrite the cross-covariance as

$$\mathbf{P}^{i,j}_{k|k} = \sum_{l=0}^{k} \mathbf{A}^{i}_{l}\mathbf{A}^{j\top}_{l} \tag{5.70}$$

$$= \mathbf{L}^{i}_{k}\mathbf{L}^{j\top}_{k}, \tag{5.71}$$

where the matrices $\mathbf{L}^{i}_{k}$ are given in block form as

$$\mathbf{L}^{i}_{k} = \left(\mathbf{A}^{i}_{0}, \mathbf{A}^{i}_{1}, \ldots, \mathbf{A}^{i}_{k}\right). \tag{5.72}$$

According to the above derivations, a recursive computation of the $\mathbf{L}^{i}_{k}$ is possible by

$$\mathbf{L}^{i}_{k} = \mathbf{M}^{i}_{k}\left(\mathbf{F}_{k|k-1}\mathbf{L}^{i}_{k-1}, \sqrt{\mathbf{Q}_{k|k-1}}\right). \tag{5.73}$$

By definition, the matrices $\mathbf{L}^{i}_{k}$ is growing linearly in time. In order to keep the required bandwidth limited, a solution with a fixed number of block entries is required. To this end, let

$$\mathbf{L}^{i}_{k,T} = \left(\mathbf{A}^{i}_{k-T+1}, \mathbf{A}^{i}_{k-T}, \ldots, \mathbf{A}^{i}_{k}\right) \tag{5.74}$$

be a sliding window square root decomposition matrix of the joint covariance. Then, a residual $\mathbf{O}_k^i$ with respect to the contribution of sensor i only is given by

$$\mathbf{P}_{k|k}^i = \mathbf{L}_k^i \mathbf{L}_k^{i\top} \tag{5.75}$$

$$= \mathbf{L}_{k,T}^i \mathbf{L}_{k,T}^{i\top} + \mathbf{O}_k^i, \tag{5.76}$$

which can be computed recursively via

$$\mathbf{O}_0^i = 0 \tag{5.77}$$

$$\mathbf{O}_k^i = \mathbf{O}_{k-1}^i + \mathbf{A}_k^i \mathbf{A}_k^{i\top}. \tag{5.78}$$

However, since parts of the correlations are ignored, the cross-covariance matrix $\mathbf{P}_{k|k}^{i,j}$ has an unknown residual $\mathbf{P}_{k,O}^{i,j}$ with

$$\mathbf{P}_{k|k}^{i,j} = \mathbf{P}_{k,T}^{i,j} + \mathbf{P}_{k,O}^{i,j} \tag{5.79}$$

where

$$\mathbf{P}_{k,T}^{i,j} = \mathbf{L}_{k,T}^i \mathbf{L}_{k,T}^{j\top}. \tag{5.80}$$

In order to circumvent that the joint error covariance becomes too small due to the missing residual of the cross-covariances, one can approximate it. The true joint covariance for sensor i and j is given by

$$\begin{pmatrix} \mathbf{P}_{k|k}^i & \mathbf{P}_{k|k}^{i,j} \\ \mathbf{P}_{k|k}^{j,i} & \mathbf{P}_{k|k}^j \end{pmatrix} = \begin{pmatrix} \mathbf{P}_{k|k}^i & \mathbf{P}_{k,T}^{i,j} + \mathbf{P}_{k,O}^{i,j} \\ \mathbf{P}_{k,T}^{j,i} + \mathbf{P}_{k,O}^{j,i} & \mathbf{P}_{k|k}^j \end{pmatrix} \tag{5.81}$$

$$= \begin{pmatrix} \mathbf{P}_{k|k}^i - \mathbf{O}_k^i & \mathbf{P}_{k,T}^{i,j} \\ \mathbf{P}_{k,T}^{j,i} & \mathbf{P}_{k|k}^j - \mathbf{O}_k^j \end{pmatrix} + \begin{pmatrix} \mathbf{O}_k^i & \mathbf{P}_{k,O}^{i,j} \\ \mathbf{P}_{k,O}^{j,i} & \mathbf{O}_k^j \end{pmatrix} \tag{5.82}$$

Since the additional term in (5.82) is a proper (yet unknown) covariance matrix, one may use the *covariance intersection* rule which will be part of the next chapter in (6.87) to find $\omega \in [0, 1]$ such that

$$\begin{pmatrix} \mathbf{O}_k^i & \mathbf{P}_{k,O}^{i,j} \\ \mathbf{P}_{k,O}^{j,i} & \mathbf{O}_k^j \end{pmatrix} \leq \begin{pmatrix} \frac{1}{\omega}\mathbf{O}_k^i & \\ & \frac{1}{1-\omega}\mathbf{O}_k^j \end{pmatrix}. \tag{5.83}$$

This naturally extends to the full joint covariance matrix such that one obtains the following approximation:

$$\begin{pmatrix} \mathbf{P}^{1}_{k|k} & \mathbf{P}^{1,2}_{k|k} & \cdots & \mathbf{P}^{1,S}_{k|k} \\ \mathbf{P}^{2,1}_{k|k} & \mathbf{P}^{2}_{k|k} & \cdots & \mathbf{P}^{2,S}_{k|k} \\ \vdots & \vdots & \ddots & \vdots \\ \mathbf{P}^{S,1}_{k|k} & \mathbf{P}^{S,S-1}_{k|k} & \cdots & \mathbf{P}^{S}_{k|k} \end{pmatrix} \approx \begin{pmatrix} \mathbf{P}^{1}_{k|k} - \mathbf{O}^{1}_{k} & \mathbf{P}^{1,2}_{k|k} & \cdots & \mathbf{P}^{1,S}_{k|k} \\ \mathbf{P}^{2,1}_{k|k} & \mathbf{P}^{2}_{k|k} - \mathbf{O}^{2}_{k} & \cdots & \mathbf{P}^{2,S}_{k|k} \\ \vdots & \vdots & \ddots & \vdots \\ \mathbf{P}^{S,1}_{k|k} & \mathbf{P}^{S,S-1}_{k|k} & \cdots & \mathbf{P}^{S}_{k|k} - \mathbf{O}^{S}_{k} \end{pmatrix}$$

$$+ \begin{pmatrix} \frac{1}{\omega_1}\mathbf{O}^{1}_{k} & & & \\ & \frac{1}{\omega_2}\mathbf{O}^{1}_{k} & & \\ & & \ddots & \\ & & & \frac{1}{\omega_S}\mathbf{O}^{S}_{k} \end{pmatrix}, \tag{5.84}$$

where the weights ω_i sum up to one and may be found by minimizing the fused error covariance.

Discussion

The distributed computation of the approximated cross-covariance requires some additional computations at the local sensor sites and bandwith in order to transmit the auxiliary parameters to a fusion center. Once the FC reconstructs all $\mathbf{P}^{i,j}_{k|k}$ for all pairs of sensors, T2TF with known cross-covariances may be applied. In practical applications, the local processing sometimes is hindered by processing nodes which are provided as a "black-box," such that local parameters may not be accessed. The upcoming chapter provides T2TF methods, where the cross-covariances are not required.

Chapter 6
Track-to-track fusion with unknown correlations

In the previous chapter, the iterative computation of cross-covariances for Kalman-tracks on a jointly observed target was derived. Due to the dependence on local parameters from each update step such as the gain matrices $\mathbf{W}^i_{k|k-1}$, these computations are often infeasible in real-world applications. Still, though, for theoretical considerations, those methods play an important role to compare approximative algorithms, which will be the focus of this chapter. By relaxing some of the conditions, non-optimal methods can be derived, which would have – in average – a higher estimation error than the "exact" algorithms, but since knowledge on the cross-covariances is not assumed, the former can easily be applied. The non-optimal estimate often is less critical than the consistency of the corresponding error covariance. An *optimistic* posterior covariances $\mathbf{P}_{k|k}$ which is smaller* than the actual squared estimation error such that

$$\mathbf{P}_{k|k} < \mathrm{E}\left[(\mathbf{x}_k - \mathbf{x}_{k|k})(\mathbf{x}_k - \mathbf{x}_{k|k})^\top\right] \tag{6.1}$$

should be avoided at all times since it would degrade the performance during the following data association and fusion process. Analogously, *pessimistic* or *conservative* covariance matrices denote the opposite whereas *consistency* is given, whenever the result is matching the actual estimation error.

However, there exist a number of methods to solve the problem of T2TF without known cross-covariances, which all have different assumptions, advantages and challenges. It is the goal of this chapter to introduce, derive, and explain them in all details such that the reader will have a sound knowledge of when to apply which for a given application.

6.1 Naïve fusion

In many applications, the cross-covariances are not available at the fusion center such that approximative solutions have to be applied. In the most simple approach for T2TF, potentially existing correlations are ignored by setting the cross-covariances $\mathbf{P}^{i,j}_{k|k} = 0$ for all $i \neq j$. Therefore, it is called *Naïve Fusion*. Since the covariances $\mathbf{P}^i_{k|k}$

*In the sense that the difference is negative definite.

for $i = 1 \ldots, S$ are provided by the sensors, the approximated joint covariance $\mathbf{R}_k$ of the pseudo data $\mathbf{z}_k$ from (5.30) is block-diagonal:

$$\mathbf{R}_k = \begin{pmatrix} \mathbf{P}^1_{k|k} & & \\ & \ddots & \\ & & \mathbf{P}^S_{k|k} \end{pmatrix} \tag{6.2}$$

The inverse of this approximation is easy to obtain by means of the inverse of the matrix block entries:

$$\begin{pmatrix} \mathbf{P}^1_{k|k} & & \\ & \ddots & \\ & & \mathbf{P}^S_{k|k} \end{pmatrix}^{-1} = \begin{pmatrix} (\mathbf{P}^1_{k|k})^{-1} & & \\ & \ddots & \\ & & (\mathbf{P}^S_{k|k})^{-1} \end{pmatrix} \tag{6.3}$$

As a consequence, the fused estimate and covariance of the naïve approach can be calculated in the closed form based on the least squares solution. They are given by

$$\mathbf{P}_{k|k} = (\mathbf{H}_k^\top \mathbf{R}_k^{-1} \mathbf{H}_k)^{-1} \tag{6.4}$$

$$= \left[(\mathbf{I}_\mathbf{x}, \ldots, \mathbf{I}_\mathbf{x}) \cdot \begin{pmatrix} (\mathbf{P}^1_{k|k})^{-1} & & \\ & \ddots & \\ & & (\mathbf{P}^S_{k|k})^{-1} \end{pmatrix} \cdot \begin{pmatrix} \mathbf{I}_\mathbf{x} \\ \vdots \\ \mathbf{I}_\mathbf{x} \end{pmatrix} \right]^{-1} \tag{6.5}$$

$$= \left(\sum_{s=1}^{S} (\mathbf{P}^s_{k|k})^{-1} \right)^{-1} \tag{6.6}$$

and

$$\mathbf{x}_{k|k} = \mathbf{P}_{k|k}(\mathbf{H}_k^\top \mathbf{R}_k^{-1} \mathbf{z}_k) \tag{6.7}$$

$$= \mathbf{P}_{k|k} \left[(\mathbf{I}_\mathbf{x}, \ldots, \mathbf{I}_\mathbf{x}) \cdot \begin{pmatrix} (\mathbf{P}^1_{k|k})^{-1} & & \\ & \ddots & \\ & & (\mathbf{P}^S_{k|k})^{-1} \end{pmatrix} \cdot \begin{pmatrix} \mathbf{x}^1_{k|k} \\ \vdots \\ \mathbf{x}^S_{k|k} \end{pmatrix} \right] \tag{6.8}$$

$$= \mathbf{P}_{k|k} \left(\sum_{s=1}^{S} (\mathbf{P}^s_{k|k})^{-1} \mathbf{x}^s_{k|k} \right) \tag{6.9}$$

Together one has for the naïve fusion approach

$$\mathbf{x}_{k|k} = \mathbf{P}_{k|k} \left(\sum_{s=1}^{S} (\mathbf{P}_{k|k}^{s})^{-1} \mathbf{x}_{k|k}^{s} \right) \tag{6.10}$$

$$\mathbf{P}_{k|k} = \left(\sum_{s=1}^{S} (\mathbf{P}_{k|k}^{s})^{-1} \right)^{-1} . \tag{6.11}$$

Those fusion rules will also be used in more sophisticated algorithms, which avoid cross-correlations by modifying the local tracks. Considering the gain matrix-based fusion rule as in (5.43), one can directly see that for the naïve fusion rule it holds that

$$\mathbf{K}^{s} = \mathbf{P}_{k|k} (\mathbf{P}_{k|k}^{s})^{-1} \tag{6.12}$$

This motivates the more general name *convex combination* for this fusion rule.

Discussion

The naïve fusion approach fuses tracks as if they were mutually independent, which is not true in most scenarios. In sections below within this chapter, we will see methods which *modify* the local track parameters in order to have them uncorrelated. One can see that the fused information matrix $\mathbf{P}_{k|k}^{-1}$ is given by the sum of the local information matrices. Also, an analogous sum can be found for the *information state* $\mathbf{P}_{k|k}^{-1}\mathbf{x}_{k|k}$. This leads us to the conclusion that

- *Independent information adds up.*

It should be remarked that the iterative computation of the cross-covariance matrices in (5.58) and (5.59), respectively, implies that they remain zero, whenever the process noise is zero such that $\mathbf{Q}_{k|k-1} = 0$ at all instants of time. As a consequence, the naïve fusion would be optimal, since no approximation would be applied. Thus, for the case of tracking a deterministic target state, the naïve fusion may well be applied.

The *area* or *volume* covered by error ellipse which corresponds to a given covariance matrix is proportional to its determinant. This is a direct consequence by the law for integration by substitution. As one can use (5.37) for the exact fusion of two tracks, it becomes obvious that volume for the density given by the fused posterior of two tracks with given cross-covariance matrix $\mathbf{P}_{k|k}^{1,2}$ is given by

$$\begin{aligned} V_{\text{exact}} &= \det \left[(\mathbf{P}_{k|k})^{-1} \right]^{-1} \\ &= \det \left[(\mathbf{S}^{-1} - \mathbf{S}^{-1}\mathbf{P}_{k|k}^{1,2}(\mathbf{P}_{k|k}^{2})^{-1} - (\mathbf{P}_{k|k}^{2})^{-1}(\mathbf{P}_{k|k}^{1,2})^{\top}\mathbf{S} + (\mathbf{P}_{k|k}^{2})^{-1}(\mathbf{P}_{k|k}^{1,2})^{\top}\mathbf{S}\mathbf{P}_{k|k}^{1,2}(\mathbf{P}_{k|k}^{2})^{-1} \right]^{-1}, \end{aligned} \tag{6.13}$$

where $\mathbf{S}$ is defined as in (5.37) by the Schur-complement [55]. Since the latter is positive definite, one can see that the volume becomes *smaller*, if the cross-covariance $\mathbf{P}_{k|k}^{1,2}$ is ignored as in the naïve fusion. This reflects the fact that setting it to zero is equivalent to treating the local tracks as if they were stochastically independent. Since this is actually not the case, the result is too optimistic. It must be said that this is to be avoided in all cases for the reason that optimistic filters converge can easily diverge and become biased in contrast to the pessimistic case where the covariance matrix is actually too large. This is of particular importance, if further information as in form of additional measurements or remote tracks are to be fused into the result.

6.2 Covariance intersection

It is generally well known that a covariance defines a n-dimensional error ellipsoid for a given level of confidence. For convenience, we define an origin centered ellipse for a given covariance matrix $\mathbf{P}$ by

$$\mathcal{E}_{\mathbf{P}} := \left\{ \mathbf{x} \in \mathbb{R}^n \;\middle|\; \mathbf{x}^\top \mathbf{P}^{-1} \mathbf{x} < 1 \right\}. \tag{6.14}$$

An example is illustrated in Figure 6.1 for a 2×2 covariance matrix $\mathbf{P} = (P_{ij})_{ij}$. One can see that $P_{11} > P_{22}$ since the x-axis is dominant. Moreover, it follows that $P_{12} \neq 0$ since the axes of the ellipse are not aligned with the coordinate system. In particular, $P_{12} > 0$ due to the counter-clock wise rotation.

The interpretation of a covariance as an ellipse sometimes help to interpret tracking results or theoretical problems constructions. But there is more, as we will see in the following, one can prove an interesting geometric statement with respect to T2TF:

Covariance intersection theorem

Let $\mathcal{E}_{\mathbf{P}_1}$ and $\mathcal{E}_{\mathbf{P}_2}$ be the centered error ellipses for the covariances $\mathbf{P}_1$ and $\mathbf{P}_2$, respectively. Then it holds for a vector $\mathbf{x} \in \mathbb{R}^n$ that

$$\mathbf{x} \in \mathcal{E}_{\mathbf{P}_1} \cap \mathcal{E}_{\mathbf{P}_2} \quad \Leftrightarrow \quad \exists\, \mathbf{P}_{1,2} : \mathbf{x} \in \mathcal{E}_{\hat{\mathbf{P}}}, \tag{6.15}$$

where $\hat{\mathbf{P}}$ denotes the optimal fusion estimate covariance of the Bar-Shalom–Campo formula given by

$$\hat{\mathbf{P}} = \mathbf{P}_1 - (\mathbf{P}_1 - \mathbf{P}_{1,2})(\mathbf{P}_1 - \mathbf{P}_{1,2} - \mathbf{P}_{2,1} + \mathbf{P}_2)^{-1}(\mathbf{P}_1 - \mathbf{P}_{1,2})^\top. \tag{6.16}$$

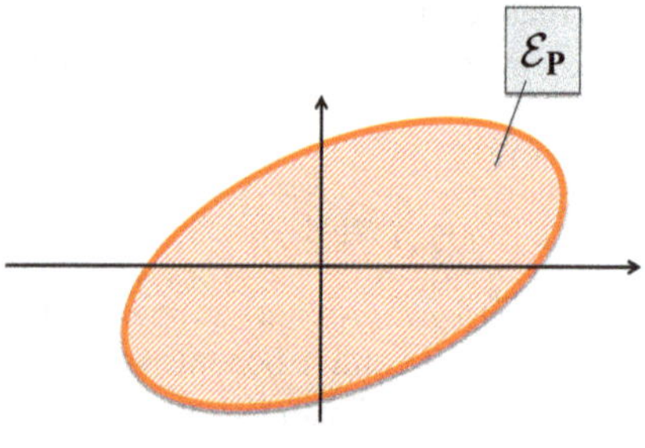

Figure 6.1 Illustration of an origin centered error ellipse

Proof.
"⇒":
W.l.o.g. we assume that $\mathbf{x}^\top \mathbf{P}_1^{-1}\mathbf{x} > \mathbf{x}^\top \mathbf{P}_2^{-1}\mathbf{x}$. Let us introduce the rescaling factor

$$\lambda := \sqrt{\frac{\mathbf{x}^\top \mathbf{P}_2^{-1}\mathbf{x}}{\mathbf{x}^\top \mathbf{P}_1^{-1}\mathbf{x}}}. \tag{6.17}$$

Obviously, we have that $\lambda < 1$. In particular, it holds that

$$\left\| \lambda \mathbf{P}_1^{-1/2}\mathbf{x} \right\| = \left\| \mathbf{P}_2^{-1/2}\mathbf{x} \right\|. \tag{6.18}$$

Therefore, a unitary rotation matrix $\mathbf{U} \in \mathbb{R}^n$ exists with $\mathbf{U}^\top \mathbf{U} = \mathbf{I}$ and

$$\lambda \mathbf{P}_1^{-1/2}\mathbf{x} = \mathbf{U}\mathbf{P}_2^{-1/2}\mathbf{x}. \tag{6.19}$$

We now define the following cross-covariance matrix $\mathbf{P}_{1,2}$:

$$\mathbf{P}_{1,2} := \lambda \mathbf{P}_1^{1/2}\mathbf{U}\mathbf{P}_2^{1/2}. \tag{6.20}$$

For the remainder of this proof, we need to show that $\mathbf{x} \in \mathcal{E}_{\hat{\mathbf{P}}}$ when $\mathbf{P}_{1,2}$ as a cross-covariance is used, that is, we need to show that

$$\mathbf{x}^\top \hat{\mathbf{P}}^{-1}\mathbf{x} < 1. \tag{6.21}$$

Because of the information filter form, the inverse of the posterior covariance $\hat{\mathbf{P}}$ is given by

$$\hat{\mathbf{P}}^{-1} = (\mathbf{I}\ \mathbf{I}) \begin{pmatrix} \mathbf{P}_1 & \mathbf{P}_{1,2} \\ \mathbf{P}_{1,2}^\top & \mathbf{P}_2 \end{pmatrix}^{-1} \begin{pmatrix} \mathbf{I} \\ \mathbf{I} \end{pmatrix} \tag{6.22}$$

According to the Inversion Lemma (see Section A in the Appendix), the closed form solution for the inverted block covariance matrix is given by

$$\begin{pmatrix} \mathbf{P}_1 & \mathbf{P}_{1,2} \\ \mathbf{P}_{1,2}^\top & \mathbf{P}_2 \end{pmatrix}^{-1} = \begin{pmatrix} \mathbf{P}_1^{-1} + \mathbf{P}_1^{-1}\mathbf{P}_{1,2}\mathbf{S}^{-1}\mathbf{P}_{1,2}^\top \mathbf{P}_1^{-1} & -\mathbf{P}_1^{-1}\mathbf{P}_{1,2}\mathbf{S}^{-1} \\ -(\mathbf{P}_1^{-1}\mathbf{P}_{1,2}\mathbf{S}^{-1})^\top & \mathbf{S}^{-1} \end{pmatrix}, \tag{6.23}$$

where the Schur-Complement $\mathbf{S}$ is given by

$$\begin{aligned} \mathbf{S} &= \mathbf{P}_2 - \mathbf{P}_{1,2}^\top \mathbf{P}_1^{-1}\mathbf{P}_{1,2} && (6.24)\\ &= \mathbf{P}_2 - \lambda^2 \mathbf{P}_2^{1/2}\mathbf{U}^\top \mathbf{P}_1^{1/2}\mathbf{P}_1^{-1}\mathbf{P}_1^{1/2}\mathbf{U}\mathbf{P}_2^{1/2} && (6.25)\\ &= (1-\lambda^2)\mathbf{P}_2 && (6.26)\\ &\quad > 0. && (6.27) \end{aligned}$$

We note that

$$\begin{aligned}
\mathbf{P}_1^{-1}\mathbf{P}_{1,2}\mathbf{S}^{-1}\mathbf{x} &= \frac{\lambda}{(1-\lambda^2)}\mathbf{P}_1^{-1}\mathbf{P}_1^{1/2}\mathbf{U}\mathbf{P}_2^{1/2}\mathbf{P}_2^{-1}\mathbf{x} \\
&= \frac{\lambda}{(1-\lambda^2)}\mathbf{P}_1^{-1}\mathbf{P}_1^{1/2}\mathbf{U}\mathbf{P}_2^{-1/2}\mathbf{x} \\
&= \frac{\lambda^2}{(1-\lambda^2)}\mathbf{P}_1^{-1}\mathbf{P}_1^{1/2}\mathbf{P}_1^{-1/2}\mathbf{x} \\
&= \frac{\lambda^2}{(1-\lambda^2)}\mathbf{P}_1^{-1}\mathbf{x} \qquad (6.28)
\end{aligned}$$

On the other hand, one can easily see that

$$\begin{aligned}
\mathbf{x}^\top\mathbf{S}^{-1}\mathbf{x} &= \frac{1}{(1-\lambda)^2}\mathbf{x}^\top\mathbf{P}_2^{-1}\mathbf{x} \\
&= \frac{\lambda^2}{(1-\lambda)^2}\mathbf{x}^\top\mathbf{P}_1^{-1}\mathbf{x} \qquad (6.29)
\end{aligned}$$

and finally, we have another simplification with

$$\begin{aligned}
\mathbf{P}_1^{-1}\mathbf{P}_{1,2}\mathbf{S}^{-1}\mathbf{P}_{1,2}^\top\mathbf{P}_1^{-1} &= \frac{\lambda^2}{(1-\lambda)^2}\mathbf{P}_1^{-1}\mathbf{P}_1^{1/2}\mathbf{U}\mathbf{P}_2^{1/2}\mathbf{P}_2^{-1}\mathbf{P}_2^{1/2}\mathbf{U}^\top\mathbf{P}_1^{1/2}\mathbf{P}_1^{-1} \\
&= \frac{\lambda^2}{(1-\lambda)^2}\mathbf{P}_1^{-1} \qquad (6.30)
\end{aligned}$$

Combining these equations yields

$$\begin{aligned}
\mathbf{x}^\top\hat{\mathbf{P}}^{-1}\mathbf{x} &= \mathbf{x}^\top\left(\mathbf{P}_1^{-1} + \mathbf{P}_1^{-1}\mathbf{P}_{1,2}\mathbf{S}^{-1}\mathbf{P}_{1,2}^\top\mathbf{P}_1^{-1} - \mathbf{P}_1^{-1}\mathbf{P}_{1,2}\mathbf{S}^{-1} - (\mathbf{P}_1^{-1}\mathbf{P}_{1,2}\mathbf{S}^{-1})^\top + \mathbf{S}^{-1}\right)\mathbf{x} \\
&= \mathbf{x}^\top\mathbf{P}_1^{-1}\mathbf{x} + \mathbf{x}^\top\frac{\lambda^2}{(1-\lambda)^2}\mathbf{P}_1^{-1}\mathbf{x} - 2\mathbf{x}^\top\frac{\lambda^2}{(1-\lambda)^2}\mathbf{P}_1^{-1}\mathbf{x} + \mathbf{x}^\top\frac{\lambda^2}{(1-\lambda)^2}\mathbf{P}_1^{-1}\mathbf{x} \\
&= \mathbf{x}^\top\mathbf{P}_1^{-1}\mathbf{x} < 1.
\end{aligned}$$

"$\Leftarrow$": Let $\mathbf{x} \in \mathcal{E}_{\hat{\mathbf{P}}}$ be given. The inverse of the posterior covariance can also be written as

$$\begin{aligned}
\hat{\mathbf{P}}^{-1} &= \mathbf{P}_1^{-1} + \mathbf{P}_1^{-1}\mathbf{P}_{1,2}\mathbf{S}^{-1}\mathbf{P}_{1,2}^\top\mathbf{P}_1^{-1} - \mathbf{P}_1^{-1}\mathbf{P}_{1,2}\mathbf{S}^{-1} - (\mathbf{P}_1^{-1}\mathbf{P}_{1,2}\mathbf{S}^{-1})^\top + \mathbf{S}^{-1} \\
&= \mathbf{P}_1^{-1} + (\mathbf{P}_1^{-1}\mathbf{P}_{1,2} - \mathbf{I})\mathbf{S}^{-1}(\mathbf{P}_1^{-1}\mathbf{P}_{1,2} - \mathbf{I})^\top
\end{aligned}$$

Since the Schur-Complement $\mathbf{S}$ of a positive definite matrix is positive definite, we may conclude that

$$1 > \mathbf{x}^\top\hat{\mathbf{P}}^{-1}\mathbf{x} > \mathbf{x}^\top\mathbf{P}_1^{-1}\mathbf{x}, \qquad (6.31)$$

and therefore it holds that $\mathbf{x} \in \mathcal{E}_{\mathbf{P}_1}$. The same argument can be applied to $\mathbf{P}_2$, which directly yields the desired conclusion. This concludes the proof.

□

A visualization of the Covariance Intersection Theorem can be found in Figure 6.2.

Discussion

The Covariance Intersection Theorem (6.15) implies that the union of all *possible* fused error covariances fills the intersection as indicated in Figure 6.3:

$$\bigcup_{\mathbf{P}_{1,2}} \mathcal{E}_{\hat{\mathbf{P}}} = \mathcal{E}_{\mathbf{P}_1} \cap \mathcal{E}_{\mathbf{P}_2} \tag{6.32}$$

Thus, one can see that the intersection of two error ellipses covers the optimally fused covariance matrix for *some arbitrary* cross-covariance $\mathbf{P}_{1,2}$. As a consequence, a conservative fused error covariance matrix is obtained, if it contains the intersection.

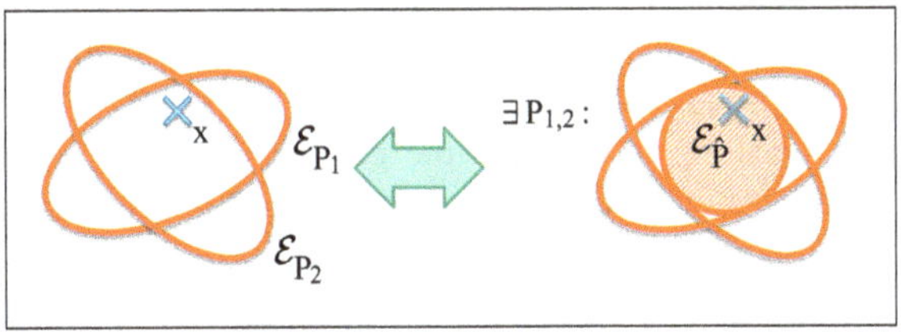

Figure 6.2 Visualization of the Covariance Intersection Theorem

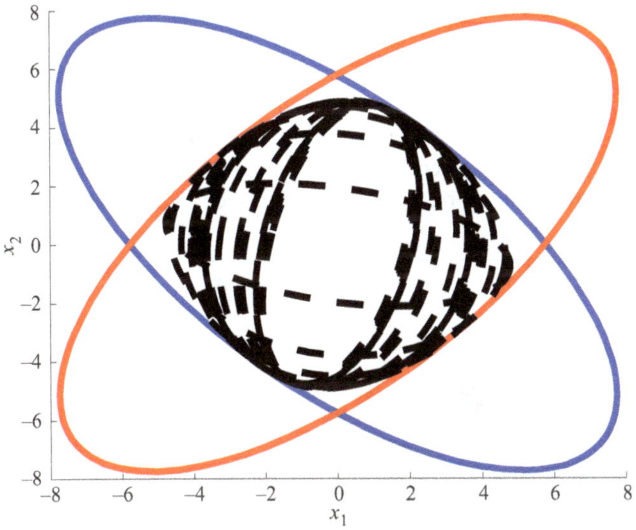

Figure 6.3 Error ellipses (black) of fused covariance matrices for two tracks (blue, red) for different cross-covariances $\mathbf{P}_{1,2}$*. It can be seen that the fused result in contained in the intersection of the original tracks.*

However, it should be avoided to be overly conservative, thus, we are seeking for the *smallest* error ellipse, which still fulfills this condition:

Definition:
A **tight bound** is defined as the *smallest* error ellipse $\mathcal{E}_{\hat{\mathbf{P}}}$ for a corresponding covariance matrix $\hat{\mathbf{P}}$ which still contains the full intersection of the error ellipses for two covariance matrices $\mathbf{P}_1$ and $\mathbf{P}_2$, respectively:
For all arbitrary symmetric positive definite matrices $\mathbf{P}$, it holds that

$$(\mathcal{E}_{\mathbf{P}_1} \cap \mathcal{E}_{\mathbf{P}_2}) \subseteq \mathcal{E}_{\mathbf{P}} \subseteq \mathcal{E}_{\hat{\mathbf{P}}} \quad \Longrightarrow \quad \mathbf{P} = \hat{\mathbf{P}}. \tag{6.33}$$

This is the formal way to express that if the intersection is contained in the error ellipse $\mathcal{E}_{\mathbf{P}}$ and $\mathbf{P}$ is "equal size or smaller" than the tight bound $\hat{\mathbf{P}}$, then $\mathbf{P}$ actually is the tight bound. The tight bound as the smallest ellipse surrounding the intersection of two given error ellipses is shown in Figure 6.4.

There is an astonishing result from the geometric algebra by Khan published in 1969, which can be used in order to construct the tight bound numerically. His theorem makes the following statement:

Tight bound theorem:
Let $\hat{\mathbf{P}}$ be the tight bound for the fusion of $\mathbf{P}_1$ and $\mathbf{P}_2$. Then:

$$\exists \lambda \in [0, 1] : \quad \hat{\mathbf{P}}^{-1} = \lambda \mathbf{P}_1^{-1} + (1 - \lambda)\mathbf{P}_2^{-1}. \tag{6.34}$$

Since a tight bound is unique, it is sufficient to prove that there exists such a λ such that $\hat{\mathbf{P}}^{-1} = \lambda \mathbf{P}_1^{-1} + (1 - \lambda)\mathbf{P}_2^{-1}$ is a tight bound. While its formula is compact and easy to remember, its proof is not. The construction is going through some smart geometric and analytic arguments [56]. The maxim of this book is to provide the full details of all proves, therefore, in the following section the construction of the tight

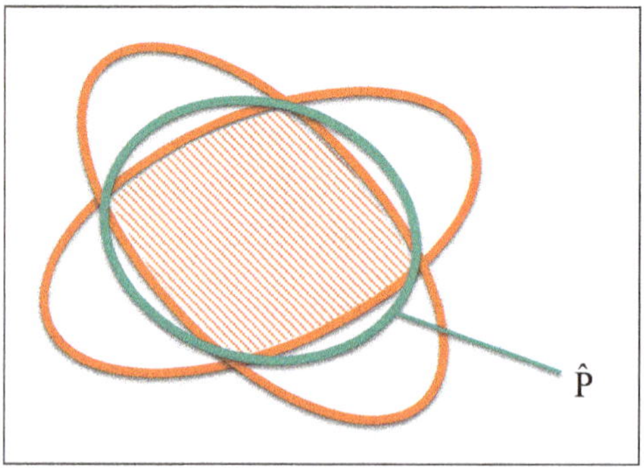

Figure 6.4 The tight bound defines the smallest ellipse enclosing the intersection of two or more error ellipses

bound (6.34) is given. The reader who is not interested in the details of the proof might prefer to skip it and continue with the next section.

The construction starts with a small lemma, which is on the quadratic norm of the full covariances beyond the intersection:

It holds that

$$(\mathcal{E}_{\mathbf{P}_1} \cap \mathcal{E}_{\mathbf{P}_2}) \subseteq \mathcal{E}_{\hat{\mathbf{P}}} \quad \Longleftrightarrow \quad \forall\, \mathbf{x} : \mathbf{x}^\top \hat{\mathbf{P}}^{-1}\mathbf{x} \le \max\left\{\mathbf{x}^\top \mathbf{P}_1^{-1}\mathbf{x},\ \mathbf{x}^\top \mathbf{P}_2^{-1}\mathbf{x}\right\} \tag{6.35}$$

To see this, define $\kappa = \max\left\{\mathbf{x}^\top \mathbf{P}_1^{-1}\mathbf{x},\ \mathbf{x}^\top \mathbf{P}_2^{-1}\mathbf{x}\right\}$ for $\mathbf{x}$. Then, if $(\mathcal{E}_{\mathbf{P}_1} \cap \mathcal{E}_{\mathbf{P}_2}) \subseteq \mathcal{E}_{\hat{\mathbf{P}}}$, then $\frac{1}{\sqrt{\kappa}}\mathbf{x} \in \mathcal{E}_{\mathbf{P}_1} \cap \mathcal{E}_{\mathbf{P}_2}$, therefore $\frac{1}{\sqrt{\kappa}}\mathbf{x} \in \mathcal{E}_{\hat{\mathbf{P}}}$, hence $\mathbf{x}^\top \hat{\mathbf{P}}^{-1}\mathbf{x} \le \kappa$. Conversely, if $\mathbf{x}^\top \hat{\mathbf{P}}^{-1}\mathbf{x} \le \kappa$ and $\mathbf{x} \in \mathcal{E}_{\mathbf{P}_1} \cap \mathcal{E}_{\mathbf{P}_2}$, then $\mathbf{x}^\top \hat{\mathbf{P}}^{-1}\mathbf{x} \le \kappa < 1$ and therefore $\mathbf{x} \in \mathcal{E}_{\hat{\mathbf{P}}}$.

As a consequence it holds in particular that

$$\mathcal{E}_{\mathbf{P}_1} \subseteq \mathcal{E}_{\hat{\mathbf{P}}} \quad \Longleftrightarrow \quad \forall\, \mathbf{x} : \mathbf{x}^\top \hat{\mathbf{P}}^{-1}\mathbf{x} \le \mathbf{x}^\top \mathbf{P}_1^{-1}\mathbf{x} \tag{6.36}$$

This can be seen by using (6.35) with $\mathbf{P}_2^{-1} \to 0$.

First of all, it is now important to see that $\hat{\mathbf{P}}$ is a conservative bound in the sense that

$$(\mathcal{E}_{\mathbf{P}_1} \cap \mathcal{E}_{\mathbf{P}_2}) \subseteq \mathcal{E}_{\hat{\mathbf{P}}} \tag{6.37}$$

For arbitrary $\mathbf{x} \in \mathcal{E}_{\mathbf{P}_1} \cap \mathcal{E}_{\mathbf{P}_2}$ we may assume w.l.o.g. that $\mathbf{x}^\top \mathbf{P}_1^{-1}\mathbf{x} \ge \mathbf{x}^\top \mathbf{P}_2^{-1}\mathbf{x}$. Then:

$$\mathbf{x}^\top \hat{\mathbf{P}}^{-1}\mathbf{x} \le \lambda \mathbf{x}^\top \mathbf{P}_1^{-1}\mathbf{x} + (1-\lambda)\mathbf{x}^\top \mathbf{P}_1^{-1}\mathbf{x} \tag{6.38}$$

$$< 1. \tag{6.39}$$

Therefore $\mathbf{x} \in \mathcal{E}_{\hat{\mathbf{P}}}$.

Now let us prove that we can construct a tight bound by means of (6.34). Thus, let a bound covariance matrix $\mathbf{H}$ be given with $\mathcal{E}_{\mathbf{P}_1} \cap \mathcal{E}_{\mathbf{P}_2} \subseteq \mathcal{E}_{\mathbf{H}}$. We need to show that there exists a $\lambda \in [0, 1]$ with

$$\hat{\mathbf{P}}_\lambda^{-1} = \lambda \mathbf{P}_1^{-1} + (1-\lambda)\mathbf{P}_2^{-1} \tag{6.40}$$

$$\mathcal{E}_{\hat{\mathbf{P}}_\lambda} \subseteq \mathcal{E}_{\mathbf{H}}. \tag{6.41}$$

Define

$$C := \inf_{\mathbf{x}\neq 0} \frac{\max\left\{\mathbf{x}^\top \mathbf{P}_1^{-1}\mathbf{x},\ \mathbf{x}^\top \mathbf{P}_2^{-1}\mathbf{x}\right\}}{\mathbf{x}^\top \mathbf{H}^{-1}\mathbf{x}} \tag{6.42}$$

It is important to note that the infimum of C is actually obtained. This can be seen as follows. For any vector $\mathbf{x}$, we may define

$$\mathbf{x}' := \frac{\mathbf{x}}{\|\mathbf{x}\|} \tag{6.43}$$

Then it holds $\|\mathbf{x}'\| = 1$ and we have

$$\frac{\max\left\{\mathbf{x}^\top \mathbf{P}_1^{-1}\mathbf{x},\ \mathbf{x}^\top \mathbf{P}_2^{-1}\mathbf{x}\right\}}{\mathbf{x}^\top \mathbf{H}^{-1}\mathbf{x}} = \frac{\max\left\{\mathbf{x}'^\top \mathbf{P}_1^{-1}\mathbf{x}',\ \mathbf{x}'^\top \mathbf{P}_2^{-1}\mathbf{x}'\right\}}{\mathbf{x}'^\top \mathbf{H}^{-1}\mathbf{x}'} \tag{6.44}$$

As a consequence, we may restrict our search to a bounded set

$$C = \inf_{\|\mathbf{x}\|=1} \frac{\max\left\{\mathbf{x}^\top \mathbf{P}_1^{-1}\mathbf{x},\ \mathbf{x}^\top \mathbf{P}_2^{-1}\mathbf{x}\right\}}{\mathbf{x}^\top \mathbf{H}^{-1}\mathbf{x}} \tag{6.45}$$

which is always achieved. We may set

$$\bar{\mathbf{x}} := \arg\inf_{\mathbf{x}\neq 0} \frac{\max\left\{\mathbf{x}^\top \mathbf{P}_1^{-1}\mathbf{x},\ \mathbf{x}^\top \mathbf{P}_2^{-1}\mathbf{x}\right\}}{\mathbf{x}^\top \mathbf{H}^{-1}\mathbf{x}} \tag{6.46}$$

Because of (6.35) we have that $C \geq 1$. Therefore, for a given vector $\mathbf{x} \in \mathcal{E}_{\mathbf{P}_1} \cap \mathcal{E}_{\mathbf{P}_2}$ according to (6.36), we need to prove that there exists a $\lambda \in [0, 1]$ with

$$\mathbf{x}^\top \hat{\mathbf{P}}_\lambda^{-1}\mathbf{x} \geq \mathbf{x}^\top \mathbf{H}^{-1}\mathbf{x} \tag{6.47}$$

Then, case A: $\bar{\mathbf{x}}^\top \mathbf{P}_1^{-1}\bar{\mathbf{x}} < \bar{\mathbf{x}}^\top \mathbf{P}_2^{-1}\bar{\mathbf{x}} = C\bar{\mathbf{x}}^\top \mathbf{H}^{-1}\bar{\mathbf{x}}$. Let

$$\mathbf{y} := \bar{\mathbf{x}} - \beta\mathbf{x} \tag{6.48}$$

For $|\beta|$ sufficiently small, it holds that

$$\mathbf{y}^\top \mathbf{P}_1^{-1}\mathbf{y} < \mathbf{y}^\top \mathbf{P}_2^{-1}\mathbf{y} \tag{6.49}$$

and according to the construction of C, it holds that

$$C\mathbf{y}^\top \mathbf{H}^{-1}\mathbf{y} \leq \mathbf{y}^\top \mathbf{P}_2^{-1}\mathbf{y} \tag{6.50}$$

We may choose the sign of β such that $\beta\mathbf{x}^\top(\mathbf{P}_2^{-1} - C\mathbf{H}^{-1})\bar{\mathbf{x}} \geq 0$. Then

$$\begin{aligned}
0 &\leq \mathbf{y}^\top(\mathbf{P}_2^{-1} - C\mathbf{H}^{-1})\mathbf{y} \\
&= (\bar{\mathbf{x}} - \beta\mathbf{x})^\top(\mathbf{P}_2^{-1} - C\mathbf{H}^{-1})(\bar{\mathbf{x}} - \beta\mathbf{x}) \\
&= \underbrace{\bar{\mathbf{x}}^\top(\mathbf{P}_2^{-1} - C\mathbf{H}^{-1})\bar{\mathbf{x}}}_{=0} - 2\beta\mathbf{x}^\top(\mathbf{P}_2^{-1} - C\mathbf{H}^{-1})\bar{\mathbf{x}} + \beta^2\mathbf{x}^\top(\mathbf{P}_2^{-1} - C\mathbf{H}^{-1})\mathbf{x} \\
&\leq \beta^2\mathbf{x}^\top(\mathbf{P}_2^{-1} - C\mathbf{H}^{-1})\mathbf{x}
\end{aligned}$$

Therefore, it holds that

$$\mathbf{x}^\top \mathbf{H}^{-1}\mathbf{x} \leq \mathbf{x}^\top C\mathbf{H}^{-1}\mathbf{x} \leq \mathbf{x}^\top \mathbf{P}_2^{-1}\mathbf{x} \tag{6.51}$$

which proves (6.47) with $\lambda = 0$ and therefore $\hat{\mathbf{P}}_\lambda = \mathbf{P}_2$.

Case B: $\bar{\mathbf{x}}^\top \mathbf{P}_1^{-1}\bar{\mathbf{x}} = \bar{\mathbf{x}}^\top \mathbf{P}_2^{-1}\bar{\mathbf{x}} = C\bar{\mathbf{x}}^\top \mathbf{H}^{-1}\bar{\mathbf{x}}$.

Define

$$\mathbf{y} := \mathbf{x} + \beta\bar{\mathbf{x}}. \tag{6.52}$$

According to the construction of C, it holds that

$$C\mathbf{y}^\top \mathbf{H}^{-1}\mathbf{y} \leq \max\left\{\mathbf{y}^\top \mathbf{P}_1^{-1}\mathbf{y},\ \mathbf{y}^\top \mathbf{P}_2^{-1}\mathbf{y}\right\} \tag{6.53}$$

Using the definition of $\mathbf{y}$ yields

$$C\mathbf{x}\mathbf{H}^{-1}\mathbf{x} + 2C\beta\mathbf{x}^\top \mathbf{H}^{-1}\bar{\mathbf{x}} \leq \max\left\{\mathbf{x}\mathbf{P}_1^{-1}\mathbf{x} + 2\beta\mathbf{x}^\top \mathbf{P}_1^{-1}\bar{\mathbf{x}},\ \mathbf{x}\mathbf{P}_2^{-1}\mathbf{x} + 2\beta\mathbf{x}^\top \mathbf{P}_2^{-1}\bar{\mathbf{x}}\right\} \tag{6.54}$$

where we used the assumption of case B.

Now, we find a scalar λ with

$$C\mathbf{H}^{-1}\bar{\mathbf{x}} = \lambda\mathbf{P}_1^{-1}\bar{\mathbf{x}} + (1-\lambda)\mathbf{P}_2^{-1}\bar{\mathbf{x}} \tag{6.55}$$

This can be obtained via LS. We define the residuum

$$\mathbf{r} := C\mathbf{H}^{-1}\bar{\mathbf{x}} - \lambda\mathbf{P}_1^{-1}\bar{\mathbf{x}} - \omega\mathbf{P}_2^{-1}\bar{\mathbf{x}} \tag{6.56}$$

and minimize the squared error $\mathbf{r}^\top\mathbf{r}$ with respect to (λ, ω). We can rewrite $\mathbf{r}$ as

$$\mathbf{r} = C\mathbf{H}^{-1}\bar{\mathbf{x}} - \mathbf{A}\begin{pmatrix}\lambda\\ \omega\end{pmatrix}, \tag{6.57}$$

where the pseudo observation matrix $\mathbf{A}$ is given by

$$\mathbf{A} := \left(\mathbf{P}_1^{-1}\bar{\mathbf{x}},\ \mathbf{P}_2^{-1}\bar{\mathbf{x}}\right) \tag{6.58}$$

The squared error function is minimized when the gradient with respect to (λ, ω) is zero:

$$0 = \nabla_{(\lambda,\omega)}\left(C\mathbf{H}^{-1}\bar{\mathbf{x}} - \mathbf{A}\begin{pmatrix}\lambda\\ \omega\end{pmatrix}\right)^\top\left(C\mathbf{H}^{-1}\bar{\mathbf{x}} - \mathbf{A}\begin{pmatrix}\lambda\\ \omega\end{pmatrix}\right) \tag{6.59}$$

$$= 2\left(C\mathbf{H}^{-1}\bar{\mathbf{x}} - \mathbf{A}\begin{pmatrix}\lambda\\ \omega\end{pmatrix}\right)^\top\mathbf{A} \tag{6.60}$$

$$= \mathbf{r}^\top\mathbf{A} \tag{6.61}$$

$$= \mathbf{r}^\top\left(\mathbf{P}_1^{-1}\bar{\mathbf{x}},\ \mathbf{P}_2^{-1}\bar{\mathbf{x}}\right) \tag{6.62}$$

Therefore, we have the two equations

$$\mathbf{r}^\top\mathbf{P}_1^{-1}\bar{\mathbf{x}} = 0 \tag{6.63}$$

$$\mathbf{r}^\top\mathbf{P}_2^{-1}\bar{\mathbf{x}} = 0 \tag{6.64}$$

Replacing $\mathbf{x}$ by $\mathbf{r}$ in (6.54) yields

$$C\mathbf{r}\mathbf{H}^{-1}\mathbf{r} + 2\beta C\mathbf{r}^\top\mathbf{H}^{-1}\bar{\mathbf{x}} \leq \max\left\{\mathbf{r}\mathbf{P}_1^{-1}\mathbf{r},\ \mathbf{r}\mathbf{P}_2^{-1}\mathbf{r}\right\} \tag{6.65}$$

By letting β go to $\pm\infty$, one obtains

$$\mathbf{r}^\top\mathbf{H}^{-1}\bar{\mathbf{x}} = 0. \tag{6.66}$$

Since this holds for all upper bounds $\mathbf{H}$, $\mathbf{r} = 0$ can be inferred. Therefore, we have found λ and ω with

$$C\mathbf{H}^{-1}\bar{\mathbf{x}} = \lambda\mathbf{P}_1^{-1}\bar{\mathbf{x}} + \omega\mathbf{P}_2^{-1}\bar{\mathbf{x}}. \tag{6.67}$$

When we multiply this equation with $\bar{\mathbf{x}}$ from the left, the result is

$$\bar{\mathbf{x}}^\top C\mathbf{H}^{-1}\bar{\mathbf{x}} = \bar{\mathbf{x}}^\top\lambda\mathbf{P}_1^{-1}\bar{\mathbf{x}} + \bar{\mathbf{x}}^\top\omega\mathbf{P}_2^{-1}\bar{\mathbf{x}} \tag{6.68}$$

$$= (\lambda + \omega)\bar{\mathbf{x}}^\top C\mathbf{H}^{-1}\bar{\mathbf{x}}. \tag{6.69}$$

As a consequence, we directly see that $\omega = 1 - \lambda$, which proves (6.55). However, we still need to show that $0 \leq \lambda \leq 1$. To this end, we choose a vector $\mathbf{v}$ with

$$\mathbf{v}^\top \mathbf{P}_1^{-1} \bar{\mathbf{x}} > 0 \tag{6.70}$$

$$\mathbf{v}^\top \mathbf{P}_2^{-1} \bar{\mathbf{x}} < 0. \tag{6.71}$$

Such a vector exists due to the law of cosines, for instance either the average $\mathbf{v} = {}^1\!/_2(\mathbf{P}_1^{-1}\bar{\mathbf{x}} + \mathbf{P}_2^{-1}\bar{\mathbf{x}})$ or its negative would do. Using (6.54) yields

$$C\mathbf{v}\mathbf{H}^{-1}\mathbf{v} + 2\beta C\mathbf{v}^\top \mathbf{H}^{-1}\bar{\mathbf{x}} \leq \max\left\{\mathbf{v}\mathbf{P}_1^{-1}\mathbf{v} + 2\beta\mathbf{v}^\top\mathbf{P}_1^{-1}\bar{\mathbf{x}},\ \mathbf{v}\mathbf{P}_2^{-1}\mathbf{v} + 2\beta\mathbf{v}^\top\mathbf{P}_2^{-1}\bar{\mathbf{x}}\right\} \tag{6.72}$$

Because of (6.55) it holds that

$$C\mathbf{v}^\top \mathbf{H}^{-1}\bar{\mathbf{x}} = \lambda \mathbf{v}^\top \mathbf{P}_1^{-1}\bar{\mathbf{x}} + (1-\lambda)\mathbf{v}^\top \mathbf{P}_2^{-1}\bar{\mathbf{x}} \tag{6.73}$$

Now letting β in (6.72) go to $\pm\infty$ yields the following two equations:

$$2\beta(\lambda\mathbf{v}^\top\mathbf{P}_1^{-1}\bar{\mathbf{x}} + (1-\lambda)\mathbf{v}^\top\mathbf{P}_2^{-1}\bar{\mathbf{x}}) \leq 2\beta\mathbf{v}^\top\mathbf{P}_1^{-1}\bar{\mathbf{x}} \tag{6.74}$$

$$2\beta(\lambda\mathbf{v}^\top\mathbf{P}_1^{-1}\bar{\mathbf{x}} + (1-\lambda)\mathbf{v}^\top\mathbf{P}_2^{-1}\bar{\mathbf{x}}) \leq 2\beta\mathbf{v}^\top\mathbf{P}_2^{-1}\bar{\mathbf{x}} \tag{6.75}$$

where $\beta > 0$ in (6.74) and $\beta < 0$ in (6.75). From (6.74), we have that

$$(1-\lambda)\mathbf{v}^\top\mathbf{P}_2^{-1}\bar{\mathbf{x}} \leq (1-\lambda)\mathbf{v}^\top\mathbf{P}_1^{-1}\bar{\mathbf{x}} \tag{6.76}$$

which implies $(1-\lambda) > 0$ due to (6.70) and (6.71), therefore $\lambda < 1$. On the other side, we have from (6.75) that

$$\lambda\mathbf{v}^\top\mathbf{P}_1^{-1}\bar{\mathbf{x}} + (1-\lambda)\mathbf{v}^\top\mathbf{P}_2^{-1}\bar{\mathbf{x}} \geq \mathbf{v}^\top\mathbf{P}_2^{-1}\bar{\mathbf{x}}, \tag{6.77}$$

therefore

$$\lambda\mathbf{v}^\top\mathbf{P}_1^{-1}\bar{\mathbf{x}} \geq \lambda\mathbf{v}^\top\mathbf{P}_2^{-1}\bar{\mathbf{x}}. \tag{6.78}$$

Again, we can employ (6.70) and (6.71) to conclude that $\lambda > 0$.

We may now set

$$\hat{\mathbf{P}}_\lambda^{-1} := \lambda\mathbf{P}_1^{-1} + (1-\lambda)\mathbf{P}_2^{-1}. \tag{6.79}$$

In order to prove that $\hat{\mathbf{P}}_\lambda$ is tight, we need to show that

$$\mathbf{x}^\top\hat{\mathbf{P}}_\lambda^{-1}\mathbf{x} \geq \mathbf{x}^\top\mathbf{H}^{-1}\mathbf{x}. \tag{6.80}$$

Indeed, let a vector $\mathbf{x}$ be given. We define

$$\beta := \frac{\mathbf{x}^\top(\mathbf{P}_1^{-1} - \mathbf{P}_2^{-1})\mathbf{x}}{\bar{\mathbf{x}}^\top(\mathbf{P}_1^{-1} - \mathbf{P}_2^{-1})\mathbf{x}} \tag{6.81}$$

$$\mathbf{y} := \mathbf{x} - {}^\beta\!/_2\bar{\mathbf{x}}. \tag{6.82}$$

For $\mathbf{y}$ one obtains that

$$\begin{aligned}
&\mathbf{y}^\top(\mathbf{P}_1^{-1} - \mathbf{P}_2^{-1})\mathbf{y} \\
&\quad = \mathbf{x}^\top(\mathbf{P}_1^{-1} - \mathbf{P}_2^{-1})\mathbf{x} - \beta\bar{\mathbf{x}}^\top(\mathbf{P}_1^{-1} - \mathbf{P}_2^{-1})\mathbf{x} + {}^{\beta^2}\!/_4\,\bar{\mathbf{x}}^\top(\mathbf{P}_1^{-1} - \mathbf{P}_2^{-1})\bar{\mathbf{x}} \\
&\quad = \mathbf{x}^\top(\mathbf{P}_1^{-1} - \mathbf{P}_2^{-1})\mathbf{x} - \mathbf{x}^\top(\mathbf{P}_1^{-1} - \mathbf{P}_2^{-1})\mathbf{x} \\
&\quad = 0,
\end{aligned}$$

which implies that

$$\mathbf{y}^\top \mathbf{P}_1^{-1} \mathbf{y} = \mathbf{y}^\top \mathbf{P}_2^{-1} \mathbf{y} = \mathbf{y}^\top \hat{\mathbf{P}}_\lambda^{-1} \mathbf{y} \tag{6.83}$$

Due to the construction of C, it holds that

$$C\mathbf{y}^\top \mathbf{H}^{-1} \mathbf{y} \leq \mathbf{y}^\top \mathbf{P}_1^{-1} \mathbf{y} = \mathbf{y}^\top \hat{\mathbf{P}}_\lambda^{-1} \mathbf{y}, \tag{6.84}$$

which translates to

$$C\mathbf{x}^\top \mathbf{H}^{-1} \mathbf{x} + C\beta \mathbf{x}^\top \mathbf{H}^{-1} \bar{\mathbf{x}} + C\frac{\beta^2}{4} \bar{\mathbf{x}}^\top \mathbf{H}^{-1} \bar{\mathbf{x}} \leq \mathbf{x}^\top \hat{\mathbf{P}}_\lambda^{-1} \mathbf{x} + \beta \mathbf{x}^\top \hat{\mathbf{P}}_\lambda^{-1} \bar{\mathbf{x}} + \frac{\beta^2}{4} \bar{\mathbf{x}}^\top \hat{\mathbf{P}}_\lambda^{-1} \bar{\mathbf{x}} \tag{6.85}$$

by using the definition of $\mathbf{y}$. Using the fact that $\bar{\mathbf{x}}^\top \mathbf{H}^{-1} \bar{\mathbf{x}} = \bar{\mathbf{x}}^\top \hat{\mathbf{P}}_\lambda^{-1} \bar{\mathbf{x}}$ and that $\mathbf{H}^{-1} \bar{\mathbf{x}} = \hat{\mathbf{P}}_\lambda^{-1} \bar{\mathbf{x}}$ due to (6.67), we obtain

$$\mathbf{x}^\top \mathbf{H}^{-1} \mathbf{x} \leq \mathbf{x}^\top \hat{\mathbf{P}}_\lambda^{-1} \mathbf{x}, \tag{6.86}$$

since $C \geq 1$. This concludes the proof for almost all cases, since we have (implicitly) assumed that $\mathbf{P}_1^{-1} \bar{\mathbf{x}}$ and $\mathbf{P}_2^{-1} \bar{\mathbf{x}}$ are linear independent to find the vector $\mathbf{v}$ for which (6.70) and (6.71) holds. Furthermore, by construction of β in (6.81) we have assumed that $\bar{\mathbf{x}}^\top (\mathbf{P}_1^{-1} - \mathbf{P}_2^{-1})\mathbf{x} \neq 0$. Both cases are given almost sure, therefore, a continuity argument actually assures that the statement of the Tight Bound Theorem holds in general.

□

Derivation of the covariance intersection fusion rule

This tedious but still elegant proof provides the astonishing simple means to construct the tight bound numerically. Since one knows that the tight bound has the smallest volume and that the volume is proportional to the determinant, it is sufficient to minimize the fusion rule

$$\hat{\mathbf{P}} = (\lambda \mathbf{P}_1^{-1} + (1 - \lambda)\mathbf{P}_2^{-1})^{-1} \tag{6.87}$$

over all $\lambda \in [0, 1]$ with respect to the determinant $|\hat{\mathbf{P}}|$. This can be solved numerically by searching for the smallest number in the discretized grid of all possible values for λ. An example is shown in Figure 6.5, where the red and the blue error ellipses correspond to track 1 and track 2. The dashed ellipses show the fused result for non-optimal lambdas 0.25 and 0.75 whereas the magenta ellipse represents the optimal fusion result for lambda 0.5. In addition, in Figure 6.6, one can see that a clear minimum exists for the determinant as a cost function and that it is in the center for the given example. Since the trace of matrix is faster to compute, the optimal lambda often is obtained by minimizing the trace.

Calculating the fused estimate

After optimizing λ for the fused error covariance according to (6.87), the question of how to compute the corresponding estimate $\hat{\mathbf{x}}$ still is to be answered. One may define

$$\hat{\mathbf{x}} = \hat{\mathbf{P}} \left(\omega \, \mathbf{P}_1^{-1} \mathbf{x}_1 + (1 - \omega) \, \mathbf{P}_2^{-1} \mathbf{x}_2 \right) \tag{6.88}$$

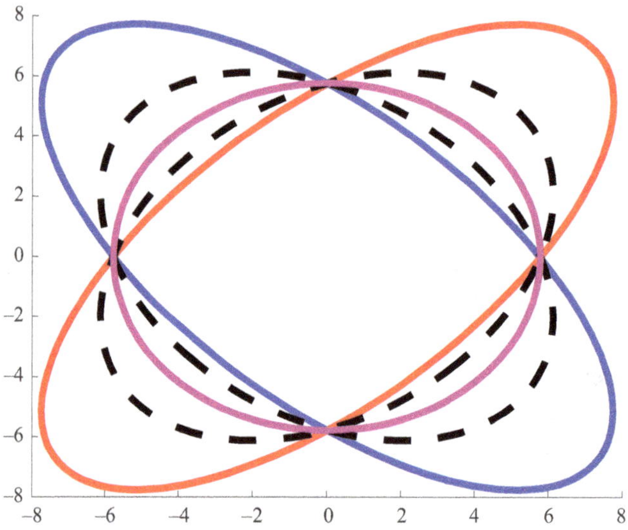

Figure 6.5 Fused results for the fused tight bound for two tracks (blue, red) for lambda = 0.25 and 0.75 (black, dashed) and for lambda = 0.5 (magenta)

for the fused estimate. The exact error covariance matrix $\mathrm{E}\left[(\hat{\mathbf{x}}-\mathbf{x})(\hat{\mathbf{x}}-\mathbf{x})^\top\right]$ cannot be calculated in the closed form without knowledge on the cross-covariance $\mathbf{P}_{1,2}$. However, one can show that $\hat{\mathbf{P}}$ is conservative in the sense that it provides an upper bound. To see this, one has to prove that

$$\begin{aligned}
& \hat{\mathbf{P}} \geq \mathrm{E}\left[(\hat{\mathbf{x}}-\mathbf{x})(\hat{\mathbf{x}}-\mathbf{x})^\top\right] \\
\Longleftrightarrow \quad & \hat{\mathbf{P}} - \mathrm{E}\left[(\hat{\mathbf{P}}\left(\omega\,\mathbf{P}_1^{-1}\mathbf{x}_1 + (1-\omega)\,\mathbf{P}_2^{-1}\mathbf{x}_2\right) - \mathbf{x})(\hat{\mathbf{P}}\left(\omega\,\mathbf{P}_1^{-1}\mathbf{x}_1 + (1-\omega)\,\mathbf{P}_2^{-1}\mathbf{x}_2\right) - \mathbf{x})^\top\right] \geq 0 \\
\Longleftrightarrow \quad & \hat{\mathbf{P}}^{-1} - \omega^2\mathbf{P}_1^{-1} - \omega(1-\omega)\mathbf{P}_1^{-1}\mathbf{P}_{1,2}\mathbf{P}_2^{-1} - (\omega(1-\omega)\mathbf{P}_1^{-1}\mathbf{P}_{1,2}\mathbf{P}_2^{-1})^\top - (1-\omega)^2\mathbf{P}_2^{-1} \geq 0,
\end{aligned} \tag{6.89}$$

where the linearity of the expectation and the fact that

$$\mathrm{E}\left[(\mathbf{P}_i^{-1}(\mathbf{x}_i-\mathbf{x}))(\mathbf{P}_i^{-1}(\mathbf{x}_i-\mathbf{x}))^\top\right] = \mathbf{P}_i^{-1}\mathrm{E}\left[(\mathbf{x}_i-\mathbf{x})(\mathbf{x}_i-\mathbf{x})^\top\right]\mathbf{P}_i^{-1} \tag{6.90}$$

$$= \mathbf{P}_i^{-1} \tag{6.91}$$

was used. Since it holds that

$$\hat{\mathbf{P}}^{-1} = \omega\,\mathbf{P}_1^{-1} + (1-\omega)\,\mathbf{P}_2^{-1} \tag{6.92}$$

and that

$$\omega\,\mathbf{P}_1^{-1} - \omega^2\mathbf{P}_1^{-1} = \omega(1-\omega)\mathbf{P}_1^{-1} \tag{6.93}$$

$$(1-\omega)\,\mathbf{P}_2^{-1} - (1-\omega)^2\mathbf{P}_2^{-1} = (1-\omega-(1-\omega)^2)\mathbf{P}_2^{-1} \tag{6.94}$$

$$= \omega(1-\omega)\mathbf{P}_2^{-1} \tag{6.95}$$

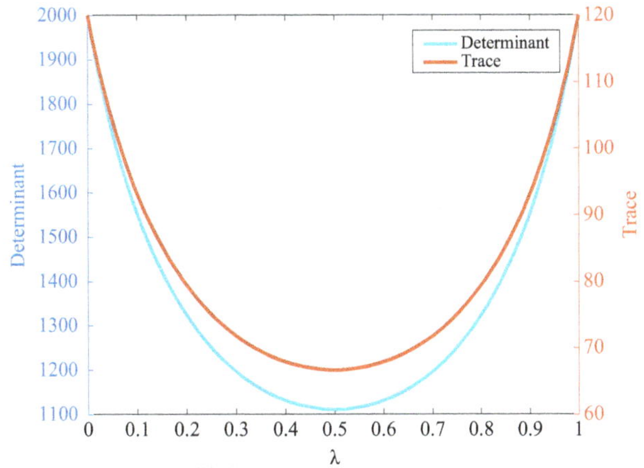

Figure 6.6 Determinant (cyan) and trace (orange) of the resulting fused tight bound covariance matrix for λ ∈ [0, 1]

(6.89) reduces to

$$\omega(1-\omega)\left(\mathbf{P}_1^{-1} - \mathbf{P}_1^{-1}\mathbf{P}_{1,2}\mathbf{P}_2^{-1} - (\mathbf{P}_1^{-1}\mathbf{P}_{1,2}\mathbf{P}_2^{-1})^\top + \mathbf{P}_2^{-1}\right) \geq 0 \tag{6.96}$$

Again with

$$\mathbf{P}_1^{-1}\mathbf{P}_{1,2}\mathbf{P}_2^{-1} = \mathrm{E}\left[(\mathbf{P}_1^{-1}(\mathbf{x}-\mathbf{x}_1))(\mathbf{P}_2^{-1}(\mathbf{x}-\mathbf{x}_2))^\top\right] \tag{6.97}$$

one obtains

$$\mathrm{E}\left[\left(\mathbf{P}_1^{-1}(\mathbf{x}-\mathbf{x}_1) - \mathbf{P}_2^{-1}(\mathbf{x}-\mathbf{x}_2)\right)\left(\mathbf{P}_1^{-1}(\mathbf{x}-\mathbf{x}_1) - \mathbf{P}_2^{-1}(\mathbf{x}-\mathbf{x}_2)\right)^\top\right] \geq 0 \tag{6.98}$$

which holds for every $\mathbf{P}_{1,2}$.

Covariance intersection for an arbitrary number of tracks

The construction of a tight bound in (6.34) provides a conservative fused error covariance for two tracks. We have proven above that this is the least pessimistic bound that exists, which still is conservative for all possible cross-covariances. The extension of the proof to an arbitrary number of tracks with error covariances $\mathbf{P}_1, \ldots, \mathbf{P}_S$ with $S > 2$ is apparently still an open research question [57]. However, a heuristic argumentation easily leads to the construction for arbitrary S. The tight bound $\hat{\mathbf{P}}$ may then be constructed as follows. The optimization to find the tight bound covariance matrix is now with respect to the parameters $\lambda_1, \ldots, \lambda_S > 0$ with

$$\sum_{s=1}^{S} \lambda_s = 1 \tag{6.99}$$

such that the determinant (or trace) of the resulting fused error covariance given by

$$\hat{\mathbf{P}} = \left(\sum_{s=1}^{S} \lambda_s \mathbf{P}_s^{-1} \right)^{-1} \tag{6.100}$$

is minimal. As a consequence $\hat{\mathbf{P}}$ is a tight bound for $\mathbf{P}_1, \ldots, \mathbf{P}_S$.

Analogously, the corresponding estimate $\hat{\mathbf{x}}$ is given by

$$\hat{\mathbf{x}} = \hat{\mathbf{P}} \left(\sum_{s=1}^{S} \lambda_s \mathbf{P}_s^{-1} \mathbf{x}_s \right). \tag{6.101}$$

Discussion

The multi-track covariance intersection fusion algorithm has been proven very useful in countless applications. This is due to various reasons. It is clear that for two tracks, the Covariance Intersection is easy to compute. The search space for $\boldsymbol{\lambda} = (\lambda_1, \ldots, \lambda_S) \in [0, 1]^S$ increases exponentially in the number of tracks S. However, the optimization process can be accelerated significantly by means of a well suited initialization. This might be either given by the equal distribution of weights with $\lambda_s = \frac{1}{S}$ or by the result of the previous fusion, if the conditions did not change drastically.

Another reason for the popularity of the covariance intersection algorithm in practical applications actually comes from its property of being overly conservative. As a matter of fact, it is inherently difficult to model all sources of noise and uncertainty in a multi-sensor fusion setup consistent in an unbiased way. Thus, unknown uncertainty is sometimes simply captured by conservative models or algorithms. That is, due to unforeseen additional sources of noise and biases, a pessimistic error covariance may become consistent.

6.3 Inverse covariance intersection

In the above section, we have seen that the intersection of two ellipses is enclosed in a unique tight bound which can be found by a numerical optimization problem (finding the best weighting factors λ_s). By construction, the *Covariance Intersection* (CI) algorithm is conservative, that is, the optimal fusion result would have a smaller covariance matrix.

In this section, the *Inverse Covariance Intersection* (ICI) will be presented, which applies the same geometrical result from Kahan (1968) to obtain a conservative bound for the *common information*. The latter is something obscure, since it fully encodes the correlation between track estimates but eventually, one does not need to compute it explicitly as the conservative approximation is used instead.

For two track estimates $\hat{\mathbf{x}}_1$, $\hat{\mathbf{x}}_2$ with covariances $\mathbf{P}_1$ and $\mathbf{P}_2$, respectively, the common information is defined implicitly via the following convex decomposition of the tracks:

Definition:
A common information estimate γ with a covariance matrix Γ obeys a joint decomposition of tracks:

$$\hat{\mathbf{x}}_1 = \mathbf{P}_1 \left(\mathbf{P}_1^{I\,-1}\hat{\mathbf{x}}_1^I + \Gamma^{-1}\gamma\right) \tag{6.102}$$

$$\hat{\mathbf{x}}_2 = \mathbf{P}_2 \left(\mathbf{P}_2^{I\,-1}\hat{\mathbf{x}}_2^I + \Gamma^{-1}\gamma\right) \tag{6.103}$$

$$\mathbf{P}_1^{-1} = \mathbf{P}_1^{I\,-1} + \Gamma^{-1} \tag{6.104}$$

$$\mathbf{P}_2^{-1} = \mathbf{P}_2^{I\,-1} + \Gamma^{-1} \tag{6.105}$$

such that the partial estimates $\hat{\mathbf{x}}_1^I$ and $\hat{\mathbf{x}}_2^I$ are conditionally independent:

$$\mathrm{E}\left[(\mathbf{x} - \hat{\mathbf{x}}_1^I)(\mathbf{x} - \hat{\mathbf{x}}_2^I)^\top\right] = \mathbf{O} \tag{6.106}$$

$$\mathrm{E}\left[(\mathbf{x} - \hat{\mathbf{x}}_i^I)(\mathbf{x} - \gamma)^\top\right] = \mathbf{O} \qquad \text{for } i = 1, 2. \tag{6.107}$$

An illustration of the shared information concept is shown in Figure 6.7.

The condition of the definition implies that the cross-covariance of the estimates can be expressed in terms of the common information:

$$\begin{aligned}
\mathbf{P}_{1,2} &= \mathrm{E}\left[(\hat{\mathbf{x}}_1 - \mathbf{x})(\hat{\mathbf{x}}_2 - \mathbf{x})^\top\right] && (6.108)\\
&= \mathrm{E}\left[(\mathbf{P}_1 \left(\mathbf{P}_1^{I\,-1}\hat{\mathbf{x}}_1^I + \Gamma^{-1}\gamma\right) - \mathbf{x})(\mathbf{P}_2 \left(\mathbf{P}_2^{I\,-1}\hat{\mathbf{x}}_2^I + \Gamma^{-1}\gamma\right) - \mathbf{x})^\top\right] && (6.109)\\
&= \mathbf{P}_1\Gamma^{-1}\mathrm{E}\left[(\gamma - \mathbf{x})(\gamma - \mathbf{x})^\top\right]\Gamma^{-1}\mathbf{P}_2 && (6.110)\\
&= \mathbf{P}_1\Gamma^{-1}\mathbf{P}_2 && (6.111)
\end{aligned}$$

This also proves the existence for the common information and the isolated estimate covariance matrices since one can set

$$\Gamma := \left(\mathbf{P}_1^{-1}\mathbf{P}_{1,2}\mathbf{P}_2^{-1}\right)^{-1}, \tag{6.112}$$

$$\mathbf{P}_i^I := \left(\mathbf{P}_i^{-1} - \Gamma^{-1}\right)^{-1} \qquad \text{for } i = 1, 2. \tag{6.113}$$

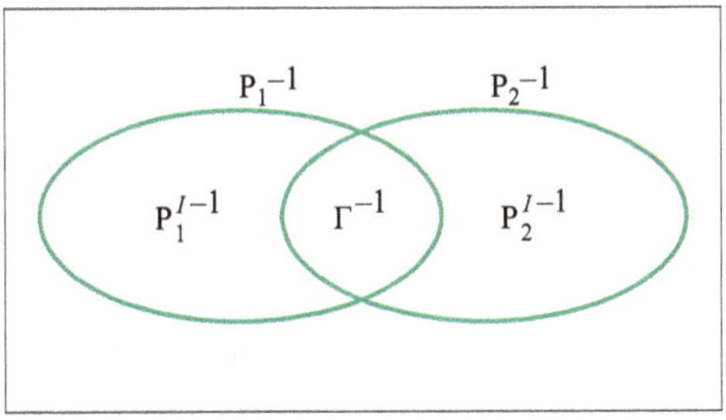

Figure 6.7 Illustration of the shared information. The shapes here represent information and not error covariances.

Optimal fusion with common information

The optimal fusion accounts for the common information exactly once, that is, information incest is avoided. The derivation is straight forward using the information filter formulas. The fused track in terms of an estimate $\hat{\mathbf{x}}$ and a covariance matrix $\hat{\mathbf{P}}$ are given by

$$\hat{\mathbf{x}} = \hat{\mathbf{P}}\left(\mathbf{P}_1^{I\,-1}\hat{\mathbf{x}}_1^I + \mathbf{P}_2^{I\,-1}\hat{\mathbf{x}}_2^I + \Gamma^{-1}\gamma\right) \tag{6.114}$$

$$= \hat{\mathbf{P}}\left(\mathbf{P}_1^{-1}\hat{\mathbf{x}}_1 + \mathbf{P}_2^{-1}\hat{\mathbf{x}}_2 - \Gamma^{-1}\gamma\right) \tag{6.115}$$

$$\hat{\mathbf{P}} = \left(\mathbf{P}_1^{I\,-1} + \mathbf{P}_2^{I\,-1} + \Gamma^{-1}\right)^{-1} \tag{6.116}$$

$$= \left(\mathbf{P}_1^{-1} + \mathbf{P}_2^{-1} - \Gamma^{-1}\right)^{-1}. \tag{6.117}$$

A generalization to an arbitrary number of sensors is obtained analogously. Since the common information is unknown in most scenarios, the next paragraph is on the construction of an approximation based on known parameters only.

Construction of the inverse covariance intersection rule

The construction follows the derivations from Noack *et al.* in [58]. At first, one observes that by construction the following inequality holds for all $\mathbf{x} \in \mathbb{R}^n$

$$\mathbf{x}^\top\mathbf{P}_i^{-1}\mathbf{x} \leq \mathbf{x}^\top\Gamma^{-1}\mathbf{x}, \qquad i = 1, 2, \tag{6.118}$$

and therefore the both error ellipses $\mathcal{E}_{\mathbf{P}_1}$ and $\mathcal{E}_{\mathbf{P}_2}$, respectively, are contained in $\mathcal{E}_\Gamma$. Vice versa it holds that

$$\mathbf{x}^\top\mathbf{P}_i\mathbf{x} \geq \mathbf{x}^\top\Gamma\mathbf{x}, \qquad i = 1, 2. \tag{6.119}$$

This statement is a direct consequence of the *Löwner–Heinz Theorem* [59] but it can also be proven by means of linear algebra. In order to prove this statement, it is first observed that w.l.o.g. it can be assumed that $\Gamma = \Gamma^{-1} = \mathbf{I}$ since otherwise there exists $\mathbf{L}$ with

$$\Gamma = \mathbf{L}^\top\mathbf{L}. \tag{6.120}$$

If now $(\mathbf{P}_i - \mathbf{I})$ is positive definite then for $\mathbf{y} = \mathbf{L}\mathbf{x}$

$$\mathbf{y}^\top\mathbf{P}_i\mathbf{y} \geq \mathbf{y}^\top\mathbf{y} \tag{6.121}$$

$$\iff \qquad \mathbf{x}^\top\mathbf{L}^\top\mathbf{P}_i\mathbf{L}\mathbf{x} \geq \mathbf{x}^\top\Gamma\mathbf{x} \tag{6.122}$$

Proving the statement for $\tilde{\mathbf{P}}_i = \mathbf{L}^\top\mathbf{P}_i\mathbf{L}$ yields the desired result.

Now for $\Gamma = \mathbf{I}$ one has that $\mathbf{P}_i$ may be diagonalized by conjugation with

$$\mathbf{V}^\top\mathbf{P}_i\mathbf{V} = \mathbf{D}, \tag{6.123}$$

where $\mathbf{V} = (\mathbf{v}_1, \ldots, \mathbf{v}_n)$ consists of the eigenvectors of $\mathbf{P}_i$ and $\mathbf{D}$ is diagonal and populated with positive eigenvalues only. Because of (6.118) it holds that $\mathbf{M} := (\mathbf{I} - \mathbf{P}_i^{-1})$ is s.p.d. and clearly $\mathbf{V}$ also diagonalizes $\mathbf{M}$ with

$$\mathbf{V}^\top\mathbf{M}\mathbf{V} = \tilde{\mathbf{D}}, \tag{6.124}$$

$$\tilde{\mathbf{D}} = \operatorname{diag}[(d_1, \ldots, d_n)], \tag{6.125}$$

where $d_i > 0$ for all i. Now since one has that $\mathbf{M} = \mathbf{P}_i^{-1}(\mathbf{P}_i - \mathbf{I})$ it follows that

$$(\mathbf{P}_i - \mathbf{I}) = \mathbf{P}_i\mathbf{V}\tilde{\mathbf{D}}\mathbf{V}^\top, \tag{6.126}$$

which is clearly s.p.d. and therefore proves (6.119).

We conclude that the ellipse

$$\mathcal{E}_{\Gamma^{-1}} = \left\{\mathbf{x} \mid \mathbf{x}^\top \Gamma \mathbf{x} < 1\right\} \tag{6.127}$$

is contained in $\mathcal{E}_{\mathbf{P}_1^{-1}}$ and $\mathcal{E}_{\mathbf{P}_2^{-1}}$. It follows that

$$\mathcal{E}_{\Gamma^{-1}} \subseteq \mathcal{E}_{\mathbf{P}_1^{-1}} \cap \mathcal{E}_{\mathbf{P}_2^{-1}} \tag{6.128}$$

According to the tight bound theorem (6.34), one can construct an upper bound for the ellipse $\mathcal{E}_{\Gamma^{-1}}$ by minimizing the term

$$\det\left(\omega\mathbf{P}_1 + (1-\omega)\mathbf{P}_2\right) \tag{6.129}$$

for $\omega \in [0, 1]$ which yields that the ellipse $\mathcal{E}_{\tilde{\Gamma}^{-1}}$ contains $\mathcal{E}_{\Gamma^{-1}}$ with

$$\tilde{\Gamma}^{-1} = (\omega\mathbf{P}_1 + (1-\omega)\mathbf{P}_2)^{-1}. \tag{6.130}$$

One can now use this upper bound $\tilde{\Gamma}^{-1}$ on the joint information matrix in order to approximate the unknown true matrix Γ^{-1} and replace it in the optimal fusion rule from (6.117).

One obtains the *inverse covariance intersection fusion rule*:

$$\hat{\mathbf{x}} = \mathbf{K}^1\hat{\mathbf{x}}_1 + \mathbf{K}^2\hat{\mathbf{x}}_2 \tag{6.131}$$

$$\hat{\mathbf{P}} = \left(\mathbf{P}_1^{-1} + \mathbf{P}_2^{-1} - (\omega\mathbf{P}_1 + (1-\omega)\mathbf{P}_2)^{-1}\right)^{-1} \tag{6.132}$$

$$\mathbf{K}^1 = \hat{\mathbf{P}}\left(\mathbf{P}_1^{-1} - \omega(\omega\mathbf{P}_1 + (1-\omega)\mathbf{P}_2)^{-1}\right) \tag{6.133}$$

$$\mathbf{K}^2 = \hat{\mathbf{P}}\left(\mathbf{P}_2^{-1} - (1-\omega)(\omega\mathbf{P}_1 + (1-\omega)\mathbf{P}_2)^{-1}\right) \tag{6.134}$$

Discussion
The ICI approach yields a good compromise between the conservative CI and too optimistic fusion rules such as Naïve Fusion. One should pay attention to the optimization step where *not* the posterior covariance $\hat{\mathbf{P}}$ is minimized but the approximated joint information matrix $\tilde{\Gamma}^{-1}$.

6.4 Tracklet fusion

All of the algorithms above can be considered as "memory-less," since there is no central track that provides a prior to which remote data is fused, that is, previous fusion results are not considered for the actual result at the current time. This is different in the *Tracklet Fusion* approach, where the fusion center maintains its own track and applies a prediction step to compute the prior parameters $\mathbf{x}_{k|k-1}$ and $\mathbf{P}_{k|k-1}$, respectively, before the new tracks received at time t_k are being processed.

The fusion step is based on the information filter, where the information parameters from all remote tracks are used:

$$\mathbf{x}_{k|k} = \mathbf{P}_{k|k}\left(\mathbf{P}_{k|k-1}^{-1}\mathbf{x}_{k|k-1} + \sum_{s=1}^{S}\mathbf{i}_k^s\right) \tag{6.135}$$

$$\mathbf{P}_{k|k} = \left(\mathbf{P}_{k|k-1}^{-1} + \sum_{s=1}^{S}\mathbf{I}_k^s\right)^{-1} \tag{6.136}$$

Obviously, with Kalman filter conditions the *measurement information parameter* $\mathbf{i}_k^s$ and *measurement information matrix* $\mathbf{I}_k^s$ are given explicitly by

$$\mathbf{I}_k^s = \mathbf{H}_k^{s\,\mathrm{T}}(\mathbf{R}_k^s)^{-1}\mathbf{H}_k^s \tag{6.137}$$

$$\mathbf{i}_k^s = \mathbf{H}_k^{s\,\mathrm{T}}(\mathbf{R}_k^s)^{-1}\mathbf{z}_k^s. \tag{6.138}$$

The tracklet fusion method, however, is very general and can be applied for all filters where the first and second moments of the prior and posterior are available. This is due to the fact that one can resolve the information update equations for the measurement contribution:

$$\mathbf{I}_k^s = (\mathbf{P}_{k|k}^s)^{-1} - (\mathbf{P}_{k|k-1}^s)^{-1} \tag{6.139}$$

$$\mathbf{i}_k^s = (\mathbf{P}_{k|k}^s)^{-1}\mathbf{x}_{k|k}^s - (\mathbf{P}_{k|k-1}^s)^{-1}\mathbf{x}_{k|k-1}^s \tag{6.140}$$

Here, it becomes obvious that the information parameter and information matrix can also be calculated based on the results of sophisticated algorithms to mitigate non-linearities, non-detections, false-alarms, etc. In particular, if the time evolution model of the stochastic model is known, the fusion center is able to apply a prediction step on the previous posterior estimate $\mathbf{x}_{k-1|k-1}^s$ and covariance $\mathbf{P}_{k-1|k-1}^s$ for each sensor s in order to obtain the local priors without additional transmissions. One obtains the following algorithm for the tracklet fusion.

At each time step t_k, the fusion center

1. calculates its own prior parameters $\mathbf{x}_{k|k-1}$ and $\mathbf{P}_{k|k-1}$ of the central track.
2. applies the prediction step for each local posterior at time t_{k-1} to obtain the local priors $\mathbf{x}_{k|k-1}^s$ and $\mathbf{P}_{k|k-1}^s$ for all $s = 1, \ldots, S$.
3. obtains the information parameters by the difference equations (6.139) and (6.140).
4. computes the central fused track parameters by using the information filter formulae (6.135) and (6.136), respectively.

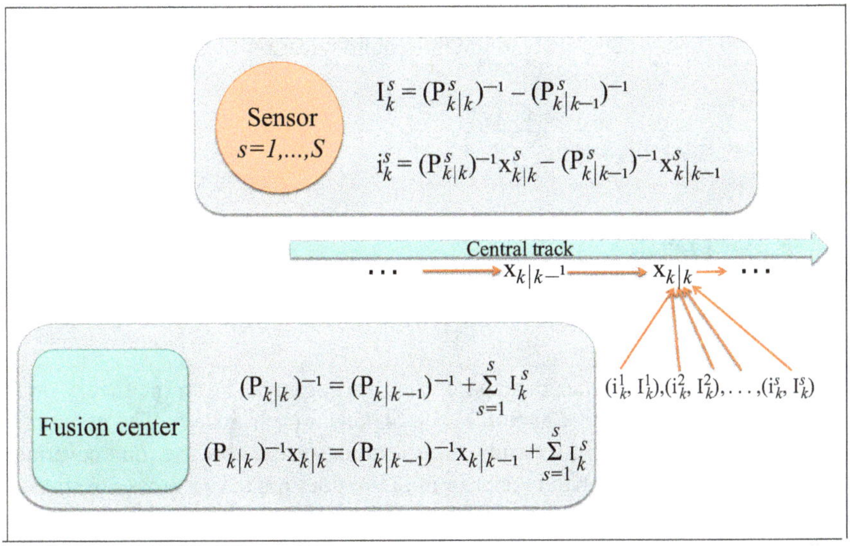

Figure 6.8 *Process flow of the tracklet fusion. From the track submissions of the local sensors, the information parameters are reconstructed and then fused into the central track of the fusion center using the information filter.*

This scheme is visualized in Figure 6.8.

Obviously, the described algorithm from above requires all sensors to send their local tracks to the fusion center after each update step – an assumption which is frequently called *"full communication"* in the literature [34]. Practical experiments, however, have shown that tracklet fusion also is possible without full communication.

If the sensors did not send their previous posterior parameters for some time, let us denote the time instant of the last transmission with t_l. By means of a multi-step prediction – or equivalently by multiple one step predictions – the local priors are obtained analogously to the case above:

$$\mathbf{x}^s_{k|l} = \mathbf{F}_{k|l}\mathbf{x}^s_{l|l} \tag{6.141}$$

$$\mathbf{P}^s_{k|l} = \mathbf{F}_{k|l}\mathbf{P}^s_{l|l}\mathbf{F}^{\top}_{k|l} + \mathbf{Q}_{k|l}. \tag{6.142}$$

When the current tracks at time t_k have been received, the fusion center can well compute the corresponding information parameters:

$$\mathbf{I}^s_{k|l} = (\mathbf{P}^s_{k|k})^{-1} - (\mathbf{P}^s_{k|l})^{-1} \tag{6.143}$$

$$\mathbf{i}^s_{k|l} = (\mathbf{P}^s_{k|k})^{-1}\mathbf{x}^s_{k|k} - (\mathbf{P}^s_{k|l})^{-1}\mathbf{x}^s_{k|l}. \tag{6.144}$$

The multi-lag update step of the central track fuses those "tracklets" after the prediction to the current time. Analogously to the full-communication case one has

$$\mathbf{x}_{k|k} = \mathbf{P}_{k|k}\left(\mathbf{P}_{k|l}^{-1}\mathbf{x}_{k|l} + \sum_{s=1}^{S}\mathbf{i}_{k|l}^{s}\right) \tag{6.145}$$

$$\mathbf{P}_{k|k} = \left(\mathbf{P}_{k|l}^{-1} + \sum_{s=1}^{S}\mathbf{I}_{k|l}^{s}\right)^{-1} \tag{6.146}$$

Discussion

The question arises, whether this approach is still optimal. From the theory on the Kalman filter equations in Chapter 3, one obtains a clear answer. The assumption for the derivation of the information filter formulas states that the joint covariance matrix of the measurement and the prior in (3.90) does not have a cross-covariance – that is, that they are mutually independent conditioned on the true state since they are Gaussian. This assumption is broken, if the pseudo measurement refers to a multi-lag, where prediction steps were yet integrated. This is due to the fact that the contained process noise in the pseudo measurement is auto-correlated with the process noise of the prior. This, however, is more relevant when it comes to theoretical considerations. Both the tracklet fusion and the multi-lag tracklet fusion are known to be robust methods for multi-sensor fusion in practical applications. Only the requirement of *regular* transmissions (with fixed frequency) implies a lack of flexibility in dynamic scenarios.

6.5 Safe fusion

The following method, which is called *safe fusion*, takes a completely different approach as cross-correlations of state estimates which are to be fused are circumvented by a joint diagonalization. Since the covariance $\mathbf{P}$ of an estimate $\hat{\mathbf{x}}$ is a symmetric, positive definite matrix, it can be diagonalized by real matrices $\mathbf{V}$ and $\mathbf{D}$ with

$$\mathbf{P} = \mathbf{V}\mathbf{D}\mathbf{V}^{\top}, \tag{6.147}$$

where $\mathbf{D} = \text{diag}[d_1, \ldots, d_n]$ and $\mathbf{V}$ is an orthogonal matrix with $\mathbf{V}^{-1} = \mathbf{V}^{\top}$. The diagonalization in (6.147) can be computed for instance by a *singular value decomposition* (SVD).

Stochastically, this implies that one obtains a transformed random variable

$$\bar{\mathbf{x}} = \mathbf{T}\mathbf{x} \tag{6.148}$$

where the linear transformation is given by $\mathbf{T} = \mathbf{V}^\top$ which has the following covariance matrix:

$$\begin{aligned}
\operatorname{cov}[\bar{\mathbf{x}}] &= \mathrm{E}\left[(\bar{\mathbf{x}} - \mathrm{E}[\mathbf{x}])(\bar{\mathbf{x}} - \mathrm{E}[\mathbf{x}])^\top\right] && (6.149)\\
&= \mathrm{E}\left[(\mathbf{T}\mathbf{x} - \mathbf{T}\hat{\mathbf{x}})(\mathbf{T}\mathbf{x} - \mathbf{T}\hat{\mathbf{x}})^\top\right] && (6.150)\\
&= \mathbf{T}\mathrm{E}\left[(\mathbf{x} - \hat{\mathbf{x}})(\mathbf{x} - \hat{\mathbf{x}})^\top\right]\mathbf{T}^\top && (6.151)\\
&= \mathbf{T}\mathbf{P}\mathbf{T}^\top && (6.152)\\
&= \mathbf{V}^\top\mathbf{P}\mathbf{V} = \mathbf{D}. && (6.153)
\end{aligned}$$

One can see that the covariance of $\bar{\mathbf{x}} = \mathbf{T}\mathbf{x}$ is diagonal by construction.

As we will see later on, one can jointly diagonalize the error covariance matrices of two given tracks $(\hat{\mathbf{x}}_1, \mathbf{P}_1)$ and $(\hat{\mathbf{x}}_2, \mathbf{P}_2)$. In the transformed state space of $\bar{\mathbf{x}} = \mathbf{T}\mathbf{x}$, the individual entries of both estimates

$$\begin{aligned}
\bar{\hat{\mathbf{x}}}_1 &= \mathbf{T}\hat{\mathbf{x}}_1 && (6.154)\\
&= (\bar{\hat{x}}_1^{(1)}, \ldots, \bar{\hat{x}}_1^{(n)})^\top && (6.155)\\
\bar{\hat{\mathbf{x}}}_2 &= \mathbf{T}\hat{\mathbf{x}}_2 && (6.156)\\
&= (\bar{\hat{x}}_2^{(1)}, \ldots, \bar{\hat{x}}_2^{(n)})^\top && (6.157)
\end{aligned}$$

are mutually independent. That is, the entries $\bar{\hat{x}}_l^{(i)}$ are independent of $\bar{\hat{x}}_l^{(j)}$ for $i \neq j$ and fixed $l = 1, 2$. Note that this does not imply that $\bar{\hat{\mathbf{x}}}_1$ and $\bar{\hat{\mathbf{x}}}_2$ are independent and can be fused optimally via convex combination. This is obvious since the transformed cross-covariance

$$\bar{\mathbf{P}}^{(1,2)} = \mathbf{T}\mathbf{P}^{(1,2)}\mathbf{T}^\top \tag{6.158}$$

does not vanish, if $\mathbf{P}^{(1,2)} \neq 0$. The idea of the safe fusion is now to choose the individual parameters from the transformed estimate, which has the higher information – or equivalently, which has the lower variance. Therefore, the fused estimate in the transformed state space is given by

$$\bar{\hat{\mathbf{x}}} = (\bar{\hat{x}}^{(1)}, \ldots, \bar{\hat{x}}^{(n)})^\top \tag{6.159}$$

$$\bar{\hat{x}}^{(i)} = \begin{cases} \bar{\hat{x}}_1^{(i)}, & \text{if } \operatorname{var}\left[\bar{\hat{x}}_1^{(i)}\right] \leq \operatorname{var}\left[\bar{\hat{x}}_2^{(i)}\right] \\ \bar{\hat{x}}_2^{(i)} & \text{else.} \end{cases} \tag{6.160}$$

But now let us see how to obtain the appropriate transformation $\mathbf{T}$, which diagonalizes both error covariance matrices $\mathbf{P}_1$ and $\mathbf{P}_2$. First of all, one starts with the SVD of $\mathbf{P}_1$, which gives us

$$\mathbf{P}_1 = \mathbf{V}_1\mathbf{D}_1\mathbf{V}_1^\top. \tag{6.161}$$

Additional to the above derivations, we note that

$$\mathbf{D}_1^{-\frac{1}{2}}\mathbf{V}_1^\top\mathbf{P}_1\mathbf{V}_1\mathbf{D}_1^{-\frac{1}{2}} = \mathbf{I}, \tag{6.162}$$

where

$$\mathbf{D}_1^{-\frac{1}{2}} = \text{diag}\left\{\frac{1}{\sqrt{d_1^{(1)}}}, \ldots, \frac{1}{\sqrt{d_1^{(n)}}}\right\}. \tag{6.163}$$

The geometrical interpretation can be seen in Figure 6.9, where the effect of this combined transformation can be seen on the corresponding error ellipses.

In the next step, one uses the fact that the unit circle does not change under additional rotations. This additional rotation is obtained by the diagonalization of the transformed covariance matrix of the second track:

$$\mathbf{D}_1^{-\frac{1}{2}}\mathbf{V}_1^\top\mathbf{P}_2\mathbf{V}_1\mathbf{D}_1^{-\frac{1}{2}} = \mathbf{V}_2\mathbf{D}_2\mathbf{V}_2^\top \tag{6.164}$$

This yields the following combined transformation matrix $\mathbf{T}$:

$$\mathbf{T} = \mathbf{V}_2^\top\mathbf{D}_1^{-\frac{1}{2}}\mathbf{V}_1^\top \tag{6.165}$$

When one defines the transformed means by

$$\bar{\hat{\mathbf{x}}}_1 = \mathbf{T}\hat{\mathbf{x}}_1 \tag{6.166}$$

$$\bar{\hat{\mathbf{x}}}_2 = \mathbf{T}\hat{\mathbf{x}}_2, \tag{6.167}$$

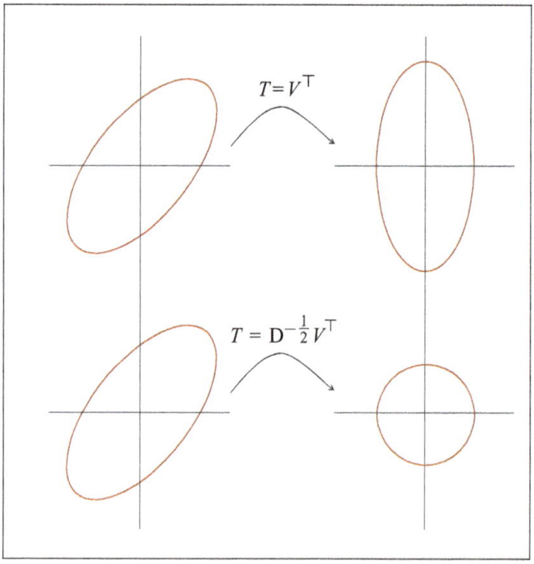

Figure 6.9 Error ellipse transformation for the rotation only based on the orthogonal matrix $\mathbf{V}$ *(above) and for the rotation followed by a dilatation matrix* $\mathbf{D}^{-\frac{1}{2}}$*. The matrices* $\mathbf{V}$ *and* $\mathbf{D}$ *diagonalize the original covariance matrix and are obtained by a singular value decomposition for instance.*

it follows that $\text{cov}\left[\hat{\bar{\mathbf{x}}}_1\right] = \mathbf{I}$ and $\text{cov}\left[\hat{\bar{\mathbf{x}}}_2\right] = \mathbf{D}_2$. Therefore we have proven that $\mathbf{T}$ jointly diagonalizes $\mathbf{P}_1$ and $\mathbf{P}_2$ and the fusion rule in (6.160) may be applied. Furthermore, one can fill in the values which just were obtained for the variances. As a last step, one only needs to apply the inverse transformation.

The fused estimate in the transformed state space is given by

$$\hat{\bar{\mathbf{x}}} = (\hat{\bar{x}}^{(1)}, \ldots, \hat{\bar{x}}^{(n)})^\top \tag{6.168}$$

$$\hat{\bar{x}}^{(i)} = \begin{cases} \hat{\bar{x}}_1^{(i)}, & \text{if } d_2^i > 1 \\ \hat{\bar{x}}_2^{(i)} & \text{else.} \end{cases} \tag{6.169}$$

The corresponding error covariance matrix also is diagonal and is given by

$$\bar{\mathbf{P}} = \text{diag}[\bar{p}^1, \ldots, \bar{p}^n] \tag{6.170}$$

$$\bar{p}^i = \begin{cases} 1, & \text{if } d_2^i > 1 \\ d_2^i & \text{else.} \end{cases} \tag{6.171}$$

An application of the inverse transformation $\mathbf{T}^{-1}$ yields the fused parameters in the original state space:

$$\hat{\mathbf{x}} = \mathbf{T}^{-1}\hat{\bar{\mathbf{x}}} \tag{6.172}$$

$$\mathbf{P} = \mathbf{T}^{-1}\bar{\mathbf{P}}(\mathbf{T}^{-1})^\top \tag{6.173}$$

Discussion

The safe fusion approach definitely has some charming aspects. It is easy to implement, the fusion can be computed fast so that it scales well in the number of sensors to be fused. The so-called *ellipsoidal intersection* [60] uses the same technique in order to compute the joint information, which is then subtracted from the naïve fusion. The computed parameters are equivalent to the safe fusion [61], but in [58] it is shown that the fused covariance matrix can slightly be too optimistic. Though, this is something which is to be avoided in most applications, the authors in [61] clearly show that the approach behaves robust and consistent after some initialization phase in a practical application.

6.6 Conclusion

The algorithms discussed in this chapter bear a lot of potential in a wide range of distributed state estimation problems. This is due to the high level of generalization of these methods, that is, no additional assumptions are made (except for the regular transmissions for the tracklet fusion in Section 6.4). On the other hand, there is no *exact* algorithm for T2TF without any assumption. One can say that the most popular approach to T2TF in practical applications still is the covariance intersection, mainly due to its simple implementation and its conservativeness. The inverse covariance intersection is a more sophisticated approach in order to enhance the degree of approximation of the exact fused covariance matrix, whereas the safe fusion is not proven to be consistent or conservative in all fusion cases. The prerequisite of regular (or "full") communication of locally fused estimates to the fusion center for the tracklet fusion clearly limits its application in spatially dynamic scenarios. Furthermore, it also should be noted that the information for the fusion with the centralized track cannot be extracted by means of a single transmission except for the case of a well-suited prior for each sensor. Therefore, the tracklet fusion is well-suited for applications with reliable communication links and a static topology.

Chapter 7
Distributed Kalman filter

In the T2TF schemes of the previous chapter, the fusion center was facing the challenge of calculating a combined estimate and estimation error covariance matrix from a set of tracks, which result from local Kalman filter processors. In contrast to that, the *Distributed Kalman Filter* (DKF) considers the data from all sensors available at (synchronized) time instants as an input to a single multi-sensor Kalman filter, for which the computation of the posterior density is being made distributed among the processing nodes. This is achieved by modifying the local tracks $(\mathbf{x}^s_{k|k}, \mathbf{P}^s_{k|k})$ such that the mutual correlations are eliminated. As a consequence, the fusion center may apply the convex combination given in (6.10) and (6.11) in order to fuse the received tracks, which is of very low computational effort. It will be shown below that the exact fusion result can be reconstructed, if Kalman filter conditions hold. However, for non-linear scenarios where the measurement function becomes data dependent, additional communication is required.

This scheme is fully based on the *product representation*, which implies stochastically independent statistics for the local densities. In the DKF, it is ensured that the product representation holds at all phases of the prediction-filtering cycle:

$$p(\mathbf{x}_k|\mathcal{Z}^k) \propto \prod_{s=1}^{S} \mathcal{N}(\mathbf{x}_k;\, \mathbf{x}^s_{k|k},\, \mathbf{P}^s_{k|k}) \tag{7.1}$$

$$p(\mathbf{x}_{k+1}|\mathcal{Z}^k) \propto \prod_{s=1}^{S} \mathcal{N}(\mathbf{x}_{k+1};\, \mathbf{x}^s_{k+1|k},\, \mathbf{P}^s_{k+1|k}). \tag{7.2}$$

One should note that if the product representation is achieved, it follows that

$$\mathcal{N}(\mathbf{x}_k;\, \mathbf{x}^s_{k|k},\, \mathbf{P}^s_{k|k}) \neq p(\mathbf{x}_k|\mathcal{Z}^k_s) \tag{7.3}$$

since in general it holds that the tracks are correlated and therefore their joint density does not factorize:

$$p(\mathbf{x}^1_{k|k}, \ldots, \mathbf{x}^S_{k|k}|\mathbf{x}_k) \neq \prod_{s=1}^{S} p(\mathbf{x}^s_{k|k}|\mathbf{x}_k). \tag{7.4}$$

It should be noted that the local tracks are sufficient statistics [62] for the sensor data produced at sensor s, therefore the global filtered density $p(\mathbf{x}_k|\mathcal{Z}^k)$ is proportional to the joint density above conditioned on the full set of data. In other words, the local optimal parameters have been replaced by some (pseudo estimate) parameters, which do not reflect the local state estimate in the Kalman filter sense. However, it is possible to modify the local parameters $\mathbf{x}^s_{k|k}$ and $\mathbf{P}^s_{k|k}$ such that the product representation holds throughout during the process of state estimation, that is, that the *global* optimal estimate can be reconstructed by fusing those local parameters.

In the following section, the prediction-filtering cycle will be derived. The derivation with its details is important to highlight the conceptual computation scheme as well as to understand the assumptions that come with the DKF solution for optimal T2TF.

7.1 Local error covariance globalization

The prediction of the distributed Kalman filter actually inserts an intermediate step, the so-called *globalization* which has to be computed first. It is assumed that the previous posterior is given in product representation as in (7.1) for time step $k-1$. Given Gaussian assumptions for the previous posterior, one can achieve that the globally fused density is given by a product representation:

$$p(\mathbf{x}_{k-1}|\mathcal{Z}^{k-1}) = \mathcal{N}(\mathbf{x}_{k-1};\, \mathbf{x}_{k-1|k-1}, \mathbf{P}_{k-1|k-1}) \tag{7.5}$$

$$\propto \prod_{s=1}^{S} \mathcal{N}(\mathbf{x}_{k-1};\, \mathbf{x}^s_{k-1|k-1}, \mathbf{P}^s_{k-1|k-1}) \tag{7.6}$$

The key of the globalization step is to compute pseudo estimates $\tilde{\mathbf{x}}^s_{k-1|k-1}$ such that

$$\mathbf{P}^1_{k-1|k-1} = \mathbf{P}^2_{k-1|k-1} = \cdots = \mathbf{P}^S_{k-1|k-1} =: \tilde{\mathbf{P}}_{k-1|k-1}. \tag{7.7}$$

Thus, all local covariance matrices are identical, such that the sensor index may be omitted.

This is achieved based on the fact that that the fused error covariance can be written as

$$\mathbf{P}_{k-1|k-1} = \left(\sum_{s=1}^{S} (S\mathbf{P}_{k-1|k-1})^{-1} \right)^{-1}. \tag{7.8}$$

Therefore, each sensor $s = 1, \ldots, S$ requires the fused covariance matrix $\mathbf{P}_{k-1|k-1}$ and then replaces its local covariance $\mathbf{P}^s_{k-1|k-1}$ by

$$\tilde{\mathbf{P}}_{k-1|k-1} := S\mathbf{P}_{k-1|k-1} \tag{7.9}$$

in the globalization step. However, in order to obtain an exact solution to the T2TF problem based on the product representation, the local estimates $\mathbf{x}^s_{k-1|k-1}$ have to be adjusted.

This process is visualized schematically in Figure 7.2. At first, the fused Gaussian density function is computed. Since the product representation holds, the local parameters are mutually independent and can be fused by using the convex combination formulas without loss of accuracy. The result of the fused error covariance matrix $\mathbf{P}_{k-1|k-1}$

is then distributed among the local processing nodes. Those replace their local covariance matrix by $\mathbf{P}_{k-1|k-1}$ multiplied with the number of sensors S. As a consequence, it holds that the global fused error covariance has not been changed during this process. However, the *local* parameters for both mean and covariance matrix, respectively, have been modified. Therefore, those are rather referenced to as pseudo parameters.

From an implementation perspective, it should be noted that this is possible completely without transmission of data between dislocated nodes under the assumption that full knowledge is available on the remote sensor models. Since the Kalman filter error covariance is data independent,* each sensor node is able to calculate the actual posterior covariances matrix $\mathbf{P}^r_{k-1|k-1}$ for all of the remote sensor nodes $r \neq s$. As a consequence, all local processors are able to calculate the current fused error covariance matrix $\mathbf{P}_{k-1|k-1}$ without any data transmission as indicated in the processing scheme given in Figure 7.1.

The following formulas describe this process in a formally exact manner. For the derivation of the globalization, the product representation is first transformed into the fused density and then retransferred into a product representation, where it has been achieved that the *globalized* local error covariances $\tilde{\mathbf{P}}_{k-1|k-1} = S\mathbf{P}_{k-1|k-1}$ hold.

According to the convex combination (6.10) the globally fused density $\mathcal{N}(\mathbf{x}_{k-1}; \mathbf{x}_{k-1|k-1}, \mathbf{P}_{k-1|k-1})$ has the mean vector

$$\mathbf{x}_{k-1|k-1} = \mathbf{P}_{k-1|k-1} \sum_{s=1}^{S} (\mathbf{P}^s_{k-1|k-1})^{-1} \mathbf{x}^s_{k-1|k-1} \tag{7.10}$$

such that it can be rewritten in the following way:

$$\begin{aligned} &\mathcal{N}(\mathbf{x}_{k-1}; \mathbf{P}_{k-1|k-1} \sum_{s=1}^{S} (\mathbf{P}^s_{k-1|k-1})^{-1} \mathbf{x}^s_{k-1|k-1}, \mathbf{P}_{k-1|k-1}) \\ &\quad = \mathcal{N}(\mathbf{x}_{k-1}; \frac{1}{S} S\mathbf{P}_{k-1|k-1} \sum_{s=1}^{S} (\mathbf{P}^s_{k-1|k-1})^{-1} \mathbf{x}^s_{k-1|k-1}, \mathbf{P}_{k-1|k-1}) \end{aligned}$$

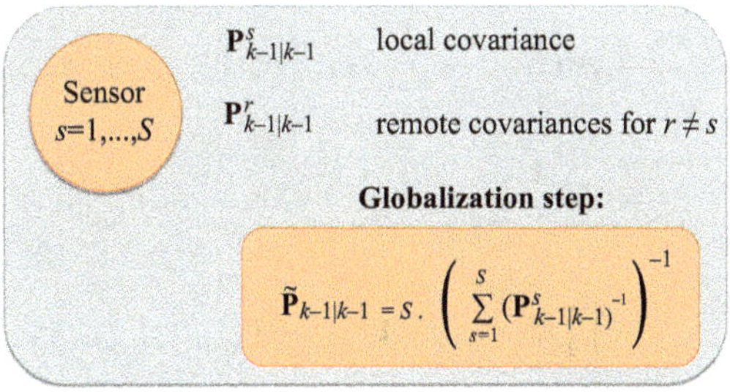

Figure 7.1 Computation scheme for the local sensors in the globalization step, where the individual error covariance matrices $\mathbf{P}^s_{k-1|k-1}$ from the previous filtering step are replaced by $\tilde{\mathbf{P}}_{k-1|k-1}$

*This excludes non-linear models and data driven hypotheses such as in the Interacting Multiple Model filter (IMM), Probabilistic Data Association (PDA), and Multi-Hypothesis Tracker (MHT).

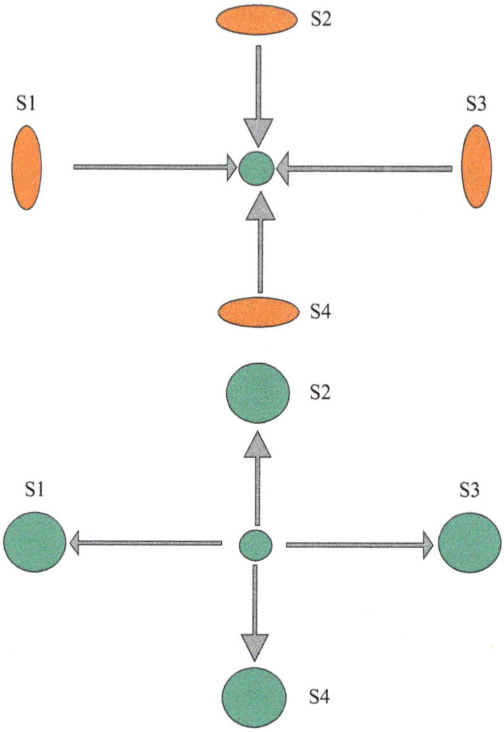

Figure 7.2 Schematic visualization of the globalization step in the DKF prediction: Local error covariances (orange) are used to compute the global posterior (above). Then, all local sensors replace their local tracks by some pseudo parameters which will yield the same global fusion result.

$$= \mathcal{N}(\mathbf{x}_{k-1}; \frac{1}{S}\sum_{s=1}^{S} S\mathbf{P}_{k-1|k-1}(\mathbf{P}^s_{k-1|k-1})^{-1}\mathbf{x}^s_{k-1|k-1}, \mathbf{P}_{k-1|k-1})$$

$$= \mathcal{N}(\frac{1}{S}\sum_{s=1}^{S}\mathbf{x}_{k-1}; \frac{1}{S}\sum_{s=1}^{S} S\mathbf{P}_{k-1|k-1}(\mathbf{P}^s_{k-1|k-1})^{-1}\mathbf{x}^s_{k-1|k-1}, \mathbf{P}_{k-1|k-1})$$

$$\propto \prod_{s=1}^{S} \mathcal{N}(\mathbf{x}_{k-1}; S\mathbf{P}_{k-1|k-1}(\mathbf{P}^s_{k-1|k-1})^{-1}\mathbf{x}^s_{k-1|k-1}, S\mathbf{P}_{k-1|k-1})$$

$$= \prod_{s=1}^{S} \mathcal{N}(\mathbf{x}_{k-1}; \tilde{\mathbf{x}}^s_{k-1|k-1}, \tilde{\mathbf{P}}_{k-1|k-1}), \tag{7.11}$$

with

$$\tilde{\mathbf{x}}^s_{k-1|k-1} := S\mathbf{P}_{k-1|k-1}(\mathbf{P}^s_{k-1|k-1})^{-1}\mathbf{x}^s_{k-1|k-1}. \tag{7.12}$$

The exact globalization only is possible, if global knowledge on the fused covariance is available locally in terms of the set of error covariances. For each instant of time, the globalized error covariance matrix $\tilde{\mathbf{P}}_{k-1|k-1}$ which is given by

$$\tilde{\mathbf{P}}_{k-1|k-1} = S\mathbf{P}_{k-1|k-1} \tag{7.13}$$

$$= S \cdot \left(\sum_{s=1}^{S} (\mathbf{P}^s_{k-1|k-1})^{-1} \right)^{-1}. \tag{7.14}$$

is introduced and used as the covariance matrix of the local track. It follows that the desired product representation holds:

$$p(\mathbf{x}_{k-1}|\mathcal{Z}^{k-1}) \propto \prod_{s=1}^{S} \mathcal{N}(\mathbf{x}_{k-1};\, \tilde{\mathbf{x}}^s_{k-1|k-1},\, \tilde{\mathbf{P}}_{k-1|k-1}). \tag{7.15}$$

In other words, the local posterior covariance is replaced by a globalized covariance matrix such that the result of the convex combination yields the optimal estimate parameters.

It should be noted that the time index refers to the instant $k-1$, since the globalization step is inserted before the prediction and is based on the previous posterior. Equivalently, one could also insert it after each filtering step.

7.2 DKF prediction

According to the globalization step, the prediction is based on a product representation at time t_{k-1} such that

$$p(\mathbf{x}_{k-1}|\mathcal{Z}^{k-1}) \propto \prod_{s=1}^{S} \mathcal{N}(\mathbf{x}_{k-1};\, \tilde{\mathbf{x}}^s_{k-1|k-1},\, \tilde{\mathbf{P}}_{k-1|k-1}) \tag{7.16}$$

This section shows how to predict the local tracks $\tilde{\mathbf{x}}^s_{k-1|k-1}$ to the present state at time t_k while keeping the product representation valid by computing a factorization of the global prior:

$$p(\mathbf{x}_{k}|\mathcal{Z}^{k-1}) \propto \prod_{s=1}^{S} \mathcal{N}(\mathbf{x}_{k};\, \tilde{\mathbf{x}}^s_{k|k-1},\, \tilde{\mathbf{P}}_{k|k-1}). \tag{7.17}$$

This can be achieved, if the transition model $p(\mathbf{x}_k|\mathbf{x}_{k-1})$ is given by a linear Gaussian density, since it can be rewritten as a product of S transition models, at least up to a constant factor of proportionality:

$$p(\mathbf{x}_k|\mathbf{x}_{k-1}) = \mathcal{N}(\mathbf{x}_k;\, \mathbf{F}_{k|k-1}\mathbf{x}_{k-1},\, \mathbf{Q}_{k|k-1}) \tag{7.18}$$

$$\propto \exp\left(-{}^1\!/\!{}_2(\mathbf{x}_k - \mathbf{F}_{k|k-1}\mathbf{x}_{k-1})^\top (\mathbf{Q}_{k|k-1})^{-1} (\mathbf{x}_k - \mathbf{F}_{k|k-1}\mathbf{x}_{k-1})\right) \tag{7.19}$$

$$\propto \exp\left(-{}^1\!/\!{}_2(\mathbf{x}_k - \mathbf{F}_{k|k-1}\mathbf{x}_{k-1})^\top (S\mathbf{Q}_{k|k-1})^{-1} (\mathbf{x}_k - \mathbf{F}_{k|k-1}\mathbf{x}_{k-1}) \cdot S\right)$$

$$\propto \mathcal{N}(\mathbf{x}_k;\, \mathbf{F}_{k|k-1}\mathbf{x}_{k-1},\, S\mathbf{Q}_{k|k-1})^S. \tag{7.20}$$

This *relaxed evolution model* can now be used in the prediction step of the Bayesian recursion. By marginalization one has

$$p(\mathbf{x}_k|\mathcal{Z}^{k-1}) = \int d\mathbf{x}_{k-1}\, p(\mathbf{x}_k|\mathbf{x}_{k-1})\, p(\mathbf{x}_{k-1}|\mathcal{Z}^{k-1}) \tag{7.21}$$

$$\propto \int d\mathbf{x}_{k-1} \prod_{s=1}^{S} \mathcal{N}(\mathbf{x}_k;\, \mathbf{F}_{k|k-1}\mathbf{x}_{k-1},\, S\mathbf{Q}_{k|k-1})\; \mathcal{N}(\mathbf{x}_{k-1};\, \tilde{\mathbf{x}}^s_{k-1|k-1},\, \tilde{\mathbf{P}}_{k-1|k-1}). \tag{7.22}$$

A repeated use of the product formula for Gaussians now yields

$$p(\mathbf{x}_k|\mathcal{Z}^{k-1}) \propto \int d\mathbf{x}_{k-1} \prod_{s=1}^{S} \mathcal{N}(\mathbf{x}_k;\, \tilde{\mathbf{x}}^s_{k|k-1},\, \tilde{\mathbf{P}}_{k|k-1})$$
$$\mathcal{N}(\mathbf{x}_{k-1};\, \tilde{\mathbf{W}}_{k-1}\mathbf{x}_k + \tilde{\mathbf{z}}^s_{k-1},\, \tilde{\mathbf{R}}_{k-1}), \tag{7.23}$$

where the following abbreviations were used:

$$\tilde{\mathbf{x}}^s_{k|k-1} = \mathbf{F}_{k|k-1}\tilde{\mathbf{x}}^s_{k-1|k-1} \tag{7.24}$$
$$\tilde{\mathbf{P}}_{k|k-1} = \mathbf{F}_{k|k-1}\tilde{\mathbf{P}}_{k-1|k-1}\mathbf{F}^\top_{k|k-1} + S\mathbf{Q}_{k|k-1} \tag{7.25}$$
$$\tilde{\mathbf{z}}^s_{k-1} = \tilde{\mathbf{x}}^s_{k-1|k-1} - \tilde{\mathbf{W}}_{k-1}\tilde{\mathbf{x}}^s_{k|k-1} \tag{7.26}$$
$$\tilde{\mathbf{W}}_{k-1} = \tilde{\mathbf{P}}_{k-1|k-1}\mathbf{F}^\top_{k|k-1}\tilde{\mathbf{R}}^{-1}_{k-1}. \tag{7.27}$$
$$\tilde{\mathbf{R}}_{k-1} = \tilde{\mathbf{P}}_{k-1|k-1} - \tilde{\mathbf{W}}_{k-1}\tilde{\mathbf{P}}_{k|k-1}\tilde{\mathbf{W}}^\top_{k-1} \tag{7.28}$$

Obviously, the prior in (7.23) can be written as

$$p(\mathbf{x}_k|\mathcal{Z}^{k-1}) = \ell_{k|k-1} \prod_{s=1}^{S} \mathcal{N}(\mathbf{x}_k;\, \tilde{\mathbf{x}}^s_{k|k-1},\, \tilde{\mathbf{P}}_{k|k-1}) \tag{7.29}$$

by introducing the function $\ell_{k|k-1}$ defined by

$$\ell_{k|k-1} = \int d\mathbf{x}_{k-1} \prod_{s=1}^{S} \mathcal{N}(\mathbf{x}_{k-1} - \tilde{\mathbf{W}}_{k-1}\mathbf{x}_k;\, \tilde{\mathbf{z}}^s_{k-1},\, \tilde{\mathbf{R}}_{k-1}). \tag{7.30}$$

It should be noted that the product representation is obtained by (7.29), if $\ell_{k|k-1}$ is independent of the random variable $\mathbf{x}_k$. This is due to the fact that the product representation yields a normalized Gaussian density, therefore factors of proportionality may be omitted.

To see this, the product formula (3.123) is applied repeatedly along the product of Gaussians for $\ell_{k|k-1}$ in (7.30). One obtains

$$\ell_{k|k-1} = \prod_{s=1}^{S-1} \mathcal{N}(\tilde{\mathbf{z}}_{k-1}^{s+1};\ \frac{1}{s}\sum_{i=1}^{s} \tilde{\mathbf{z}}_{k-1}^{i},\ \left(1+\frac{1}{s}\right)\tilde{\mathbf{R}}_{k-1})$$

$$\cdot \underbrace{\int d\mathbf{x}_{k-1}\, \mathcal{N}(\mathbf{x}_{k-1} - \tilde{\mathbf{W}}_{k-1}\mathbf{x}_k;\ \frac{1}{S}\sum_{i=1}^{S} \tilde{\mathbf{z}}_{k-1}^{i},\ \frac{1}{S}\tilde{\mathbf{R}}_{k-1})}_{=1}. \tag{7.31}$$

As one can see, the integration term becomes trivial, as it is over a single Gaussian. As a result, the function $\ell_{k|k-1}$ is independent of the current state variable $\mathbf{x}_k$ and we obtain the desired product representation:

$$p(\mathbf{x}_k|\mathcal{Z}^{k-1}) \propto \prod_{s=1}^{S} \mathcal{N}(\mathbf{x}_k;\ \tilde{\mathbf{x}}_{k|k-1}^{s},\ \tilde{\mathbf{P}}_{k|k-1}). \tag{7.32}$$

The DKF prediction can be summarized by the following formulas:

$$\tilde{\mathbf{P}}_{k-1|k-1} = S \cdot \left(\sum_{s=1}^{S} (\mathbf{P}_{k-1|k-1}^{s})^{-1}\right)^{-1} \tag{7.33}$$

$$\tilde{\mathbf{x}}_{k-1|k-1}^{s} = \tilde{\mathbf{P}}_{k-1|k-1}(\mathbf{P}_{k-1|k-1}^{s})^{-1}\mathbf{x}_{k-1|k-1}^{s}. \tag{7.34}$$

$$\tilde{\mathbf{x}}_{k|k-1}^{s} = \mathbf{F}_{k|k-1}\tilde{\mathbf{x}}_{k-1|k-1}^{s} \tag{7.35}$$

$$\tilde{\mathbf{P}}_{k|k-1} = \mathbf{F}_{k|k-1}\tilde{\mathbf{P}}_{k-1|k-1}\mathbf{F}_{k|k-1}^{\top} + S\mathbf{Q}_{k|k-1}. \tag{7.36}$$

7.3 DKF filtering

In this section, the posterior density $p(\mathbf{x}_k|\mathcal{Z}^k)$ is computed which adds the information of the set of measurements $Z_k = \{\mathbf{z}_k^s\}_{s=1,\dots,S}$ from all sensors at time t_k. To this end, it is important to notice that the sensors measurement errors are mutually independently and therefore their joint density function factorizes:

$$p(Z_k|\mathbf{x}_k) = \prod_{s=1}^{S} p(\mathbf{z}_k^s|\mathbf{x}_k), \tag{7.37}$$

where $p(\mathbf{z}_k^s|\mathbf{x}_k) = \mathcal{N}(\mathbf{z}_k^s;\ \mathbf{H}_k^s\mathbf{x}_k,\ \mathbf{R}_k^s)$ is the likelihood function of sensor s. Obviously, this perfectly fits to the product representation of the DKF. Using Bayes' rule, we obtain

$$p(\mathbf{x}_k|\mathcal{Z}^k) = \frac{p(Z_k|\mathbf{x}_k)\,p(\mathbf{x}_k|\mathcal{Z}^{k-1})}{\int \mathrm{d}\mathbf{x}_k\, p(Z_k|\mathbf{x}_k)\,p(\mathbf{x}_k|\mathcal{Z}^{k-1})}. \tag{7.38}$$

We can now fill in the prediction (7.32) and the likelihood function (7.37) and then have

$$p(\mathbf{x}_k|\mathcal{Z}^k) \propto \prod_{s=1}^{S} \mathcal{N}(\mathbf{z}_k^s;\, \mathbf{H}_k^s\mathbf{x}_k,\, \mathbf{R}_k^s)\; \mathcal{N}(\mathbf{x}_k;\, \tilde{\mathbf{x}}_{k|k-1}^s,\, \tilde{\mathbf{P}}_{k|k-1}). \tag{7.39}$$

A final use of the product formula now yields the product representation for the filtering update

$$p(\mathbf{x}_k|\mathcal{Z}^k) \propto \prod_{s=1}^{S} \mathcal{N}(\mathbf{x}_k;\, \mathbf{x}_{k|k}^s,\, \mathbf{P}_{k|k}^s). \tag{7.40}$$

Thus, the filtering parameters are obtained via Kalman filter formulas:

$$\mathbf{x}_{k|k}^s = \tilde{\mathbf{x}}_{k|k-1}^s + \mathbf{W}_{k|k-1}^s\left(\mathbf{z}_k^s - \mathbf{H}_k^s\tilde{\mathbf{x}}_{k|k-1}^s\right) \tag{7.41}$$

$$\mathbf{W}_{k|k-1}^s = \tilde{\mathbf{P}}_{k|k-1}\mathbf{H}_k^{s\top}\mathbf{S}_{k|k-1}^{s\,-1} \tag{7.42}$$

$$\mathbf{S}_{k|k-1}^s = \mathbf{H}_k^s\tilde{\mathbf{P}}_{k|k-1}\mathbf{H}_k^{s\top} + \mathbf{R}_k^s \tag{7.43}$$

$$\mathbf{P}_{k|k}^s = \tilde{\mathbf{P}}_{k|k-1} - \mathbf{W}_{k|k-1}^s\mathbf{S}_{k|k-1}^s\mathbf{W}_{k|k-1}^{s\top}. \tag{7.44}$$

Globalized likelihood function

As indicated above, it is equivalent to insert the globalization step before the prediction or after the filtering. Actually, it is possible to reconstruct a local likelihood function for each sensor s such that the global knowledge is already included. This, however, does not change any of the assumptions made for calculating the global error covariance. That is, the full knowledge of the remote sensor models is required, too. But since the globalization step is included in the likelihood, a standard update using the Kalman filter formulas directly yields the desired representation. In other words, the globalization step may be omitted using the globalized likelihood for every sensor when processing the current measurement $\mathbf{z}_k^s$.

The derivation of the globalized likelihood function $\ell_k^s(\mathbf{x}_k; \mathbf{z}_k)$ is backwards in a way. Based on the formulae above, the closed form solution for the local posterior parameters $\mathbf{x}_{k|k}^s$ and $\mathbf{P}_{k|k}^s$ is available. An application of the globalization step then would yield

$$\tilde{\mathbf{x}}_{k|k}^s = \tilde{\mathbf{P}}_{k|k}(\mathbf{P}_{k|k}^s)^{-1}\mathbf{x}_{k|k}^s \tag{7.45}$$

$$\tilde{\mathbf{P}}_{k|k} = S\left(\sum_{i=1}^{S}(\mathbf{P}_{k|k}^i)^{-1}\right)^{-1}. \tag{7.46}$$

Using the information filter formulae, one can rewrite $\tilde{\mathbf{x}}_{k|k}^s$ from (7.45) in the following way:

$$\tilde{\mathbf{x}}_{k|k}^s = \tilde{\mathbf{P}}_{k|k}\left(\tilde{\mathbf{P}}_{k|k-1}^{-1}\mathbf{x}_{k|k-1}^s + \mathbf{H}_k^{s\top}(\mathbf{R}_k^s)^{-1}\mathbf{z}_k^s\right) \tag{7.47}$$

and the globalized posterior covariance in (7.46) as

$$\tilde{\mathbf{P}}_{k|k} = S\left(\sum_{i=1}^{S}(\tilde{\mathbf{P}}_{k|k-1})^{-1} + \mathbf{H}_k^{i\top}(\mathbf{R}_k^i)^{-1}\mathbf{H}_k^i\right)^{-1}. \tag{7.48}$$

It should be noted that the prior covariance $\tilde{\mathbf{P}}_{k|k-1})$ may be assumed to be globalized, since the applied evolution model $p(\mathbf{x}_k|\mathbf{x}_{k-1})$ is identical for all sensors.

Using the above derivations for the globalized *local* posterior, one can see that it can be written as a product of the prior and a Gaussian-likelihood function a factor of proportionality:

$$\mathcal{N}(\mathbf{x}_k;\ \tilde{\mathbf{x}}_{k|k}^s,\ \tilde{\mathbf{P}}_{k|k}) \propto \mathcal{N}(\mathbf{x}_k;\ \mathbf{x}_{k|k-1}^s,\ \tilde{\mathbf{P}}_{k|k-1}) \cdot \mathcal{N}(\mathbf{x}_k;\ \tilde{\mathbf{R}}_k\mathbf{H}_k^{s\top}(\mathbf{R}_k^s)^{-1}\mathbf{z}_k^s,\ \tilde{\mathbf{R}}_k) \tag{7.49}$$

where the global measurement error covariance $\tilde{\mathbf{R}}_k$ is given by

$$\tilde{\mathbf{R}}_k = \left(\frac{1}{S}\sum_{i=1}^{S}\mathbf{H}_k^{i\top}(\mathbf{R}_k^i)^{-1}\mathbf{H}_k^i\right)^{-1} \tag{7.50}$$

As a consequence, a globalized likelihood is found by

$$\ell_k^s(\mathbf{x}_k;\mathbf{z}_k) = \mathcal{N}(\mathbf{x}_k;\ \tilde{\mathbf{R}}_k\mathbf{H}_k^{s\top}(\mathbf{R}_k^s)^{-1}\mathbf{z}_k^s,\ \tilde{\mathbf{R}}_k). \tag{7.51}$$

An application of the information filter formulas directly yields the desired globalized filtering parameters:

$$\tilde{\mathbf{x}}_{k|k}^s = \tilde{\mathbf{P}}_{k|k}\left(\tilde{\mathbf{P}}_{k|k-1}^{-1}\tilde{\mathbf{x}}_{k|k-1} + \mathbf{H}_k^{s\top}(\mathbf{R}_k^s)^{-1}\mathbf{z}_k^s\right) \tag{7.52}$$

$$\tilde{\mathbf{P}}_{k|k} = \left(\tilde{\mathbf{P}}_{k|k-1}^{-1} + \tilde{\mathbf{R}}_k^{-1}\right)^{-1} \tag{7.53}$$

$$\tilde{\mathbf{R}}_k = S\left(\sum_{s=1}^{S}\mathbf{H}_k^{s\top}(\mathbf{R}_k^s)^{-1}\mathbf{H}_k^s\right)^{-1} \tag{7.54}$$

7.4 Federated Kalman filter

The federated Kalman filter (FKF) is also based on the product representation (7.1) and the relaxed evolution model in (7.20). The globalization step of the DKF is circumvented by means of an approximation in the prediction. Its derivation is therefore quite straightforward:

$$p(\mathbf{x}_k|\mathcal{Z}^{k-1}) = \int d\mathbf{x}_{k-1}\, p(\mathbf{x}_k|\mathbf{x}_{k-1})\, p(\mathbf{x}_{k-1}|\mathcal{Z}^{k-1}) \tag{7.55}$$

$$\propto \int d\mathbf{x}_{k-1} \prod_{s=1}^{S} \mathcal{N}(\mathbf{x}_k;\ \mathbf{F}_{k|k-1}\mathbf{x}_{k-1},\ S\mathbf{Q}_{k|k-1})\ \mathcal{N}(\mathbf{x}_{k-1};\ \mathbf{x}_{k-1|k-1}^s,\ \mathbf{P}_{k-1|k-1}^s). \tag{7.56}$$

As in the DKF prediction, an application of the product formula yields that the prior is given by

$$p(\mathbf{x}_k|\mathcal{Z}^{k-1}) = \ell_{k|k-1} \prod_{s=1}^{S} \mathcal{N}(\mathbf{x}_k;\, \mathbf{x}^s_{k|k-1},\, \mathbf{P}^s_{k|k-1}). \tag{7.57}$$

We have seen above that the factor $\ell_{k|k-1}$ is independent of the random variable $\mathbf{x}_k$, if the previous posterior has been globalized such that the local error covariances $\mathbf{P}^s_{k-1|k-1}$ are identical for all s. In the FKF, the globalization is omitted for the sake of less strict assumptions for the knowledge on the remote sensor models. As a consequence, the factor $\ell_{k|k-1}$ remains state-dependent.

One obtains for the FKF prediction

$$p(\mathbf{x}_k|\mathcal{Z}^{k-1}) \approx \prod_{s=1}^{S} \mathcal{N}(\mathbf{x}_k;\, \mathbf{x}^s_{k|k-1},\, \mathbf{P}^s_{k|k-1}) \tag{7.58}$$

where the following abbreviations were used:

$$\mathbf{x}^s_{k|k-1} = \mathbf{F}_{k|k-1}\mathbf{x}^s_{k-1|k-1} \tag{7.59}$$

$$\tilde{\mathbf{P}}^s_{k|k-1} = \mathbf{F}_{k|k-1}\mathbf{P}^s_{k-1|k-1}\mathbf{F}^{\top}_{k|k-1} + S\mathbf{Q}_{k|k-1}. \tag{7.60}$$

The normalization constant (7.30) is neglected. Since it actually depends on the state variable $\mathbf{x}_k$, this is an approximation and will not yield the optimal fusion result in general. The loss, however, is compensated by the fact that the FKF is easy to implement and does not require knowledge of the remote sensor models.

The filtering equations are easily derived since the joint likelihood of observation factorizes as described in the DKF filtering section. One has:

$$p(\mathbf{x}_k|\mathcal{Z}^{k}) \propto \prod_{s=1}^{S} \mathcal{N}(\mathbf{z}^s_k;\, \mathbf{H}^s_k\mathbf{x}_k,\, \mathbf{R}^s_k)\, \mathcal{N}(\mathbf{x}_k;\, \mathbf{x}^s_{k|k-1},\, \mathbf{P}^s_{k|k-1}) \tag{7.61}$$

This yields for the FKF update step according to the Kalman filter update formulae

$$\mathbf{x}^s_{k|k} = \mathbf{x}^s_{k|k-1} + \mathbf{W}^s_{k|k-1}\left(\mathbf{z}^s_k - \mathbf{H}^s_k\mathbf{x}^s_{k|k-1}\right) \tag{7.62}$$

$$\mathbf{W}^s_{k|k-1} = \mathbf{P}^s_{k|k-1}\mathbf{H}^{s\top}_k\mathbf{S}^{s\,-1}_{k|k-1} \tag{7.63}$$

$$\mathbf{S}^s_{k|k-1} = \mathbf{H}^s_k\mathbf{P}^s_{k|k-1}\mathbf{H}^{s\top}_k + \mathbf{R}^s_k \tag{7.64}$$

$$\mathbf{P}^s_{k|k} = \mathbf{P}^s_{k|k-1} - \mathbf{W}^s_{k|k-1}\mathbf{S}^s_{k|k-1}\mathbf{W}^{s\top}_{k|k-1}. \tag{7.65}$$

7.5 DKF sum formulation

By rescaling all local state estimate parameters of the DKF $\mathbf{x}^s_{k|k}$ by a factor of $^1/_S$, the convex combination for the track fusion becomes

$$\mathbf{x}_{k|k} = \sum_{s=1}^{S} \mathbf{x}^s_{k|k}. \tag{7.66}$$

As seen above, this fusion rule is exact, if the product representation holds (before the rescaling) and is called the *sum representation*. Since the same rescaling of the globalized covariance yields the global error covariance:

$$\frac{1}{S}\tilde{\mathbf{P}}_{k|k} = \mathbf{P}_{k|k} \tag{7.67}$$

it follows that the local sensors store the latter $\mathbf{P}_{k|k}$ directly for each step in the prediction-update cycle. The Bayesian recursion can well be expressed in the sum representation, as shown in this section.

A corresponding prediction formulation is easily found since it holds that

$$\mathbf{x}_{k|k-1} = \mathbf{F}_{k|k-1}\mathbf{x}_{k-1|k-1} \tag{7.68}$$

$$= \mathbf{F}_{k|k-1}\sum_{s=1}^{S} \mathbf{x}^s_{k-1|k-1} \tag{7.69}$$

$$= \sum_{s=1}^{S} \mathbf{F}_{k|k-1}\mathbf{x}^s_{k-1|k-1}. \tag{7.70}$$

The prior parameters in the rescaled version of the DKF are given by

$$\mathbf{x}^s_{k|k-1} = \mathbf{F}_{k|k-1}\mathbf{x}^s_{k-1|k-1}. \tag{7.71}$$

Since the local sensor nodes always need to be aware of the global error covariance, each processor also computes its prior matrix:

$$\mathbf{P}_{k|k-1} = \mathbf{F}_{k|k-1}\mathbf{P}_{k|k}\mathbf{F}^{\top}_{k|k-1} + \mathbf{Q}_{k|k-1}. \tag{7.72}$$

It should be noted that in this case the relaxed evolution model is *not* used, since the prediction of the global track covariance matrix is considered on a globally fused level instead of local parameters. This, however, is equivalent to the application of the relaxed evolution model in the product representation.

In order to derive the filtering equations, one also considers the Kalman update for multiple measurements at first and then defines the corresponding local parameters. Using the information filter, one has that

$$\mathbf{x}_{k|k} = \mathbf{P}_{k|k}\left(\mathbf{P}^{-1}_{k|k-1}\mathbf{x}_{k|k-1} + \sum_{s=1}^{S} \mathbf{H}^{s\top}_k(\mathbf{R}^s_k)^{-1}\mathbf{z}^s_k\right) \tag{7.73}$$

$$= \mathbf{P}_{k|k} \left(\mathbf{P}_{k|k-1}^{-1} \sum_{s=1}^{S} \mathbf{x}_{k|k-1}^{s} + \sum_{s=1}^{S} \mathbf{H}_k^{s\top} (\mathbf{R}_k^s)^{-1} \mathbf{z}_k^s \right) \tag{7.74}$$

$$= \mathbf{P}_{k|k} \left(\sum_{s=1}^{S} \mathbf{P}_{k|k-1}^{-1} \mathbf{x}_{k|k-1}^{s} + \mathbf{H}_k^{s\top} (\mathbf{R}_k^s)^{-1} \mathbf{z}_k^s \right) \tag{7.75}$$

$$= \sum_{s=1}^{S} \mathbf{P}_{k|k} \left(\mathbf{P}_{k|k-1}^{-1} \mathbf{x}_{k|k-1}^{s} + \mathbf{H}_k^{s\top} (\mathbf{R}_k^s)^{-1} \mathbf{z}_k^s \right) \tag{7.76}$$

As a consequence, one can see that the local fusion is equivalent to a Kalman filter update where the *global* (fused) error covariance is used, as if it would refer to a local track (which is not the case).

Since the Kalman filter update equations are equivalent, the local parameters may well be computed by

$$\mathbf{x}_{k|k}^{s} = \mathbf{x}_{k|k-1}^{s} + \mathbf{W}_{k|k-1}^{s} \left(\mathbf{z}_k^s - \mathbf{H}_k^s \mathbf{x}_{k|k-1}^{s} \right) \tag{7.77}$$

$$\mathbf{W}_{k|k-1}^{s} = \mathbf{P}_{k|k-1} \mathbf{H}_k^{s\top} \mathbf{S}_{k|k-1}^{s\,-1} \tag{7.78}$$

$$\mathbf{S}_{k|k-1}^{s} = \mathbf{H}_k^s \mathbf{P}_{k|k-1} \mathbf{H}_k^{s\top} + \mathbf{R}_k^s \tag{7.79}$$

$$\mathbf{P}_{k|k} = \left(\mathbf{P}_{k|k-1}^{-1} + \sum_{s=1}^{S} \mathbf{H}_k^{s\top} (\mathbf{R}_k^s)^{-1} \mathbf{H}_k^s \right)^{-1}. \tag{7.80}$$

Here, the information filter equation for the fused covariance matrix update is proposed to avoid a successive computation.

Obviously, the global knowledge also applies here for each local sensor node, since each processor is fully aware of the fused error covariance at each instant of time.

Discussion

The DKF is an exact reconstruction of the Kalman filter which is optimal in the linear-Gaussian case but with a distributed computation of intermediate parameters. It is of theoretical interest, since for quite some decades it remained unclear, whether such formulae exist. Its practical use, however, is limited due to the fact that the remote sensor models must be known. This excludes in particular non-linear scenarios, where the observation matrix $\mathbf{H}_k^s$ for each local measurement must be approximated based on some linearization technique. Those in turn use the local state estimate as a first order sampling point such that the approximation becomes data dependent. Investigations to use the local state estimate as a sampling point in order to approximate the remote observation matrices showed that this approach was not superior to the much simpler approach of the FKF. However, the DKF shows its potential in multi-sensor applications with sufficient network capacity, such as on-site data fusion-based transmissions via cables. In this case, the global prior may be transmitted to the local processing nodes for the processing of the current measurements. As a consequence, the data association and filtering process are solved in a distributed manner such that global fusion at the fusion center remains feasible in real-time applications.

7.6 Consensus of agents in graphs

Consensus algorithms for multi-agent networked systems address the problem of achieving an identical state for processing platforms with limited connectivity. An agent at node $i \in V$ in a graph $\mathcal{G} = (V, E)$ with $E \subseteq V \times V$ has a scalar state variable x_i. At each instant of time, it may send its current state to all its neighbors $j \in N_i$ with

$$N_i = \{j \in V \quad | \quad (i,j) \in E\}. \tag{7.81}$$

The collective dynamics of the global system with n nodes can then be written as a matrix formulation:

$$\dot{\mathbf{x}} = -\mathbf{L}\mathbf{x}, \tag{7.82}$$

where $\mathbf{x} = (x_1, \ldots, x_n)^\top$ and

$$(\mathbf{L})_{i,j} = \begin{cases} -a_{i,j}, & \text{if } i \neq j \\ \sum_{\substack{k=1 \\ k \neq i}}^{n} a_{i,k} & \text{else.} \end{cases} \tag{7.83}$$

The matrix $\mathbf{A} = (a_{i,j})$ is a weighting matrix, which models the information flow within the graph. As an example, consider the case that $a_{i,j} = 1$ for all $j \in N_i$. Then the dynamics of the estimates in (7.82) becomes

$$\dot{x}_i = \sum_{j \in N_i} (x_j - x_i) \tag{7.84}$$

such that

$$x_i(k) = x_i(k-1) + \epsilon \sum_{j \in N_i} (x_j - x_i), \tag{7.85}$$

where ϵ is the time increment at time step $k - 1$. In other words, this weighting matrix $\mathbf{A}$ models equal information for all nodes.

An example for a multi-agent networked system is shown in Figure 7.3.

Obviously, the dynamics equation has the time discrete solution

$$\mathbf{x}_k = (\mathbf{I} - \epsilon\mathbf{L})\mathbf{x} \tag{7.86}$$

and for the time-continuous case, one has that

$$\mathbf{x}_t = \exp(-t\mathbf{L})\mathbf{x}_0. \tag{7.87}$$

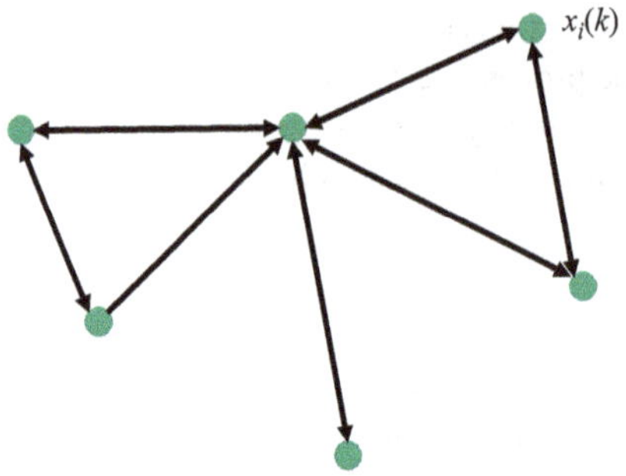

Figure 7.3 Illustration of a small connected graph, where each node may send its current scalar state variable x_i to its neighbors. The goal is to achieve consensus throughout the network such that $x_i = x_j$ for all pairs i,j.

For $\epsilon > 0$ small enough, $P = \mathbf{I} - \epsilon\mathbf{L}$ is a Markov transition matrix as introduced in Section 3.9 with

$$\sum_{j=1}^{n} p_{i,j} = 1 \tag{7.88}$$

$$p_{i,j} \geq 0. \tag{7.89}$$

In a static scenario, where only the initial states $x_i = z^i$ are based on local observations z^i, the convergence of the system $\mathbf{x}$ to a joint estimate

$$\mathbf{x} = (\alpha, \ldots, \alpha)^\top \tag{7.90}$$

with $\alpha = {}^1\!/_n \sum_i z^i$ is easily verified. In the literature, one can find detailed studies of the conversion in complex scenarios [53], for instance, if the graph topology changes dynamically over time or if new information from local measurements is to be considered.

The scalar consensus may well be extended to multivariate estimates by an application of multiple processes in parallel. By means of the consensus, the *global* information on available information may be propagated at each instant of time. Therefore, if global knowledge is achieved, the distributed Kalman filter can be applied such that the state estimate at each sensor is equal to a centralized fusion [63]:

$$\mathbf{x}_{k|k} = \mathbf{x}^s_{k|k} \qquad \forall\, s \tag{7.91}$$

$$\mathbf{P}_{k|k} = \mathbf{P}^s_{k|k} \qquad \forall\, s \tag{7.92}$$

This can be seen as follows. Since the prediction $\mathbf{P}_{k|k-1} = \mathbf{F}_{k|k-1}\mathbf{P}_{k-1|k-1}\mathbf{F}^\top_{k|k-1} + \mathbf{Q}_{k|k-1}$ is data independent, it can be applied at each sensor node based on the result of the previous time step.

The filtering update in information filter in (3.104) and (3.105) form is given by

$$\mathbf{P}_{k|k}^{-1} = \mathbf{P}_{k|k-1}^{-1} + \sum_{s=1}^{S} \mathbf{H}_k^\top \mathbf{R}_k^{s\,-1} \mathbf{H}_k \tag{7.93}$$

$$\mathbf{P}_{k|k}^{-1}\mathbf{x}_{k|k} = \mathbf{P}_{k|k-1}^{-1}\mathbf{x}_{k|k-1} + \sum_{s=1}^{S} \mathbf{H}_k^\top \mathbf{R}_k^{s\,-1} \mathbf{z}_k^s. \tag{7.94}$$

Therefore, it becomes obvious that the central fused estimate $\mathbf{x}_{k|k}$ and its error covariance matrix $\mathbf{P}_{k|k}$ can be computed by each sensor node, if the gossip algorithm for propagating parameters achieves consensus across the network on the measurement information parameters

$$\mathbf{I}_k = \sum_{s=1}^{S} \mathbf{H}_k^\top \mathbf{R}_k^{s\,-1} \mathbf{H}_k \tag{7.95}$$

$$\mathbf{i}_k = \sum_{s=1}^{S} \mathbf{H}_k^\top \mathbf{R}_k^{s\,-1} \mathbf{z}_k^s. \tag{7.96}$$

The process of alternating consensus and filtering cycles is shown in Figure 7.4.

For the consensus DKF, there exist different versions. For instance, one may modify the interaction between the consensus and the Bayesian update cycle such

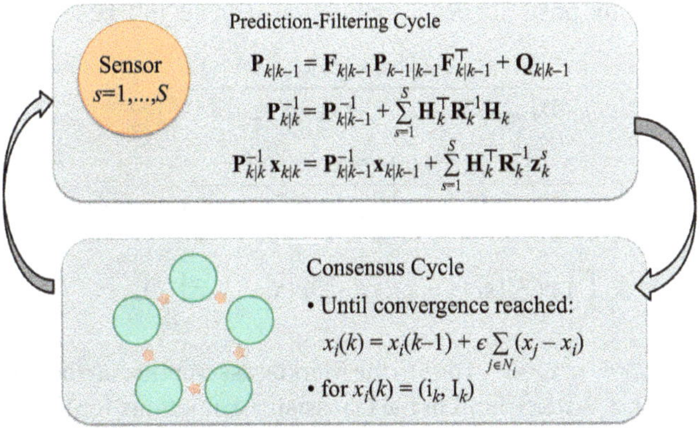

Figure 7.4 Schematic illustration of the consensus DKF, where global knowledge is achieved in each time step by propagating the information contribution within the network until consensus is achieved. Then, the filtering update of a global Kalman filter can be computed by each sensor node.

that they do not run separately. To this end, the transmission at each instant of time reduces to

$$\mathbf{I}_k^{N_i} = \sum_{s \in N_i} \mathbf{H}_k^\top \mathbf{R}_k^{s\,-1} \mathbf{H}_k \tag{7.97}$$

$$\mathbf{i}_k^{N_i} = \sum_{s \in N_i} \mathbf{H}_k^\top \mathbf{R}_k^{s\,-1} \mathbf{z}_k^s. \tag{7.98}$$

Obviously, this does not reconstruct the optimal centralized fusion estimate, but still convergence can be observed and may even be proven depending on the network topology.

7.7 Distributed accumulated state density filter

In this section, we will come back to the accumulated state densities (ASD) as introduced in Section 4.2. It turns out that besides trajectory estimation, and out-of-sequence processing ASDs can also be the means to decorrelate local estimate densities, that is, to achieve the product representation. In contrast to the DKF, the *distributed accumulated state density* (DASD) filter does not require knowledge on the remote sensor models and still can recover the optimal fused parameters. As a consequence, this approach may well be applied in non-linear scenarios and where measurement ambiguities such as false alarms and non-detections are present. There is a close relation between the DASD and the FKF, which will be highlighted at the end of the section.

At first, the global ASD for the multi-sensor case is derived. In the multi-sensor case, the posterior density can be decomposed as follows. An application of the Bayes' theorem yields for the posterior density conditioned on the full data set of all S sensors

$$p(\mathbf{x}_{k:n}|\mathcal{Z}^k) = \frac{p(Z_k^1, \ldots, Z_k^S|\mathbf{x}_k) \cdot p(\mathbf{x}_{k:n}|\mathcal{Z}^{k-1})}{\int \mathrm{d}\mathbf{x}_{k:n} p(Z_k^1, \ldots, Z_k^S|\mathbf{x}_k) \cdot p(\mathbf{x}_{k:n}|\mathcal{Z}^{k-1})} \tag{7.99}$$

$$= \frac{\prod_{s=1}^S p(Z_k^s|\mathbf{x}_k) \cdot p(\mathbf{x}_{k:n}|\mathcal{Z}^{k-1})}{\int \mathrm{d}\mathbf{x}_{k:n} \prod_{s=1}^S p(Z_k^s|\mathbf{x}_k) \cdot p(\mathbf{x}_{k:n}|\mathcal{Z}^{k-1})} \tag{7.100}$$

$$\propto \prod_{s=1}^S p(Z_k^s|\mathbf{x}_k) \cdot p(\mathbf{x}_k|\mathbf{x}_{k-1:n}) p(\mathbf{x}_{k-1:n}|\mathcal{Z}^{k-1}), \tag{7.101}$$

where $p(\mathbf{x}_k|\mathbf{x}_{k-1:n}) = p(\mathbf{x}_k|\mathbf{x}_{k-1})$ due to the Markov assumption on the random process of the state variable. After filling in the Gaussian assumptions for the sensor and the dynamics model as well as for the posterior of the previous time step, the following terms are obtained:

$$p(\mathbf{x}_{k:n}|\mathcal{Z}^k) \propto \prod_{s=1}^S \mathcal{N}(\mathbf{z}_k^s;\, \mathbf{H}_k^s \mathbf{x}_k,\, \mathbf{R}_k^s) \cdot \mathcal{N}(\mathbf{x}_k;\, \mathbf{F}_{k|k-1}\mathbf{x}_{k-1},\, \mathbf{Q}_{k|k-1})$$
$$\cdot\ \mathcal{N}(\mathbf{x}_{k-1:n};\, \mathbf{x}_{k-1:n|k-1},\, \mathbf{P}_{k-1:n|k-1}). \tag{7.102}$$

As in the distributed Kalman filter in (7.20), one may use the fact that the transition density can be factorized, if the relaxed evolution model is applied:

$$\mathcal{N}(\mathbf{x}_k;\ \mathbf{F}_{k|k-1}\mathbf{x}_{k-1},\ \mathbf{Q}_{k|k-1}) \propto \mathcal{N}(\mathbf{x}_k;\ \mathbf{F}_{k|k-1}\mathbf{x}_{k-1},\ S\mathbf{Q}_{k|k-1})^S. \tag{7.103}$$

As a consequence, the product in (7.102) is over both, the sensor and the evolution model:

$$p(\mathbf{x}_{k:n}|\mathcal{Z}^k) \propto \prod_{s=1}^{S} \{\mathcal{N}(\mathbf{z}_k^s;\ \mathbf{H}_k^s\mathbf{x}_k,\ \mathbf{R}_k^s)\,\mathcal{N}(\mathbf{x}_k;\ \mathbf{F}_{k|k-1}\mathbf{x}_{k-1},\ S\mathbf{Q}_{k|k-1})\} \\ \cdot\ \mathcal{N}(\mathbf{x}_{k-1:n};\ \mathbf{x}_{k-1:n|k-1},\ \mathbf{P}_{k-1:n|k-1}). \tag{7.104}$$

An iterative application of the same line of arguments on the remaining posterior posterior $\mathcal{N}(\mathbf{x}_{k-1:n};\ \mathbf{x}_{k-1:n|k-1},\ \mathbf{P}_{k-1:n|k-1})$ at time t_{k-1} yields

$$p(\mathbf{x}_{k:n}|\mathcal{Z}^k) \propto \prod_{l=n+1}^{k} \Big[\prod_{s=1}^{S} \{\mathcal{N}(\mathbf{z}_l^s;\ \mathbf{H}_l^s\mathbf{x}_l,\ \mathbf{R}_l^s)\,\mathcal{N}(\mathbf{x}_l;\ \mathbf{F}_{l|l-1}\mathbf{x}_{l-1},\ S\mathbf{Q}_{l|l-1})\}\Big] \cdot \\ \mathcal{N}(\mathbf{x}_n;\ \mathbf{x}_{n|n},\ \mathbf{P}_{n|n}), \tag{7.105}$$

where $\mathbf{x}_{n|n}$ and $\mathbf{P}_{n|n}$ refer to the result of a standard Kalman filter after processing all data up to time t_n or are the initialization parameters, respectively. Based on the factorization rule for the relaxed evolution model in (7.20), one has that

$$\mathcal{N}(\mathbf{x}_n;\ \mathbf{x}_{n|n},\ \mathbf{P}_{n|n}) \propto \mathcal{N}(\mathbf{x}_n;\ \mathbf{x}_{n|n},\ S\mathbf{P}_{n|n})^S, \tag{7.106}$$

which yields

$$p(\mathbf{x}_{k:n}|\mathcal{Z}^k) \propto \prod_{s=1}^{S} \Big\{ \prod_{l=n+1}^{k} \mathcal{N}(\mathbf{z}_l^s;\ \mathbf{H}_l^s\mathbf{x}_l,\ \mathbf{R}_l^s)\,\mathcal{N}(\mathbf{x}_l;\ \mathbf{F}_{l|l-1}\mathbf{x}_{l-1},\ S\mathbf{Q}_{l|l-1}) \\ \mathcal{N}(\mathbf{x}_n;\ \mathbf{x}_{n|n},\ S\mathbf{P}_{n|n})\Big\}. \tag{7.107}$$

At this point, one can see that (7.107) is a product of local ASDs since one has the same terms as in (4.20) in Section 4.3 except for the spread initial covariance matrix $S\mathbf{P}_{n|n}$ instead of $\mathbf{P}_{n|n}$ and the relaxed evolution model.

Following the calculation of an ASD in Section 4.3, one obtains that the global posterior is factorized into local ASD densities (up to a constant factor of proportionality), which are multivariate Gaussians with means $\mathbf{x}^s_{k:n|k}$ and covariance matrices $\mathbf{P}^s_{k:n|k}$:

$$p(\mathbf{x}_{k:n}|\mathcal{Z}^k) \propto \prod_{s=1}^{S} \mathcal{N}(\mathbf{x}_{k:n};\, \mathbf{x}^s_{k:n|k},\, \mathbf{P}^s_{k:n|k}) \tag{7.108}$$

Analogously to the distributed Kalman filter this directly implies that the *fused* ASD estimate and covariance matrix are given by the convex combination

$$\mathbf{x}_{k:n|k} = \mathbf{P}_{k:n|k} \sum_{s=1}^{S} \mathbf{P}^{s\,-1}_{k:n|k} \mathbf{x}^s_{k:n|k} \tag{7.109}$$

$$\mathbf{P}_{k:n|k} = \left(\sum_{s=1}^{S} \mathbf{P}^{s\,-1}_{k:n|k} \right)^{-1} . \tag{7.110}$$

This is quite an interesting observation since we have used the temporal correlations encoded within the ASDs in order to avoid spatial correlations between the local tracks. The required ingredients are local ASDs with the relaxed evolution model and the inflated initial covariance matrix.

In order to make the computation of the local parameters more explicit, we are going to derive an iterative solution of the local ASD tracks given by $\mathbf{x}^s_{k:n|k}$ and $\mathbf{P}^s_{k:n|k}$.

Prediction

Assume the local posterior ASD at time t_{k-1} for sensor s is given in terms of $\mathbf{x}^s_{k-1:n|k-1}$ and $\mathbf{P}^s_{k-1:n|k-1}$. The prediction of the state is straight forward due to the Markov proposition:

$$\mathbf{x}^s_{k:n|k-1} = \begin{pmatrix} \mathbf{x}^s_{k|k-1} \\ \mathbf{x}^s_{k-1|k-1} \\ \vdots \\ \mathbf{x}^s_{n|k-1} \end{pmatrix}, \tag{7.111}$$

where $\mathbf{x}^s_{k|k-1} = \mathbf{F}_{k|k-1}\mathbf{x}^s_{k-1|k-1}$ is equivalent to a local Kalman filter prediction. For the ASD covariance prediction, a recursive formulation as in (4.38) is used:

$$\mathbf{P}^s_{k:n|k-1} = \begin{pmatrix} \mathbf{P}^s_{k|k-1} & \mathbf{P}^s_{k|k-1}\mathbf{W}^{s\,\top}_{k-1:n} \\ \mathbf{W}^s_{k-1:n}\mathbf{P}^s_{k|k-1} & \mathbf{P}^s_{k-1:n|k-1} \end{pmatrix}, \tag{7.112}$$

where

$$\mathbf{P}^s_{k|k-1} = \mathbf{F}_{k|k-1}\mathbf{P}^s_{k-1|k-1}\mathbf{F}^\top_{k|k-1} + S\mathbf{Q}_{k|k-1}, \tag{7.113}$$

$$\mathbf{W}^s_{k-1:n} = \begin{pmatrix} \mathbf{W}^s_{k-1|k} \\ \mathbf{W}^s_{k-2:n}\mathbf{W}^s_{k-1|k} \end{pmatrix}, \tag{7.114}$$

$$\mathbf{W}^s_{l|l+1} = \mathbf{P}^s_{l|l}\mathbf{F}^\top_{l+1|l}\left(\mathbf{F}_{l+1|l}\mathbf{P}^s_{l|l}\mathbf{F}^\top_{l+1|l} + S\mathbf{Q}_{l+1|l}\right)^{-1}. \tag{7.115}$$

The notations above are an abbreviation for the result in (4.36) which states that the prior cross-correlation of two state estimates $\mathbf{x}_l$ and $\mathbf{x}_k$ referring to time t_l and t_k (which is the present time) respectively is given by

$$\operatorname{cov}\left[\mathbf{x}_l, \mathbf{x}_k | \mathcal{Z}^{k-1}\right] = \mathbf{W}_{l|k}\mathbf{P}_{k|k-1}, \tag{7.116}$$

where

$$\mathbf{W}_{l|k} = \prod_{i=l}^{k-1} \mathbf{W}_{i|i+1}. \tag{7.117}$$

Filtering

For the filtering step, it is assumed that the prior parameters $\mathbf{x}^s_{k:n|k-1}$ and $\mathbf{P}^s_{k:n|k-1}$ for sensor s are given. As the measurement error is assumed to be independent from the past, the sensor likelihood function can be expressed by an application of projections $\mathbf{\Pi}_k$ onto the current state:

$$p(\mathbf{z}^s_k|\mathbf{x}_k) = p(\mathbf{z}^s_k|\mathbf{\Pi}_k\mathbf{x}_{k:n}) \tag{7.118}$$

$$= \mathcal{N}(\mathbf{z}^s_k;\ \mathbf{H}^s_k\mathbf{\Pi}_k\mathbf{x}_{k:n},\ \mathbf{R}^s_k), \tag{7.119}$$

where

$$\mathbf{\Pi}_k = (\mathbf{1}, \mathbf{O}, \dots, \mathbf{O}) \tag{7.120}$$

is a projection matrix to the current state $\mathbf{x}_k$. Then, the posterior parameters are obtained by the multiplication of the local prior density and the likelihood function. An application of the product formula (3.123) yields

$$\mathbf{x}^s_{k:n|k} = \mathbf{x}^s_{k:n|k-1} + \mathbf{W}^s_{k:n|k-1}(\mathbf{z}^s_k - \mathbf{H}^s_k\mathbf{\Pi}_k\mathbf{x}^s_{k:n|k-1}), \tag{7.121}$$

$$\mathbf{W}^s_{k:n|k-1} = \mathbf{P}^s_{k:n|k-1}\mathbf{\Pi}^\top_k\mathbf{H}^{s\top}_k\mathbf{S}^{s\,-1}_k, \tag{7.122}$$

$$\mathbf{S}^s_k = \mathbf{H}^s_k\mathbf{\Pi}_k\mathbf{P}^s_{k:n|k-1}\mathbf{\Pi}^\top_k\mathbf{H}^{s\top}_k + \mathbf{R}^s_k, \tag{7.123}$$

$$\mathbf{P}^s_{k:n|k} = \mathbf{P}^s_{k:n|k-1} - \mathbf{W}^s_{k:n|k-1}\mathbf{S}^s_k\mathbf{W}^{s\top}_{k:n|k-1}. \tag{7.124}$$

Sliding window mechanism

After the filtering step of the multi-sensor ASD, the local parameters have the following structure:

$$\mathbf{x}^s_{k:n|k} = \begin{pmatrix} \mathbf{x}^s_{k|k} \\ \mathbf{x}^s_{k-1|k} \\ \vdots \\ \mathbf{x}^s_{n|k} \end{pmatrix}, \tag{7.125}$$

$$\mathbf{P}^s_{k:n|k} = \begin{pmatrix} \mathbf{P}^s_{k|k} & \mathbf{P}^s_{k|k}\mathbf{W}^{s\top}_{k-1:n} \\ \mathbf{W}^s_{k-1:n}\mathbf{P}^s_{k|k} & \mathbf{P}^s_{k-1:n|k} \end{pmatrix}. \tag{7.126}$$

If $\dim_{\mathbf{x}}$ denotes the dimension of the state $\mathbf{x}_k$, both the ASD parameter $\mathbf{x}^s_{k:n|k}$ and the square matrix $\mathbf{P}^s_{k:n|k}$ have the dimension $(k-n+1)\cdot\dim_{\mathbf{x}}$. In order to keep the dimension fixed during the prediction–filtering iterations, the block row entry which refers to time t_n can be skipped since it has the least impact on the estimate of the current time t_k. This can be seen by the off-diagonal entries of the posterior ASD covariance matrix since these sub-matrices describe the correlations of the estimates. As given in (7.116), the cross-covariance of $\mathbf{x}_k$ and $\mathbf{x}_l$ for some $l < k$ is given by

$$\text{cov}\left[\mathbf{x}_n, \mathbf{x}_k | \mathcal{Z}^k\right] = \mathbf{W}_{n|k} \cdot \mathbf{P}_{k|k} \tag{7.127}$$

$$= \prod_{l=n}^{k-1} \mathbf{W}_{l|l+1} \cdot \mathbf{P}_{k|k} \tag{7.128}$$

Using the definition of $\mathbf{W}_{l|l+1}$ and the fact that $\det(\mathbf{F}_{l+1|l}) = 1$ for most tracking applications, one can see that

$$\det(\mathbf{W}_{l|l+1}) = \frac{\det(\mathbf{P}_{l|l})}{\det(\mathbf{P}_{l|l} + \mathbf{Q}_{l+1|l})} < 1. \tag{7.129}$$

This implies the following relation:

$$\det(\mathbf{W}_{n|k}) < \det(\mathbf{W}_{l|k}) < \det(\mathbf{W}_{k-1|k}) \text{ for } n < l < k, \tag{7.130}$$

which clearly shows that the estimate for time t_n has the least impact on time t_k. As a result, the corresponding block entries are omitted and one obtains the fixed-size parameters:

$$\mathbf{x}^s_{k:n+1|k} = \begin{pmatrix} \mathbf{x}^s_{k|k} \\ \mathbf{x}^s_{k-1|k} \\ \vdots \\ \mathbf{x}^s_{n+1|k} \end{pmatrix}, \tag{7.131}$$

$$\mathbf{P}^s_{k:n+1|k} = \begin{pmatrix} \mathbf{P}^s_{k|k} & \mathbf{P}^s_{k|k}\mathbf{W}^{s\top}_{k-1:n+1} \\ \mathbf{W}^s_{k-1:n+1}\mathbf{P}^s_{k|k} & \mathbf{P}^s_{k-1:n+1|k} \end{pmatrix}. \tag{7.132}$$

It should be noted that the fused estimate density is an approximation to the exact result which is obtained by transmitting the full ASD length. However, it can be seen by a numerical evaluation that already for ASDs with only few states as a sliding window hardly any difference to the optimal centralized solution can be observed for the estimation error of the current time t_k. Figure 7.6 shows the normalized† *root mean square error* (RMSE) in position with respect to the RMSE of a centralized Kalman filter (CKF) averaged over 100 Monte Carlo runs for different ASD lengths in a typical target tracking scenario with four sensors. It can be seen that in this scenario, an ASD length of 10 states is already sufficient to provide a result which matches the estimate of the CKF very close.

The limiting case of an ASD length 1 is equivalent to a federated Kalman filter as derived in Section 7.4. The numerical comparison of this scenario is shown in Figure 7.5. This is obvious from the formulae since the ASD with a window size of one is a single state Kalman filter with the relaxed evolution model. This is exactly how the federated Kalman filter was defined.

Discussion

The DASD filter provides the means for optimal or approximative distributed fusion without the requirements of the distributed Kalman filter. In particular, it is of great

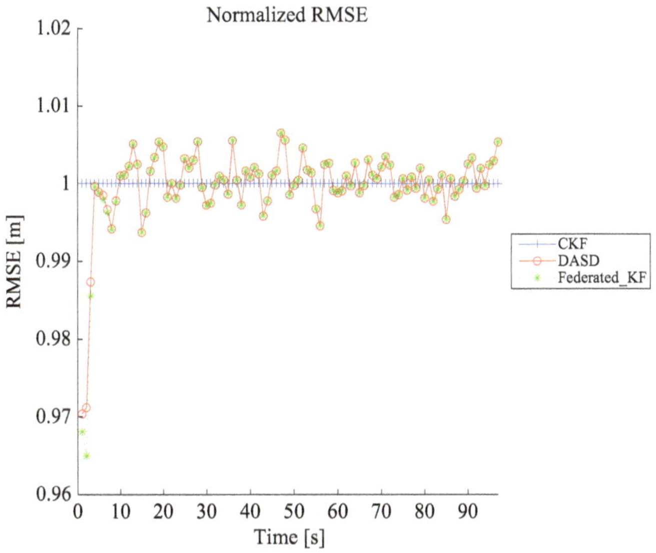

Figure 7.5 Normalized RMSE in position for the same algorithms as in Figure 7.6 for a sliding window of length 1. Here, the DASD result is equivalent to the federated Kalman filter.

†That is, a normalized RMSE of 1 implies that the estimate is identical to the CKF.

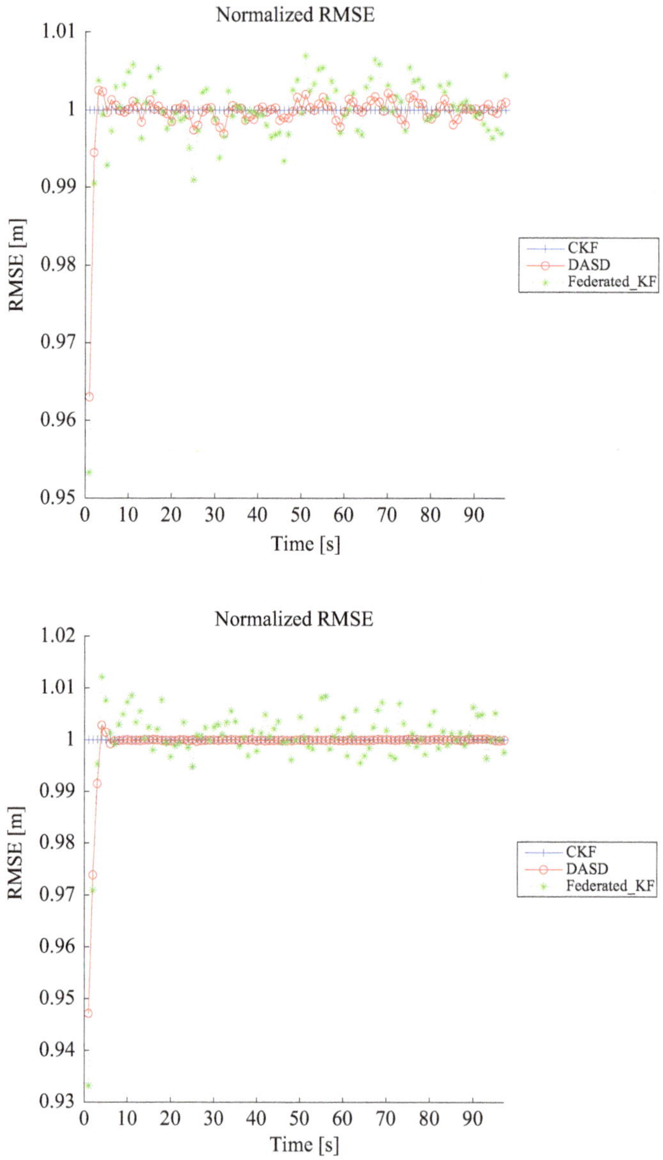

Figure 7.6 Normalized Root Mean Square Error *(RMSE) in position for a centralized Kalman filter (CKF) and the Distributed Accumulated State Density (DASD) and a Federated Kalman filter. The ASD length was fixed to include 5 (top), and 10 (bottom) states, respectively.*

practical importance that the measurement model of the remote platforms may well be unknown. This opens the approach to all kinds of non-linear scenarios, where the local measurement matrix $\mathbf{H}_k^s$ must be approximated by some linearization technique. This comes, of cause, at a cost. In this case, the additional transmission cannot be neglected. The transmission of the local ASD posterior covariance requires to send $(k - n + 1) \cdot (k - n + 2)/2$ matrices[‡] in the dimension of the state.

[‡]The number of independent elements in a symmetric matrix of dimension n is $\frac{n(n+1)}{2}$.

Chapter 8
Track-to-track association and distributed detection

8.1 Track-to-track association

Within the methods of the previous chapters, it was assumed that there is a jointly observed object or system, whose state is of interest and shall be estimated based on the obtained measurements from multiple sensors. In multi-target tracking, track association conflicts may arise in closely spaced scenarios, that is, the assignment of mutual tracks from the sensors such that they refer to the same target can be ambiguous. This problem is analogous to the measurement-to-track association problem for the single sensor case, except the fact that the measurements are not correlated to the prior estimates of the target, whereas we have seen previously that locally optimal tracks are mutually correlated. *Track-to-track-association* (T2TA) therefore is an important issue in dense scenarios, that is, if multiple targets are close to each other, that the association of tracks which refer to the same object is not obvious. An exemplary scenario is depicted in Figure 8.1.

Let n be the number of tracks given by each of two sensors ($s \in \{1, 2\}$), which are to be associated. The two sensor scenario is sufficient to discuss the procedure, since an iterative argument can be applied to obtain the general case.

> Note that the time synchronization which is assumed here is not given in practical applications due to the heterogeneous nature of the sensor architectures, data caching, and transmission link qualities. This however may easily be solved by choosing some joint current time for both sensors
>
> $$\hat{t}_k := \max\left\{t_k^1, t_k^2\right\}. \tag{8.1}$$
>
> The application of a prediction step on all tracks to this joint current time yields the corresponding statistics. Those "priors" may well be treated as some synchronized posteriors for the application of T2TA and T2TF.

For the sake of notational simplicity, the sensor index is omitted and instead index variable i is used for the track candidate from sensor 1 and j is used for sensor 2,

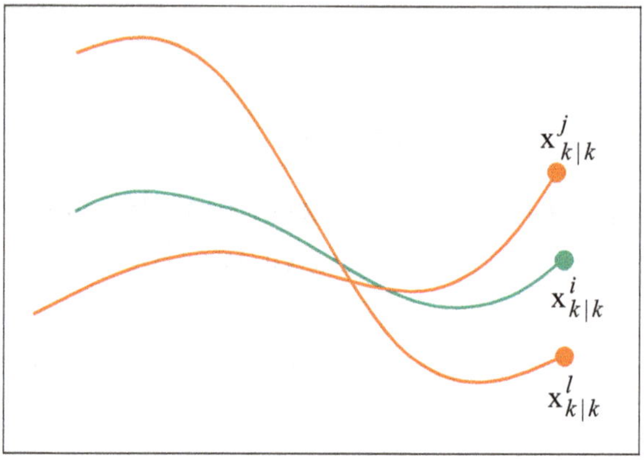

Figure 8.1 The problem of track-to-track association may be ambiguous, if targets are closely spaced. In this figure, the green track is from sensor 1 and the orange tracks are from sensor 2.

respectively. The objective for stochastically optimal association is to maximize the association likelihood given by

$$P(\mathbf{a}) = \prod_{i=1}^{n} \ell(\mathbf{a}_{i,j}), \tag{8.2}$$

where $\mathbf{a}_{i,j}$ is the association matrix with

$$\mathbf{a}_{i,j} = \begin{cases} 1, & \text{if track } i \text{ from sensor 1 is associated with track } j \text{ from sensor 2} \\ 0, & \text{else.} \end{cases}$$

Instead of the association likelihood density function $\ell(\mathbf{a}_{i,j})$, it is convenient to derive a solution based on an additive cost functions given by the negative logarithm:

$$C(\mathbf{a}) = \sum_{i,j} c(\mathbf{a}_{i,j}) \tag{8.3}$$

$$c(\mathbf{a}_{i,j}) := -2\ln(\ell(\mathbf{a}_{i,j})). \tag{8.4}$$

Factors of proportionality can be neglected such that one obtains metrics for Chi-square tests.

During the past 50 years, various metrics have been proposed for T2TA. A pioneering work by Singer and Kanyuck used

$$c(\mathbf{a}_{i,j}) = (\mathbf{x}^i_{k|k} - \mathbf{x}^j_{k|k})^\top (\mathbf{P}^i_{k|k} + \mathbf{P}^j_{k|k})^{-1} (\mathbf{x}^i_{k|k} - \mathbf{x}^j_{k|k}), \tag{8.5}$$

which comes from the likelihood

$$\ell(\mathbf{a}_{i,j}) = \int \mathrm{d}\mathbf{x}\ \mathcal{N}(\mathbf{x};\ \mathbf{x}^i_{k|k},\ \mathbf{P}^i_{k|k})\ \mathcal{N}(\mathbf{x};\ \mathbf{x}^j_{k|k},\ \mathbf{P}^j_{k|k}) \tag{8.6}$$

$$= \mathcal{N}(\mathbf{x}^i_{k|k};\ \mathbf{x}^j_{k|k},\ (\mathbf{P}^i_{k|k} + \mathbf{P}^j_{k|k})). \tag{8.7}$$

Obviously, this metric is equivalent to the gating or measurement-to-track association problem for single sensor target tracking with $\mathbf{H}_k = \mathbf{I}$ and $\mathbf{R}_k = \mathbf{P}^j_{k|k}$. Thus, the potential cross-covariance matrix is ignored.

If the cross-covariances are known, the Bar–Shalom metric given by

$$c(\mathbf{a}_{i,j}) = (\mathbf{x}^i_{k|k} - \mathbf{x}^j_{k|k})^\top (\mathbf{P}^i_{k|k} + \mathbf{P}^j_{k|k} - \mathbf{P}^{i,j}_{k|k} - \mathbf{P}^{i,j\,\top}_{k|k})^{-1} (\mathbf{x}^i_{k|k} - \mathbf{x}^j_{k|k}) \tag{8.8}$$

can be used in order to incorporate the correlated estimation error.

In the case that a common prior exists, which was used by both tracks and is denoted by $\bar{\mathbf{x}}_{k|k-1}, \bar{\mathbf{P}}_{k|k-1}$, then this joint information should be subtracted in order to avoid overconfidence of the fused error covariance. For the association metric, this yields

$$\begin{aligned} c(\mathbf{a}_{i,j}) &= (\mathbf{x}_{k|k} - \mathbf{x}^i_{k|k})^\top (\mathbf{P}^i_{k|k})^{-1} (\mathbf{x}_{k|k} - \mathbf{x}^i_{k|k}) \\ &\quad + (\mathbf{x}_{k|k} - \mathbf{x}^j_{k|k})^\top (\mathbf{P}^j_{k|k})^{-1} (\mathbf{x}_{k|k} - \mathbf{x}^j_{k|k}) \\ &\quad - (\mathbf{x}_{k|k} - \bar{\mathbf{x}}_{k|k-1})^\top (\bar{\mathbf{P}}_{k|k-1})^{-1} (\mathbf{x}_{k|k} - \bar{\mathbf{x}}_{k|k-1}). \end{aligned} \tag{8.9}$$

Here $\mathbf{x}_{k|k}$ denotes the fused estimate with its error covariance matrix $\mathbf{P}_{k|k}$ from both tracks.

In many applications, the tracks are sufficiently separated such that the *Mahalanobis*-distance in (8.5) performs well. In closely spaced scenarios with high track density, it is highly beneficial to incorporate the track history in order to avoid false association. The simple example shown in Figure 8.1 already indicates that track j has a higher similarity to track i than track l, which only becomes evident when the history is considered, too. Based on the ASD estimates as derived in Chapter 4, the Mahalanobis-distance easily expands to either the full trajectory or parts of it in terms of a sliding window:

$$c(\mathbf{a}_{i,j}) = (\mathbf{x}^i_{k:n|k} - \mathbf{x}^j_{k:n|k})^\top (\mathbf{P}^i_{k:n|k} + \mathbf{P}^j_{k:n|k})^{-1} (\mathbf{x}^i_{k:n|k} - \mathbf{x}^j_{k:n|k}). \tag{8.10}$$

This assumes that the ASD estimates from both sensors $\mathbf{x}^i_{k:n|k}$ and $\mathbf{x}^j_{k:n|k}$, respectively, have equal dimension and that all intermediate states are synchronized, too. However, the full ASD states are not required, if only the individual (smoothed) states and their error covariances as a result from the retrodiction formulae in (4.12) and (4.13) are available. By ignoring the cross-covariances between state estimates at different instants of time, one obtains

$$c(\mathbf{a}_{i,j}) = \sum_{l=n}^{k} (\mathbf{x}^i_{l|k} - \mathbf{x}^j_{l|k})^\top (\mathbf{P}^i_{l|k} + \mathbf{P}^j_{l|k})^{-1} (\mathbf{x}^i_{l|k} - \mathbf{x}^j_{l|k}). \tag{8.11}$$

For a fair comparison of different association costs, it is again important that the considered trajectory length is equal for all pairs of tracks.

8.2 Exact distributed sequential likelihood ratio test

Track extraction, confirmation, and deletion from a time series of data is challenging if non-detections and false alarms from the sensors are present. In Section 3.10, the *sequential likelihood ratio test* (SLRT) was presented, which calculates the likelihood ratio (LR) in order to test the existence based on two thresholds A and B, respectively. One of the great merits of the DKF is a distributed computation of the LR-score for the statistical test on track existence. The problem of the multi-sensor case is illustrated in Figure 8.2.

An approximative distributed computation of the decision is easily obtained by fusing local decisions based on local data. One could for instance use the mean of binary decision variables from each sensor. A more sophisticated approach incorporates a change detection on the local data distribution such as [64] for instance. In other decentralized decision schemes, local tracking filters on spatially distributed platforms $s = 1, \ldots, S$ compute their LR score to make a decision based on own sensor data $\mathcal{Z}_s^k = \{Z_1^s, \ldots, Z_k^s\}$ only. The set of decisions may then be exchanged via communication links and the global decision is made by means of maximum voting among all agents.

This is different in the distributed decision setup, where it is the goal to reconstruct the global LR-score based on the full information contained in the set of produced measurements $\mathcal{Z}^k = \{\mathcal{Z}_1^k, \ldots, \mathcal{Z}_S^k\}$. The optimal result is therefore equivalent to a *centralized Kalman filter* (CKF), which processes the full set of measurements from all sensors at each time step. In practical applications, the centralized approach often is hindered by limitations in the communication bandwidth between the processing units at the sensor site and the fusion center as well as by limited processing capabilities.

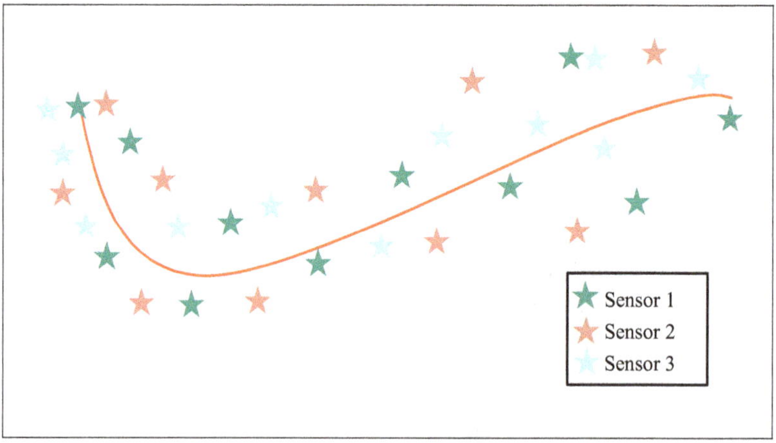

Figure 8.2 Problem of track existence based on measurements from multiple sensors

In particular in conditions with high number of false alarms and non-detections, a pre-processing for in order "dense" information is required. The approach described in this section proposes a distributed computation of LR(k) based on the product representation and the derivations of the DKF. Eventually, the fusion center is able to reconstruct the exact global LR-score based on scalar parameters transmitted by each sensor along with the tentative track in terms of its estimate and covariance matrix.

For the problem of track extraction, a decision based on the LR-score has to be computed. In Section 3.10, a recursive computation for the single sensor case was derived:

$$\mathrm{LR}(k) = \mathrm{LR}(k-1) \cdot \Lambda(k) \tag{8.12}$$

$$\Lambda(k) = \frac{\int \mathrm{d}\mathbf{x}_k \, p(Z_k|\mathbf{x}_k, h_1)\, p(\mathbf{x}_k|\mathcal{Z}^{k-1}, h_1)}{p(Z_k|h_0)} \tag{8.13}$$

$$= \int \mathrm{d}\mathbf{x}_k \Big((1-p_D) + \frac{p_D}{\rho_F} \sum_{j=1}^{m_k} \mathcal{N}(\mathbf{z}_j;\, \mathbf{H}_k\mathbf{x}_k,\, \mathbf{R}_k)\Big) p(\mathbf{x}_k|\mathcal{Z}^{k-1}). \tag{8.14}$$

Application of the DKF

The DKF is based on the product representation where the prior $p(\mathbf{x}_k|\mathcal{Z}^{k-1})$ is given by

$$p(\mathbf{x}_k|\mathcal{Z}^{k-1}) = \frac{1}{c_{k|k-1}} \prod_{s=1}^{S} \mathcal{N}(\mathbf{x}_k;\, \mathbf{x}^s_{k|k-1},\, \mathbf{P}^s_{k|k-1}) \tag{8.15}$$

For the computation of the resulting Gaussian density $p(\mathbf{x}_k|\mathcal{Z}^{k-1}) = \mathcal{N}(\mathbf{x}_k;\, \mathbf{x}_{k|k-1},\, \mathbf{P}_{k|k-1})$, the normalization constant $c_{k|k-1}$ was not required. However, this is different for calculating the factor $\Lambda(k)$ in (8.14). The normalization constant $c_{k|k-1}$ can be obtained by solving the integral:

$$c_{k|k-1} = \int \mathrm{d}\mathbf{x}_k \prod_{s=1}^{S} \mathcal{N}(\mathbf{x}_k;\, \mathbf{x}^s_{k|k-1},\, \mathbf{P}^s_{k|k-1}). \tag{8.16}$$

In case of a two sensor scenario ($S = 2$), a single application of the product formula directly yields

$$c_{k|k-1} = \mathcal{N}(\mathbf{x}^1_{k|k-1};\, \mathbf{x}^2_{k|k-1},\, \mathbf{P}^1_{k|k-1} + \mathbf{P}^2_{k|k-1}) \cdot \int \mathrm{d}\mathbf{x}_k \; \mathcal{N}(\mathbf{x}_k;\, \mathbf{x}_{k|k-1},\, \mathbf{P}_{k|k-1}) \tag{8.17}$$

$$= \mathcal{N}(\mathbf{x}^1_{k|k-1};\, \mathbf{x}^2_{k|k-1},\, \mathbf{P}^1_{k|k-1} + \mathbf{P}^2_{k|k-1}), \tag{8.18}$$

where

$$\mathbf{x}_{k|k-1} = \mathbf{P}_{k|k-1}((\mathbf{P}^1_{k|k-1})^{-1}\mathbf{x}^1_{k|k-1} + (\mathbf{P}^2_{k|k-1})^{-1}\mathbf{x}^2_{k|k-1}) \tag{8.19}$$

$$\mathbf{P}_{k|k-1} = ((\mathbf{P}^1_{k|k-1})^{-1} + (\mathbf{P}^2_{k|k-1})^{-1})^{-1} \tag{8.20}$$

are the fused parameters, which in this case are integrated away.

In the general case ($S > 2$), one obtains by an iterative application of the very same formula

$$c_{k|k-1} = \prod_{s=1}^{S-1} \mathcal{N}(\mathbf{x}_{k|k-1}^{s+1};\, \mathbf{x}_{k|k-1}^{(1:s)},\, \mathbf{P}_{k|k-1}^{(1:s)} + \mathbf{P}_{k|k-1}^{s+1}), \tag{8.21}$$

where $\mathbf{x}_{k|k-1}^{(1:s)}$ and $\mathbf{P}_{k|k-1}^{(1:s)}$ are the fused estimates and covariances resulting from the parameters from all sensors from 1 to s given by

$$\mathbf{x}_{k|k-1}^{(1:s)} = \mathbf{P}_{k|k-1}^{(1:s)} \sum_{i=1}^{s} (\mathbf{P}_{k|k-1}^{i})^{-1} \mathbf{x}_{k|k-1}^{i}, \tag{8.22}$$

$$\mathbf{P}_{k|k-1}^{(1:s)} = \left(\sum_{i=1}^{s} (\mathbf{P}_{k|k-1}^{i})^{-1} \right)^{-1}. \tag{8.23}$$

The product in (8.21) is initialized with the parameters

$$\mathbf{x}_{k|k-1}^{(1:1)} = \mathbf{x}_{k|k-1}^{1}, \tag{8.24}$$

$$\mathbf{P}_{k|k-1}^{(1:1)} = \mathbf{P}_{k|k-1}^{1}. \tag{8.25}$$

For the application on the DKF, one has to consider that at each time step, S mutually independent measurement processes are realized. Therefore, the density $p(Z_k|\mathbf{x}_k, h_1)$ which describes the joint likelihood for the set $Z_k = \{Z_k^1, \ldots, Z_k^S\}$ of current measurements from all sensors given the assumption that the track actually exists (h_1) factorizes. As a consequence, $\Lambda(k)$ in (8.14) for the DKF is given by

$$\Lambda(k) = \frac{1}{c_{k|k-1}} \int \mathrm{d}\mathbf{x}_k \prod_{s=1}^{S} \Bigg\{ \left((1 - p_D) + \frac{p_D}{\rho_F} \sum_{j=1}^{m_k^s} \mathcal{N}(\mathbf{z}_k^{j,s};\, \mathbf{H}_k^s \mathbf{x}_k,\, \mathbf{R}_k^s) \right) \cdot \mathcal{N}(\mathbf{x}_k; \mathbf{x}_{k|k-1}^s, \mathbf{P}_{k|k-1}^s) \Bigg\}. \tag{8.26}$$

For the sake of simplicity, it was assumed that the probability of detection p_D and the intensity of false alarms ρ_F is identical for all sensors. A generalization to the case of individual sensor parameters is straight forward. The total number of measurements is given by

$$m_k = \sum_{s=1}^{S} m_k^s, \tag{8.27}$$

where m_k^s is the number of observations from sensor s at time t_k.

By introducing the unnormalized weights

$$p^{\star j,s} = \begin{cases} (1 - p_D) & \text{if } j = 0 \\ \frac{p_D}{\rho_F} \mathcal{N}(\mathbf{z}_k^{j,s};\, \mathbf{H}_k^s \mathbf{x}_{k|k-1}^s,\, \mathbf{S}_k^s) & \text{else} \end{cases} \tag{8.28}$$

and applying the product formula, the factor is given by

$$\Lambda(k) = \frac{1}{c_{k|k-1}} \int \mathrm{d}\mathbf{x}_k \Big\{ \prod_{s=1}^{S} \sum_{j=0}^{m_k^s} p^{\star j,s} \mathcal{N}(\mathbf{x}_k;\, \mathbf{x}_{k|k}^{j,s},\, \mathbf{P}_{k|k}^{j,s}) \Big\} \tag{8.29}$$

where the following abbreviations were used

$$\mathbf{S}_k^s = \mathbf{H}_k^s \mathbf{P}_{k|k-1}^s \mathbf{H}_k^{s\top} + \mathbf{R}_k^s, \tag{8.30}$$

$$\mathbf{x}_{k|k}^{j,s} = \mathbf{x}_{k|k-1}^s + \mathbf{W}_{k|k-1}^s (\mathbf{z}_k^{j,s} - \mathbf{H}_k^s \mathbf{x}_{k|k-1}^s), \tag{8.31}$$

$$\mathbf{W}_{k|k-1}^s = \mathbf{P}_{k|k-1}^s \mathbf{H}_k^{s\top} (\mathbf{S}_k^s)^{-1}, \tag{8.32}$$

$$\mathbf{P}_{k|k}^{j,s} = \mathbf{P}_{k|k-1}^s - \mathbf{W}_{k|k-1}^s \mathbf{S}_k^s \mathbf{W}_{k|k-1}^{s\top}, \tag{8.33}$$

$$\mathbf{x}_{k|k}^{0,s} = \mathbf{x}_{k|k-1}^s, \tag{8.34}$$

$$\mathbf{P}_{k|k}^{0,s} = \mathbf{P}_{k|k-1}^s. \tag{8.35}$$

Now, the normalized weights are given by

$$p^{j,s} = \frac{p^{\star j,s}}{\bar{p}^s} \tag{8.36}$$

where

$$\bar{p}^s = \sum_{j=0}^{m_s} p^{\star j,s}. \tag{8.37}$$

Therefore, one has that

$$\sum_{j=0}^{m_s} p^{\star j,s} \mathcal{N}(\mathbf{x}_k;\, \mathbf{x}_{k|k}^{j,s},\, \mathbf{P}_{k|k}^{j,s}) = \bar{p}^s \sum_{j=0}^{m_s} p^{j,s} \mathcal{N}(\mathbf{x}_k;\, \mathbf{x}_{k|k}^{j,s},\, \mathbf{P}_{k|k}^{j,s}). \tag{8.38}$$

Since $\sum_{j=0}^{m_s} p^{j,s} \mathcal{N}(\mathbf{x}_k;\, \mathbf{x}_{k|k}^{j,s},\, \mathbf{P}_{k|k}^{j,s})$ is a normalized Gaussian mixture, moment matching can be applied to compute an approximation in terms of a single Gaussian:

$$\sum_{j=0}^{m_s} p^{j,s} \mathcal{N}(\mathbf{x}_k;\, \mathbf{x}_{k|k}^{j,s},\, \mathbf{P}_{k|k}^{j,s}) \approx \mathcal{N}(\mathbf{x}_k;\, \mathbf{x}_{k|k}^{s},\, \mathbf{P}_{k|k}^{s}) \tag{8.39}$$

where

$$\mathbf{x}_{k|k}^s = \sum_{j=0}^{m_s} p^{j,s} \mathbf{x}_{k|k}^{j,s} \tag{8.40}$$

$$\mathbf{P}_{k|k}^s = \sum_{j=0}^{m_s} p^{j,s} \Big\{ \mathbf{P}_{k|k}^{j,s} + (\mathbf{x}_{k|k}^{j,s} - \mathbf{x}_{k|k}^s)(\mathbf{x}_{k|k}^{j,s} - \mathbf{x}_{k|k}^s)^\top \Big\}. \tag{8.41}$$

Using this approximation in the factor $\Lambda(k)$ case yields

$$\Lambda(k) = \frac{1}{c_{k|k-1}} \int \mathrm{d}\mathbf{x}_k \prod_{s=1}^{S} \bar{p}^s \mathcal{N}(\mathbf{x}_k;\, \mathbf{x}_{k|k}^{s},\, \mathbf{P}_{k|k}^{s}). \tag{8.42}$$

An iterative application of the product formula as in (8.21) yields

$$\Lambda(k) = \frac{c_{k|k}}{c_{k|k-1}} \prod_{s=1}^{S} \bar{p}^s \int \mathrm{d}\mathbf{x}_k \; \mathcal{N}(\mathbf{x}_k; \mathbf{x}_{k|k}, \mathbf{P}_{k|k}) \tag{8.43}$$

$$= \frac{c_{k|k}}{c_{k|k-1}} \prod_{s=1}^{S} \bar{p}^s \tag{8.44}$$

where the posterior normalization constant is given by

$$c_{k|k} = \prod_{s=1}^{S-1} \mathcal{N}(\mathbf{x}_{k|k}^{s+1}; \mathbf{x}_{k|k}^{(1:s)}, \mathbf{P}_{k|k}^{(1:s)} + \mathbf{P}_{k|k}^{s+1}), \tag{8.45}$$

$$\mathbf{x}_{k|k}^{(1:s)} = \mathbf{P}_{k|k}^{(1:s)} \sum_{i=1}^{s} (\mathbf{P}_{k|k}^{i})^{-1} \mathbf{x}_{k|k}^{i}, \tag{8.46}$$

$$\mathbf{P}_{k|k}^{(1:s)} = \left(\sum_{i=1}^{s} (\mathbf{P}_{k|k}^{i})^{-1} \right)^{-1}. \tag{8.47}$$

The processing scheme is concluded in Figure 8.3.

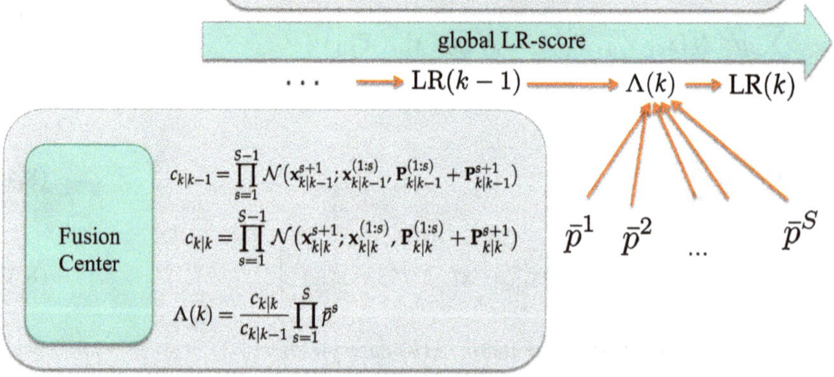

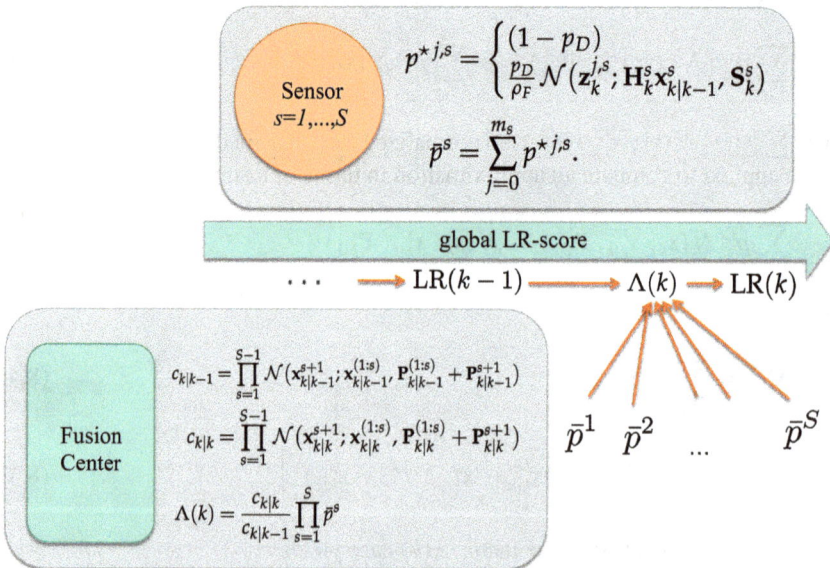

Figure 8.3 Processing scheme of the distributed computation of the likelihood ratio (LR) score for optimal decisions on track existence

Discussion

The distributed computation of the likelihood ratio score allows a fast and easy reconstruction of the exact value under conditions where the distributed Kalman filter works. In particular, additional communication to the fusion center is limited to sharing an additional real valued $\bar{p}^s$ number for each processing platform s. The normalization constants $c_{k|k-1}$ and $c_{k|k}$ are obtained from the prior and posterior parameters, respectively, where the local priors can be computed by a prediction step from the previous posterior without additional transmissions, if full communication is available.

8.3 Approximative distributed sequential likelihood ratio test

In Chapter 7, it was discussed that the practical applications of the DKF are limited since it requires either communication of the fused covariance matrix or knowledge on the remote sensor models in order to compute the globalized local tracks, which are fully decorrelated. Approximative versions such as the *federated Kalman filter* (FKF) or *the naïve fusion* (NF) circumvent this requirement and therefore enable a distributed computation in non-linear scenarios with measurement ambiguity present without full communication. A distributed non-linear scenario is illustrated in Figure 8.4 since the sensor measurements from radars, passive emitter bearing antennas, and imaging systems have a non-linear relationship with respect to the Cartesian coordinates and the kinematic parameters of a tracked air fighter. In addition, false alarms and non-detections may appear such that the local estimation error covariance matrices become highly data dependent. In this section, these relaxations on the requirements are applied to the distributed computation of the LR-score for a distributed computation

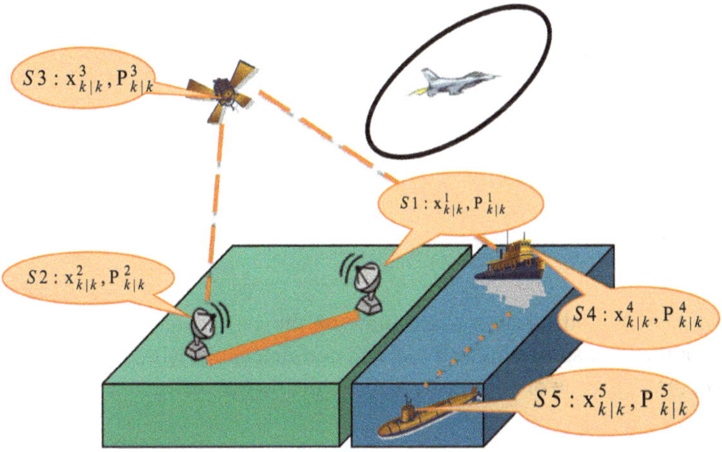

Figure 8.4 Illustration of a distributed non-linear scenario for target existence decision and tracking

of decisions on track existence. The approximation required to relax the assumptions often is insignificant relative to the LR-score, such that the resulting algorithm can well be used in practical applications.

To this end, one may differentiate between the two cases

(a) The local trackers can be modified.
(b) The local trackers are readable.

The first case (a) refers to the FKF, where the prediction step of the local tracks is modified* in order to account for the cross-correlations of the estimation errors, which come from the process noise of the jointly observed targets as discussed in Chapter 5. In other applications of T2TF, the local tracks are computed in proprietary black-box systems such that the corresponding estimate and covariance matrix cannot be altered. This is what the second case (b) refers to, in which the local tracks remain locally optimal but a mutual independence is assumed which also yields a product representation based on the NF.

In both cases, it is assumed that the global prior on the current state $\mathbf{x}_k$ can be approximated by mutual independent local densities, such that the product representation holds:

$$p(\mathbf{x}_k|\mathcal{Z}^{k-1}) \propto \prod_{s=1}^{S} \mathcal{N}(\mathbf{x}_k;\, \mathbf{x}^s_{k|k-1},\, \mathbf{P}^s_{k|k-1}). \tag{8.48}$$

Since the measurement error of the sensors is mutually independent, the filtering methodologies of Kalman-based algorithms can be applied also to the FKF/NF as in the DKF. An application of the Bayes' theorem directly yields

$$p(\mathbf{x}_k|\mathcal{Z}^{k}) \propto p(\mathbf{x}_k|\mathcal{Z}^{k-1}) \cdot \ell(Z_k;\mathbf{x}_k) \tag{8.49}$$

$$\propto \prod_{s=1}^{S} \mathcal{N}(\mathbf{x}_k;\, \mathbf{x}^s_{k|k-1},\, \mathbf{P}^s_{k|k-1})\ell(Z^s_k;\mathbf{x}_k), \tag{8.50}$$

where

$$\ell(Z_k;\mathbf{x}_k) = \prod_{s=1}^{S} \ell(Z^s_k;\mathbf{x}_k) \tag{8.51}$$

is the joint likelihood function of the state on the current data at time t_k. With the models for the false alarms intensity ρ_F and non-detections with probability $(1 - p_D)$ as above, the same line of arguments leads to the formulae:

*The modifications refer to a local optimal Kalman-based processor.

$$\Lambda(k) \approx \frac{1}{c_{k|k-1}} \int \mathrm{d}\mathbf{x}_k \Big\{ \prod_{s=1}^{S} \sum_{j=0}^{m_s} p^{\star j,s} \mathcal{N}(\mathbf{x}_k;\, \mathbf{x}_{k|k}^{j,s},\, \mathbf{P}_{k|k}^{j,s}) \Big\} \tag{8.52}$$

$$= \frac{c_{k|k}}{c_{k|k-1}} \prod_{s=1}^{S} \bar{p}^s \tag{8.53}$$

with

$$\bar{p}^s = \sum_{j=0}^{m_s} p^{\star j,s} \tag{8.54}$$

$$p^{\star j,s} = \begin{cases} (1 - p_D) & \text{if}\, j = 0 \\ \frac{p_D}{\rho_F} \mathcal{N}(\mathbf{z}_k^{j,s};\, \mathbf{h}_k^s(\mathbf{x}_{k|k-1}^s),\, \tilde{\mathbf{S}}_k^s) & \text{else.} \end{cases} \tag{8.55}$$

Here, $\mathbf{h}_k^s$ is the non-linear measurement function, which maps a state to its corresponding measurement coordinates. One may use a linearization of this function at the prior estimate $\mathbf{x}_{k|k-1}^s$ in order to obtain the approximative measurement matrix $\tilde{\mathbf{H}}_k^s$, which is required to compute the innovation covariance matrix

$$\tilde{\mathbf{S}}_k^s = \tilde{\mathbf{H}}_k^s \mathbf{P}_{k|k-1}^s \tilde{\mathbf{H}}_k^{s\top} + \mathbf{R}_k^s. \tag{8.56}$$

Discussion

The approximative computation of the likelihood ratio score LR(k) either does not compensate for the cross-covariances (NF) or uses the relaxed evolution model as in the FKF. The computation of the parameters is analogous to the exact version based on the DKF but since the approximative version does not require the globalization step, assumptions on knowledge regarding the remote sensor models are relaxed. Since the LR-score often has changes in higher orders of magnitudes, the degree of approximation may be considered low such that a reliable fused decision can be computed.

Appendix A

A.1 Inversion lemma

Assume that we are looking for the inverse of a block matrix consisting of four matrices **A**, **B**, **C**, and **D**. Its inverse has the same quadratic shape and can also be divided into sub-blocks **E**, **F**, **G**, and **H**. Since they commute, one has that

$$\begin{pmatrix} \mathbf{A} & \mathbf{B} \\ \mathbf{C} & \mathbf{D} \end{pmatrix} \cdot \begin{pmatrix} \mathbf{E} & \mathbf{F} \\ \mathbf{G} & \mathbf{H} \end{pmatrix} = \begin{pmatrix} \mathbf{I} & \mathbf{O} \\ \mathbf{O} & \mathbf{I} \end{pmatrix} \tag{A.1}$$

and

$$\begin{pmatrix} \mathbf{E} & \mathbf{F} \\ \mathbf{G} & \mathbf{H} \end{pmatrix} \cdot \begin{pmatrix} \mathbf{A} & \mathbf{B} \\ \mathbf{C} & \mathbf{D} \end{pmatrix} = \begin{pmatrix} \mathbf{I} & \mathbf{O} \\ \mathbf{O} & \mathbf{I} \end{pmatrix}. \tag{A.2}$$

From (A.1), a block of four equations can be derived:

$$\mathbf{AE} + \mathbf{BG} = \mathbf{I} \tag{A.3}$$

$$\mathbf{AF} + \mathbf{BH} = \mathbf{O} \tag{A.4}$$

$$\mathbf{CE} + \mathbf{DG} = \mathbf{O} \tag{A.5}$$

$$\mathbf{CF} + \mathbf{DH} = \mathbf{I} \tag{A.6}$$

Solving for the four unknowns gives us

$$\mathbf{BG} = -\mathbf{BD}^{-1}\mathbf{CE} \tag{A.7}$$

$$\mathbf{AE} - \mathbf{BD}^{-1}\mathbf{CE} = \mathbf{I} \tag{A.8}$$

$$\mathbf{E} = (\mathbf{A} - \mathbf{BD}^{-1}\mathbf{C})^{-1} \tag{A.9}$$

$$\mathbf{G} = -\mathbf{D}^{-1}\mathbf{CE} \tag{A.10}$$

$$= -\mathbf{D}^{-1}\mathbf{C}(\mathbf{A} - \mathbf{BD}^{-1}\mathbf{C})^{-1} \tag{A.11}$$

$$\mathbf{H} = \mathbf{D}^{-1} - \mathbf{GBD}^{-1} \tag{A.12}$$

$$= \mathbf{D}^{-1} + \mathbf{D}^{-1}\mathbf{C}(\mathbf{A} - \mathbf{BD}^{-1}\mathbf{C})^{-1}\mathbf{BD}^{-1} \tag{A.13}$$

$$\mathbf{CF} = -\mathbf{CA}^{-1}\mathbf{BH} \tag{A.14}$$

$$-\mathbf{CA}^{-1}\mathbf{BH} + \mathbf{DH} = \mathbf{I} \tag{A.15}$$

$$\mathbf{F} = -\mathbf{A}^{-1}\mathbf{B}(\mathbf{D} - \mathbf{CA}^{-1}\mathbf{B})^{-1} \tag{A.16}$$

Analogously, (A.2) yields another set of four equations:

$$\mathbf{EA} + \mathbf{FC} = \mathbf{I} \tag{A.17}$$

$$\mathbf{EB} + \mathbf{FD} = \mathbf{O} \tag{A.18}$$

$$\mathbf{GA} + \mathbf{HC} = \mathbf{O} \tag{A.19}$$

$$\mathbf{GB} + \mathbf{HD} = \mathbf{I} \tag{A.20}$$

which result in

$$\mathbf{E} = \mathbf{A}^{-1} - \mathbf{FCA}^{-1} \tag{A.21}$$

$$= \mathbf{A}^{-1} + \mathbf{A}^{-1}\mathbf{B}(\mathbf{D} - \mathbf{CA}^{-1}\mathbf{B})^{-1}\mathbf{CA}^{-1} \tag{A.22}$$

$$\mathbf{F} = -(\mathbf{A} - \mathbf{BD}^{-1}\mathbf{C})^{-1}\mathbf{BD}^{-1} \tag{A.23}$$

$$\mathbf{GB} = -\mathbf{HCA}^{-1}\mathbf{B} \tag{A.24}$$

$$-\mathbf{HCA}^{-1}\mathbf{B} + \mathbf{HD} = \mathbf{I} \tag{A.25}$$

$$\mathbf{H} = (\mathbf{D} - \mathbf{CA}^{-1}\mathbf{B})^{-1} \tag{A.26}$$

$$\mathbf{G} = -(\mathbf{D} - \mathbf{CA}^{-1}\mathbf{B})^{-1}\mathbf{CA}^{-1} \tag{A.27}$$

Since the inverse matrix is unique, both equations must be equivalent.

Therefore, we have that

$$\begin{pmatrix} \mathbf{A} & \mathbf{B} \\ \mathbf{C} & \mathbf{D} \end{pmatrix}^{-1} = \begin{pmatrix} (\mathbf{A} - \mathbf{BD}^{-1}\mathbf{C})^{-1} & -(\mathbf{A} - \mathbf{BD}^{-1}\mathbf{C})^{-1}\mathbf{BD}^{-1} \\ -\mathbf{D}^{-1}\mathbf{C}(\mathbf{A} - \mathbf{BD}^{-1}\mathbf{C})^{-1} & \mathbf{D}^{-1} + \mathbf{D}^{-1}\mathbf{C}(\mathbf{A} - \mathbf{BD}^{-1}\mathbf{C})^{-1}\mathbf{BD}^{-1} \end{pmatrix} \tag{A.28}$$

$$= \begin{pmatrix} \mathbf{A}^{-1} + \mathbf{A}^{-1}\mathbf{B}(\mathbf{D} - \mathbf{CA}^{-1}\mathbf{B})^{-1}\mathbf{CA}^{-1} & -\mathbf{A}^{-1}\mathbf{B}(\mathbf{D} - \mathbf{CA}^{-1}\mathbf{B})^{-1} \\ -(\mathbf{D} - \mathbf{CA}^{-1}\mathbf{B})^{-1}\mathbf{CA}^{-1} & (\mathbf{D} - \mathbf{CA}^{-1}\mathbf{B})^{-1} \end{pmatrix}. \tag{A.29}$$

In particular, one can see that the following four equations hold:

$$(\mathbf{A} - \mathbf{BD}^{-1}\mathbf{C})^{-1} = \mathbf{A}^{-1} + \mathbf{A}^{-1}\mathbf{B}(\mathbf{D} - \mathbf{CA}^{-1}\mathbf{B})^{-1}\mathbf{CA}^{-1} \tag{A.30}$$

$$-(\mathbf{A} - \mathbf{BD}^{-1}\mathbf{C})^{-1}\mathbf{BD}^{-1} = -\mathbf{A}^{-1}\mathbf{B}(\mathbf{D} - \mathbf{CA}^{-1}\mathbf{B})^{-1} \tag{A.31}$$

$$-\mathbf{D}^{-1}\mathbf{C}(\mathbf{A} - \mathbf{BD}^{-1}\mathbf{C})^{-1} = -(\mathbf{D} - \mathbf{CA}^{-1}\mathbf{B})^{-1}\mathbf{CA}^{-1} \tag{A.32}$$

$$\mathbf{D}^{-1} + \mathbf{D}^{-1}\mathbf{C}(\mathbf{A} - \mathbf{BD}^{-1}\mathbf{C})^{-1}\mathbf{BD}^{-1} = (\mathbf{D} - \mathbf{CA}^{-1}\mathbf{B})^{-1} \tag{A.33}$$

These equations are known as the *matrix inversion lemma.*

It is obvious that the inverse matrices of **A**, **D**, and of the "*Schur-complements*" $(\mathbf{A} - \mathbf{BD}^{-1}\mathbf{C})$ and $(\mathbf{D} - \mathbf{CA}^{-1}\mathbf{B})$ must exist.

Appendix B

B.1 Test code for linear—Gaussian Feynman Kernel

This code runs a test simulation on the linear Gaussian Feynman kernel for an initial Gaussian density $p(\mathbf{x}_n|\mathcal{Z}^n) = \mathcal{N}(\mathbf{x}_n; \mathbf{x}, \mathbf{P})$.

In the first section, initial values are set. The second section runs a single Monte Carlo simulation and iteratively calculates the Kalman filter result x_kf (mean) and P_kf (covariance). In the third section, the Feynman kernel update is applied and the result x_kk (mean) and P_kk (covariance) is compared to the Kalman filter.

```
% initial values for x\_n (truth)
P = eye(4) * 10;
x = [100 100 10 10]';

T = 1; % time interval between subsequent steps
q = 1; % process noise

[F,Q] = dynamics(T,q); % dynamics model matrices

R = eye(2) * 25; % measurement error covariance
H = [eye(2) zeros(2)]; % measurement matrix

% Initial values for Kalman filter
x_kf = x;
P_kf = P;

% Initial values for Feynman kernel update
x_n = x;
P_n = P;

steps = 50; % number of steps
z_history = zeros(2,steps); % array to store all measurements


%%
for t=1:steps
    x = F * x + mvnrnd(zeros(4,1),Q)'; % prediction of truth
    z = x(1:2) + mvnrnd([0 0], R)'; % generation of a measurement
    z_history(:,t) = z; % save the measurement

```

```
    % Kalman filter prediction
    x_kf = F * x_kf;
    P_kf = F * P_kf * F' + Q;

    % Kalman filter update
    S_kf = H * P_kf * H' + R;
    W_kf = P_kf * H' * S_kf^-1;

    x_kf = x_kf + W_kf * (z - H * x_kf);
    P_kf = P_kf - W_kf * S_kf * W_kf';

end


%%
% align accumulated measurment vector
bz = zeros(2*steps,1);
i=1;
for k=steps:-1:1
    bz(i:i+1) = z_history(:,k);
    i = i + 2;
end
% align accumulated matrices for measurement error,
% process noise, measuremnt matrix and transition matrix
R_kn = kron(eye(steps),R);
Q_kn = [];
Q_tmp = Q; % only used for alignmnent
for k=1:steps
    if k > 1
        G = F * Q_kn(1:4,:);
    else
        G = [];
    end
    Q_kn = [Q_tmp, G; G', Q_kn]; %#ok<*AGROW>
    Q_tmp = F * Q_tmp * F' + Q;
end
H_kn = kron(eye(steps),H);
F_kn = [];
for i=steps:-1:1
    F_kn = [F_kn; F^i];
end

% Feynman kernel update
S_kn = H_kn * F_kn * P_n * F_kn' * H_kn' + H_kn * Q_kn * H_kn' + R_kn;
W_kn = P_n * F_kn' * H_kn' * S_kn^-1;
m_kn = x_n + W_kn * (bz - H_kn * F_kn * x_n);
M_kn = P_n - W_kn * S_kn * W_kn';

Pi_k = [eye(4) zeros(4,(steps-1)*4)];
U = H_kn * Q_kn * H_kn' + R_kn;
J = Q_kn * H_kn' * U^-1;
G = Q_kn - J * U * J';
T = F_kn - J * H_kn * F_kn;

```

```
x_kk = Pi_k * (F_kn * m_kn + J * (bz - H_kn * F_kn * m_kn));
P_kk = Pi_k * (T * M_kn * T' + G) * Pi_k';

% compare the results
disp('x diff:')
disp(x_kf - x_kk)

disp('P diff:')
disp(P_kf - P_kk)
```

```
x diff:
   1.0e-07 *

    0.0632
    0.1512
    0.0019
    0.0046

P diff:
   1.0e-05 *

   -0.6500         0   -0.0198         0
         0   -0.6479         0   -0.0198
   -0.0198         0   -0.0006         0
         0   -0.0198         0   -0.0006
```

Index

accumulated state density (ASD) 82–7, 168
notion of 81–2
recursive measurement fusion using 87–92
see also distributed accumulated state density (DASD) filter
Advanced Driver Assistance Systems (ADAS) 2, 113
Airborne Early Warning and Control System (AWACS) aircrafts 1
approximative distributed sequential likelihood ratio test 185–7
arbitrary gains, fusion with 119–21
Autonomous Driving (AD) 2

Bar-Shalom–Campo formula 115–17
batch processing 96–8
Bayes formula 3, 30, 58–9, 62, 66
Bayesian prediction-filtering recursion 113
Bayesian state estimation and target tracking 29
data association 52
global nearest neighbor 54–5
nearest neighbor 53
probabilistic data association 55–8
Fundamental Equations of Linear Estimation 32–5
interacting multiple model filter 69–72
Kalman filter 35–8
filtering step 36–8
as least squares solution 40–3
as orthogonal projection 43–52
prediction step 35–6
least squares 38–9
multi-hypotheses-tracker 63–9
Probabilistic Data Association Filter (PDAF) 59
moment matching 60–3
sequential likelihood ratio test (SLRT) 72–7
Bayes' theorem 82–3, 94, 96, 98, 105, 168, 186
Bernoulli distributed random variable 77
Bernoulli process 12
block-line combined estimator and smoother 98–103
Borell–Cantelli lemma 10
Brownian motion 4, 10, 16–18, 25
and diffusion equation 8–16

centralized fusion 113–14
centralized Kalman filter (CKF) 173–4, 180
Central Limit Theorem (CLT) 17
Chapman–Kolmogorov equation 103
Cholesky decomposition 43, 124
conjugate prior 32, 55–6, 60
continuous time retrodiction 89
Counter Unmanned Air Systems (C-UAS) 2
covariance, evolution of 23–4
covariance intersection (CI) 130–42
see also Inverse Covariance Intersection (ICI)
covariance intersection rule 125
covariance matrix 29, 32, 36, 53, 61. 84, 86, 144

cross-covariances
  distributed calculation of 123–6
  iterative calculation of 121–3

data association 52
  global nearest neighbor 54–5
  nearest neighbor 53
  probabilistic data association 55–8
decentralized fusion 114
diffusion equation 9, 18, 20, 24, 105
distributed accumulated state density (DASD) filter 168–75
distributed calculation of cross-covariances using square root decomposition 123–6
distributed fusion 1, 114
distributed Kalman filter (DKF) 153, 180–2
  federated Kalman filter (FKF) 161–2, 174, 185
  filtering 159–61
  graphs, consensus of agents in 165–8
  local error covariance globalization 154–7
  prediction 157–9
  sum formulation 163–5
distributed sequential likelihood ratio test
  approximative 185–7
  exact 180–5

ellipsoidal intersection 151
error covariance matrix 2, 5, 32, 77
exact distributed sequential likelihood ratio test 180–5

federated Kalman filter (FKF) 161–2, 174, 185
Feynman kernel 103–6, 191
Field of View (FoV) 55, 58, 73, 75
filtering step 36–8, 66, 77, 89, 91
Fisher Information Matrix (FIM) 41
Fokker–Planck equation 21, 25
free dynamic particle 106–8
free quantum particle 103
  free dynamic particle 106–8
  linear-Gaussian measurements 109–10
  state estimation for macroscopic objects with path integrals 105–6
full communication 3, 147–8
Fundamental Equations of Linear Estimation 32–5
fusion center (FC) node 113
Future Combat Air System (FCAS) 1

Gamma distribution 30
Gaussian density 32, 56, 59, 80, 89, 96
  application on 110–11
Gaussian distributed stochastic process 22
Gaussian-likelihood function 105, 161
Gaussian mixture 60–3, 65–7, 69
Gaussian probability density functions 8
generalized smoothing and out-of-sequence processing 92
  batch processing 96–8
  block-line combined estimator and smoother 98–103
globalized likelihood function 160–1
global nearest neighbor 54–5
graphs, consensus of agents in 165–8

Heisenberg's uncertainty principle 24, 103, 106
hypotheses-based target tracking 63

IMM algorithm 71
information aging 90
information parameters 41, 146
innovation error covariance 38
interacting multiple model filter 69–72
Inverse Covariance Intersection (ICI) 142–5
  optimal fusion with common information 144
  rule 144–5

Inversion Lemma (IL) 42, 189–90
Itô integral 19

Joint Directors of Laboratories (JDL) 25
Joseph form 51

Kalman-based algorithms 186
Kalman filter 3, 25, 35–8, 55–6, 81, 90, 111, 115, 117, 121, 191
  centralized Kalman filter (CKF) 173–4, 180
  distributed Kalman filter (DKF) 153, 180–2
    filtering 159–61
    graphs, consensus of agents in 165–8
    local error covariance globalization 154–7
    prediction 157–9
    sum formulation 163–5
  federated Kalman filter (FKF) 161–2, 174, 185
  filtering step 36–8
  as least squares solution 40–3
  as orthogonal projection 43–52
  prediction step 35–6
Kalman filter equations 42, 48, 51, 53, 65, 87, 93, 148
Kalman gain 37, 90
Kronecker–Delta function 18

Lagrangian function 103
laws of statistical mechanics 25
least squares 38–9
  Kalman filter as least squares solution 40–3
Least Squares (LS) estimate 118
likelihood function 3, 30–1, 65, 75–6, 90–1
  Gaussian-likelihood function 105, 161
  globalized likelihood function 160–1
likelihood ratio (LR) 73, 180
  distributed sequential likelihood ratio test
    approximative 185–7
    exact 180–5
  sequential likelihood ratio test (SLRT) 72–7, 180
    approximative distributed 185–7
linear Gaussian Feynman kernel 191–3
linear-Gaussian measurements 109–10
local error covariance globalization 154–7
Löwner–Heinz Theorem 144

Mahalanobis distance 53, 67–8, 179
Markov Chain 8, 69, 70, 77
Markovian transition kernel 8
Markov kernel 103
Markov process 77
Markov property 83–4, 105
Markov proposition 91, 170
maximum likelihood estimator (MLE) 39
mean, evolution of 23–4
mean drift velocity 13
measurement information matrix 146
measurement matrix 32, 39, 88, 175
moment matching 60–3
multi-hypotheses-tracker 63–9
multiple-hypotheses-tracker (MHT) 63–5, 72
multiple sensors 1–2, 5, 113, 177, 180
multivariate Brownian motion 16, 18

Naïve Fusion (NF) 127–30, 185
nearest neighbor (NN) 53
  global 54–5
Newton's laws of motion 7
normalization constant 30, 33, 162

orthogonal projection, Kalman filter as 43–52
Out-of-Sequence (OoS) data 88, 90

path integral 24–5
in quantum physics 25–7
for stochastic processes 27–8
posterior covariance 37, 40, 52, 127, 131, 133, 145
prediction 29
prediction step 35–6, 66, 77, 146
principle of uncertainty 24–5
probabilistic data association 55–8
probabilistic data association filter (PDAF) 59
moment matching 60–3
product formula 38

quantum physical interpretation 103
quantum physics, path integral in 25–7

Rauch–Tung–Striebel (RTS) equations 81
recursive measurement fusion using ASDs 87–92
relaxed evolution model 158
retrodiction 79–81
retrodiction gain matrix 80–1
Riemann integral 19
root mean square error (RMSE) 173–4

safe fusion 148–51
Schrödinger equation 24–5
Schur-Complement 129, 131–2
sensor model 30, 32
sequential likelihood ratio test (SLRT) 72–7, 180
singular value decomposition (SVD) 148
spread term 61
square root decomposition of a symmetric positive definite matrix 123–6
state estimation for macroscopic objects with path integrals 105–6
stochastic differential equation 19
mean and covariance, evolution of 23–4
stochastic motion 7
path integral 24
in quantum physics 25–7
for stochastic processes 27–8
Wiener process 16–18
stochastic part 19, 21–2
stochastic processes 8
Brownian motion and diffusion equation 8–16
path integral for 27–8
superposition 8, 19

Theorem of Orthogonal Projection 47
Theorem of the Orthogonal Estimate 46–7
tight bound, defined 134
Tight Bound Theorem 134, 139, 145
tracklet fusion 3, 145–8
track-to-track-association (T2TA) 177–9
track-to-track fusion (T2TF) 1, 113
arbitrary gains, fusion with 119–21
for arbitrary number of sensors with known cross-covariances 118–19
Bar-Shalom–Campo formula 115–17
cross-covariances, iterative calculation of 121–3
square root decomposition, distributed calculation of cross-covariances using 123–6
with unknown correlations 127
covariance intersection 130–42
Inverse Covariance Intersection (ICI) 142–5
Naïve Fusion 127–30
safe fusion 148–51
Tracklet Fusion 145–8
trajectory estimation 79
accumulated state density (ASD) 82–7
notion of 81–2
recursive measurement fusion using 87–92

free quantum particle 103
  free dynamic particle 106–8
  linear-Gaussian measurements 109–10
  state estimation for macroscopic objects with path integrals 105–6
Gaussian density, application on 110–11
generalized smoothing and out-of-sequence processing 92
  batch processing 96–8
  block-line combined estimator and smoother 98–103
  quantum physical interpretation 103
  retrodiction 79–81
transformation theory 8

Wick-rotated Schrödiger equation 106
Wick-rotation 25
Wiener process 16–18, 20–1

zero knowledge initialization 41–2

www.ingramcontent.com/pod-product-compliance
Lightning Source LLC
Chambersburg PA
CBHW050517190326
41458CB00005B/1569